Unraveling DNA

Molecular Biology for the Laboratory

Unraveling DNA

Molecular Biology for the Laboratory

Michael R. Winfrey
University of Wisconsin–La Crosse

Marc A. Rott
University of Wisconsin–La Crosse

Alan T. Wortman
Applied Biosystems
Foster City, California

PRENTICE HALL, UPPER SADDLE RIVER, NJ 07458

Library of Congress Cataloging-in-Publication Data

Winfrey, Michael R , 1952–
    Unraveling DNA : molecular biology for the laboratory / Michael R
Winfrey, Marc A. Rott, Alan T Wortman
        p   cm.
    ISBN 0-13-270034-4 (pbk.)
    1 Molecular biology—Laboratory manuals   2. Recombinant DNA—
Laboratory manuals. I Rott, Marc A., 1956-   II Wortman,
Alan T , 1951-   . III Title
    [DNLM: 1 DNA—analysis—laboratory manuals   2. DNA, Recombinant—
laboratory manuals. QU 25 W768u 1997]
    QH506.W54   1997
    574.87'3282—dc20
    DNLM/DLC
    for Library of Congress   96-36610
                                                                          CIP

To Jane for showing us how to teach with high standards and
to Chuck for showing us how to keep our sanity while maintaining them
                                                        Mike and Mark

Executive Editor: *Sheri Snavely*
Editorial Director: *Tim Bozik*
Editor in Chief: *Paul F. Corey*
Assistant Vice President and Director of Production: *David W. Riccardi*
Special Projects Manager: *Barbara A. Murray*
Production Editorial/Composition: *Jane Judge Bonassar*
Cover Design: *Heather Scott*
Interior Design: Tamara Newnam-Cavallo/Judith Matz-Coniglio
Cover Photo Credits: Cover photograph: William Ormerod, CEMMA-Department of Biological Sciences,
                University of Southern California
                Used with permission of: Margaret McFall-Ngai, Pacific Biomedical Research
                Center, University of Hawaii

                Cover Illustration: Patrice Van Acker

                Back cover photo provided by: Michael Winfrey, Department of Biology and
                Microbiology, University of Wisconsin-La Crosse

                Interior photographs: Jerry Davis, Department of Biology and Microbiology,
                University of Wisconsin-La Crosse

        © 1997, Prentice-Hall, Inc
        A Pearson Education Company
        Upper Saddle River, NJ 07458

Printed in the United States of America

10 9 8 7 6 5 4 3 2 1

ISBN 0-13-270034-4

Prentice-Hall International (UK) Limited,London
Prentice-Hall of Australia Pty. Limited, Sydney
Prentice-Hall Canada Inc., Toronto
Prentice-Hall Hispanoamericana, S.A., Mexico
Prentice-Hall of India Private Limited, New Delhi
Prentice-Hall of Japan, Inc., Tokyo
Pearson Education Asia Pte. Ltd., Singapore
Editora Prentice-Hall do Brasil, Ltda., Rio de Janeiro

CONTENTS

Laboratory Exercises

I. Introductory Techniques

II. DNA Isolation and Analysis

III. Cloning the *lux* Operon

IV. Restriction Mapping and Southern Blotting

V. Subcloning the *lux*A Gene

VI. Advanced Techniques

Appendices

FOREWORD

Some of us have been fortunate to have worked with bioluminescent systems for part or all of our careers. For some, it was a matter of fate: being handed a project by a professor. For others, it was the attraction of studying something so totally fascinating that it was almost impossible not to study. I confess to being in the latter group—drawn to the study of bacterial bioluminescence like an insect to a flame after hearing a lecture presented by Professor J. W. Hastings at the University of Chicago in the early 1960s. This led to a thesis project on the subject of bacterial bioluminescence, and I have continued this work for more than 30 years!

During the ensuing years, I have used the bacterial bioluminescence system as a teaching tool, using its beauty and simplicity to capture the interest and imagination of many a student, and have often mused about the possibility of designing a whole course in bacteriology (or some part of it) using a bioluminescent bacterium and bioluminescence as the major organism and system of study. However, as often happens, due to the limitations of time and energy, such endeavors languish on the shelf as "great ideas" but fail to mature into working systems.

It was thus with some excitement that I have observed Mike Winfrey, Marc Rott, and Al Wortman design, implement, and now publish, in manual format, a series of exercises in molecular biology using the bacterial luminescence (*lux*) genes. Since the original cloning of these genes in 1982 by Dan Cohn, and the resulting excitement of seeing *Escherichia coli* glowing in the darkroom, it has seemed obvious that this could and should be a powerful teaching tool. Yet until now there was no procedural manual that allowed this to be done in the classroom format.

It has been my great pleasure to follow the development of this manual, to see summer students struggle with and complete the experiments, and to watch it mature into a usable manual for teaching at many levels. The feedback from students has been uniformly positive—and the exercises are enjoyable and intellectually challenging. Winfrey, Rott, and Wortman have produced an excellent manual with a selection of experiments that goes far beyond just molecular biology.

They step beyond molecular biology in experiments that examine the ecology and distribution of the luminous bacteria using molecular probes. The ecology and distribution of the luminous bacteria remains one of the unexplored horizons of marine (and soil) microbiology, and when such discussions are added to the molecular biology of bioluminescence, the usefulness of these techniques and approaches for studying organisms, populations, and ecology becomes apparent. I believe that this manual has tremendous potential for educating, exciting, and challenging minds at all levels—from beginning students to jaded veterans—and I am pleased to see its publication.

Kenneth H. Nealson
Distinguished Professor of Biological Sciences
University of Wisconsin-Milwaukee
Center for Great Lakes Studies

PREFACE

In slightly less than two decades, the biological sciences have been revolutionized by new procedures used to manipulate and study genetic material. At the heart of this "biological revolution" are procedures known as **recombinant DNA techniques**— the ability to splice pieces of foreign DNA into a vector and transfer these recombinant DNA molecules into a living organism. In the early 1980s, like many university faculty, I realized that this revolution in how biological systems are studied also required changes in how the biological sciences are taught. The applications of recombinant DNA techniques were so widespread and were having such a universal impact that it was important for young scientists to have training in the principles, methods, and applications of these procedures. Although it was relatively easy to incorporate this novel information into lectures, performing recombinant DNA techniques in a teaching lab seemed almost overwhelming.

However, I had always been inspired by the Chinese proverb that states:

> I hear and I forget,
> I see and I remember,
> I do and I understand

and felt that for students to truly comprehend the elegance and power of these techniques, they should actually perform gene cloning experiments in a laboratory setting. Thus, in the mid-1980s, I began developing a curriculum to bring the relatively new procedures in recombinant DNA techniques into undergraduate teaching laboratories. My first attempts in 1985 involved the use of homemade gel boxes put together with scraps of plexiglass and nicrome wire, a single micropipet capable of measuring 2–10 μl (for a class of 20 students), and a handheld UV light.

During these early attempts at teaching recombinant DNA techniques, I was intrigued by the possibility of having students clone a gene from an entire genome, even though this seemed far beyond the scope of undergraduate teaching laboratories. In 1988, while pondering how to develop teaching labs that would allow such a cloning, I heard a talk by Ken Nealson on bacterial bioluminescence. I had been fascinated with these organisms since graduate school, where I had isolated them from fresh shrimp in a dark (and rather smelly) room. Ken described, amid beautiful slides of bacteria producing a soothing blue luminescence, how they had cloned the genes. The usually extremely time-consuming process of screening the thousands of clones from a genomic library for the clone of interest was done in mere minutes in a dark room! This launched the idea of developing a laboratory teaching curriculum centered around the cloning of the bioluminescence genes.

In 1990, Al Wortman, a new molecular biologist at University of Wisconsin-La Crosse, and I wrote an Undergraduate Faculty Enhancement Program grant through the National Science Foundation to offer a two-week summer laboratory workshop in molecular biology for college faculty. We were funded for three years, and these workshops, along with a new undergraduate course in microbial genetics, resulted in the birth of this manual. We wanted to create a curriculum that not only allowed the molecular biology research techniques to be done within the limitations of college teaching laboratories, but also provided students with a sense of completing a major research project rather than just presenting a collection of seemingly unrelated techniques.

In the first year of the NSF–funded workshop, we compiled an integrated series of exercises involving the cloning, mapping, subcloning, and sequencing of the bioluminescence (*lux*) genes from the marine bacterium *Vibrio fischeri*. Following the first year of the course, Al Wortman resigned and was replaced by Marc Rott. Marc had extensive experience in bacterial molecular biology, assisted in part of the 1992 NSF workshop, and was a co-instructor in the 1993 course. Marc and I received an additional two years of funding from NSF to offer an expanded two-and-a-half-week workshop for college faculty in the summers of 1994 and 1995. Through the NSF-sponsored summer courses (attended by 100 college faculty) and our undergraduate microbiology courses, the exercises were thoroughly and repeatedly tested. We also extensively revised the manual each year based on the experiences and valuable comments of college faculty and our students.

So, more than a decade after the inception of an idea, and after over seven years of development, we are pleased to publish this manual—an integrated series of laboratory exercises based on the cloning and analysis of genes encoding bacterial bioluminescence. In *Unraveling DNA: Molecular Biology for the Laboratory,* we guide students and instructors alike through the tangled maze of modern protocols in molecular biology and make them feasible in the time constraints of undergraduate laboratories. Although in many cases we have pushed the envelope in determining how short one can make incubations, these exercises still yield excellent results in the hands of students. We have not varied conceptually from **how** these fundamental procedures in molecular biology are done, and the protocols in this manual are suitable for use in research labs as well.

At first glance, cloning a set of genes from an entire genome in an undergraduate course seems overly ambitious. We received some criticism from reviewers in our NSF proposals that it was not possible to clone a gene in a two-week course. What they failed to realize was that we cloned the genes in the first week and spent the second week analyzing the clones! Conducting a series of integrated exercises where subsequent exercises depend on the successful completion of previous labs may also seem problematic. Students, like all scientists, will make mistakes and not all will be successful in completing each exercise. However, the techniques provided here have been exhaustively tested by thousands of undergraduate students and more than a hundred faculty, and have proven to work exceptionally well. In addition, we have designed the exercises such that each group generates a large excess of the material needed for subsequent labs. Thus, student groups that encounter problems in any exercise will be able to borrow materials, DNA, or strains from other groups. This approach allows everyone to complete the entire cloning project successfully.

Besides providing basic experience and understanding in the principles and practice of modern molecular techniques, it is also our intention to present this material in a format that is exciting and fun for students. With this objective in mind, we have based the entire series of exercises on one of the world's most fascinating biological systems: the biological production of light. We have found that this adds an additional level of interest and biological relevance to the already intriguing study of molecular biology.

One limitation of laboratory manuals in rapidly advancing areas such as molecular biology is that they often become out of date even before they are published. This is particularly true with the use of computers in molecular biology. To circumvent this, we have established a Molecular Biology Home Page on the World Wide Web:

http://www.uwlax.edu/MoBio

The computer analysis of DNA (Exercise 26) is linked to our home page to allow students to use the most recent and relevant Internet sites to conduct the exercise. Our home page also provides links to numerous other sites on the World Wide Web of relevance to many of the exercises in the manual.

In addition to the 28 exercises in the manual, 19 appendices provide a wealth of information on basic procedures, principles and precautions in molecular biology, recipes for media and reagents, lists of suppliers of equipment and materials used in the course, and current references. These provide a valuable resource for student and instructor alike and are used in many of the exercises. An instructor's manual is also available with detailed instructions on how to prepare the materials for each exercise, tips on interpreting results, troubleshooting potential errors, and answers to the questions following each exercise. Instructors are also referred to our home page for updated preparation tips, suppliers, and so on.

Finally, it is easy to get lost in the details of molecular biology and forget the big picture. Molecular biology is an elegant science in its own right, but today it is most frequently used as a powerful set of tools to study a myriad of biological processes. In this manual, we have attempted not to lose the connection with the organisms we are examining and the environment from which they came. Thus, we hope you come away from using this manual with not only an appreciation of the power of molecular biology but also an interest in a unique group of luminous bacteria and the often bizarre deep-sea creatures that provide these organisms a home.

Mike Winfrey
(with Marc Rott and Al Wortman)

ACKNOWLEDGMENTS

We would like to thank the Japanese pinecone fish, *Monocentris japonicus,* for harboring and caring for its symbiont *Vibrio fischeri* since before the dawn of humanity. We would also like to thank *Vibrio fischeri* for providing illumination for its caretaker through all those millennia. We also express a deep appreciation for the process of evolution for providing all the genes required for bioluminescence on a single 9 kb *Sal* I restriction fragment. This manual would not be possible without this extraordinary bond between these two species and the fortuitous location of the *lux* genes.

We would also like to thank Ken Nealson for his continued assistance throughout the development of this manual, providing the MJ1 strain of *Vibrio fischeri* and numerous ideas, and presenting the opening seminars at each of our NSF workshops. Ken has an infectious enthusiasm and love of science that have been an inspiration to us and all who have used this manual.

We are deeply indebted to the Instrumentation and Laboratory Improvement Program (formerly the College Science Instrumentation Program) and the Undergraduate Faculty Enhancement Program of the National Science Foundation for their support in the development of this manual. We and all of higher education in the United States are fortunate to have such an organization dedicated to furthering science education. This manual would not have been possible without grants from the NSF (CSI-8750784; USE-9054261; DUE-9353970).

The Japanese pinecone fish (*Monocentris japonicus*). This fish, commonly referred to as the "port and starboard" fish, has two light organs under its lower jaw. The bioluminescent bacterium used in this manual, *Vibrio fischeri* MJ1, was isolated from this fish by Ken Nealson. Bar = 1.0 cm (from Hastings and Nealson, 1981, with permission).

Many people contributed in many ways to the development of this manual. We thank Chuck Whimpee and Lisa Van Ert for providing the PCR primers used in the PCR exercise. Dave Mead, Dave Essar, Ford *Lux*, Pat Singer, and Paul Barney assisted in several of the NSF-sponsored workshops and contributed valuable suggestions to the manual. We thank Jerry Davis for the photography in the manual, and Jill Rouselle, Zac Triemert, and Lindsay Dunnum for editing and assisting with the photography.

The NSF-sponsored workshops would not have been possible without the assistance of Kathie Self and the herculean efforts of our student assistants (affectionately referred to as "elves") who spent many hours before, during, and after each day's labs preparing the media, reagents, and materials. They were able to troubleshoot the preparations for the labs in this manual to ensure that they run smoothly. They are Tony Bladl, Todd Clark, Cindy Doriot, Darin Ellingson, Meredith Jones, Josie Murphy, Servet Ozcan, Sasikala Perumal, Heather Pryer, Polly Quiram, Patrick Splinter, Dave Steffes, Steve Titus, Beth Yanke, and Dave Young.

The quality of this manual is due in part to the extensive feedback from the 100 college and university faculty who participated in the NSF workshops from 1991 to 1995. They have been ideal reviewers because they not only read and evaluated the manual, but also used it. We owe each of them a debt of gratitude:

Kemi Adewusi
Ferris State University

Oluwatoyin Akinwunmi
Muskingum College

Les Albin
Austin Community College-Rio Grande Campus

William Andresen
Missouri Western State College

Kathleen Angel
University of Mary

Josephine Arogyasami
Teiko Westmar University

Judy Awong-Taylor
Armstrong State College

Davinderjit Bagga
University of Monevallo

Paul Barney, Jr.
Pennsylvania State University-Erie

Marvin Bartell
Concordia University

Steven Berg
Winona State University

Dean Bishop
Pittsburgh State University

Shelly Bock
University of Pittsburgh at Johnstown

Paul Boehlke
Martin Luther College

Robert Boomsma
Trinity Christian College

Michael Bowes
Humboldt University

Bonnie Boyle
University of Minnesota-Crookston

John Brauner
Jamestown College

Carolyn Brooks
University of Maryland Eastern Shore

Attila Buday
Houston Community College System

Michael Bunch
Amarillo College

Clara Carrasco
University of Puerto Rico-Ponce

Arthur Carroll
College of Santa Fe

John Clausz
Carroll College

Lorenzo Coats
Shaw University

Charlene Cole
Tarrant County Junior College

Brian Coughlin
University of Northern Florida

Garry Duncan
Northeast Wesleyan University

Deborah Dunn
Bethany College

Lenore Durkee
Grinnell College

Jan Ely
Fort Hays State University

Wayne England
Salem-Teikyo University

Robert Evans
Illinois College

Samuel Fan
Bradley University

Beth Ferro Mitchell
LeMoyne College

Scott Figdore
Upper Iowa University

Cynthia Fitch
Seattle Pacific University

Lorita Gaffney
Silver Lake College

Stephen Gallik
Mary Washington College

Geoff Gearner
Morehead State University

Tom Glover
Hobart & William Smith Colleges

Rodney Hagley
St. Mary's College

Richard Halliburton
Western Connecticut State University

Betty Harris
Westfield State College

Jean Helgeson
Collin County Community College

Robert Herforth
Augsburg College

Deborah Hettinger
Texas Lutheran College

William Hixon
St. Ambrose University

Angela Hoffman
University of Portland

Shelley Jansky
University of Wisconsin-Stevens Point

Philip Jardim
City College of San Francisco

Warren Johnson
University of Wisconsin-Green Bay

Peggy Jean Jones
William Jewell College

Elizabeth Juergensmeyer
Judson College

Gerhard Kalmus
East Carolina University

Samuel Kent
Vermont Technical College

John Kiss
Miami University

Patricia Knopp
Southeastern Community College

Joan Kosan
New York City Technical College

Barbara Krumhardt
Des Moines Area Community College

Catherine Kwan
Mount St. Mary's College

Mary Leida
Morningside College

James Leslie
Adrian College

Michael Lockhart
Northeast Missouri State University

Fordyce G. Lux III
Lander University

Vicky McKinley
Roosevelt University

Ronald Meyers
Grand Rapids Baptist College

Byron Noordewier
Northwestern College

Ted Nuttall
Lock Haven University

G. Dale Orkney
George Fox College

Patricia G. S. Pagni
Knoxville College

Robin Pettit
Western Oregon State College

Onkar Phalora
Anderson University

Jan Phelps
University of Wisconsin Center Baraboo-Sauk County

Jo Anne Quinlivan
Holy Names College

William Quinn
St. Edwards University

Peggy Redshaw
Austin College

Richard Renner
Laredo Community College

William Roberts (deceased)
College of Eastern Utah

Lori Rynd
Pacific University

Lauri Sammartano
St. Olaf College

Eduardo Schroder
University of Puerto Rico

Emeric Schultz
Bloomsburg University

David Scupham
Valparaiso University

Bruce Simat
Northwestern College

Patricia Singer
Simpson College

Mary Spratt
William Woods University

Richard Stark
Heritage College

Ronald Stephens
Ferrum College

David Stock
Stetson University

James Stukes
South Carolina State University

Phil Stukus
Denison University

Robert Turner
Western Oregon State College

Anthony Udeogalanya
Medgar Evers College

Charissa Urbano
Delta College

William Vail
Frostburg State University

Frank Verley
Northern Michigan University

Sandra Whelan
Bridgewater State College

Andrew Whipple
Taylor University

Bambi Wilson
University of Wisconsin-Madison

We would also like to thank the following reviewers of the manual for their careful and extensive review of a lengthy manuscript: Marsha Altschuler, Williams College; Clifford F. Brunk, University of California–Los Angeles; Bruce Chase, University of Nebraska at Omaha; Cynthia Cooper, Northeast Missouri State University; Elliot S. Goldstein, Arizona State University; and Mike Lockhart, Northeast Missouri State University. We are especially indebted to Bruce Chase, whose objective criticisms resulted in significant improvements in introducing student independence and critical thinking aspects in the final version.

We thank Bio-Rad Laboratories, Inc.; FOTODYNE, Inc.; the Genetics Computer Group (GCG); Gibco-BRL; Perkin Elmer; and Promega, Inc., for their sponsorship of the NSF workshops and supplying much of the enzymes, DNAs, reagents, and expertise needed to develop this manual. We would also like to offer thanks and a toast to the late Bernie Meyer, who passed away in September, 1995, for his support and sponsorship of the NSF workshops and showing us that a nonscientist can appreciate molecular biology.

We also thank Carrie Mattaini and Kelly McDonald of Prentice Hall for stumbling into Mike Winfrey's office one day when he was desperately looking for a publisher for this manual. Our production editor, Jane Judge Bonassar, has been very helpful in assisting us in getting the manual to print and we appreciate her patient help. We are also indebted to our production manager, Barbara Murray, and our art director, Heather Scott. Finally, we thank Sheri Snavely, our editor at Prentice Hall, for accelerating the publication of this manual, for her patience and support of three naive authors, and for her guidance in our first experience through the maze of commercial publishing.

INTRODUCTION TO BACTERIAL BIOLUMINESCENCE

The biological production of light, or bioluminescence, has intrigued and fascinated humans for thousands of years, and scientific studies on bioluminescence date back more than 300 years (Meighen 1988). Numerous organisms have the ability to emit biologically produced light, including fireflies, fish, clams, worms, algae, and bacteria (Harvey 1952). Perhaps some of the most bizarre and fascinating of these are marine fish and squid, which have a variety of unique light organs (Goode and Bean 1895, McCosker 1977, Robison 1995, Ruby and McFall-Ngai 1992). However, most of the luminescent marine animals do not produce bioluminescence themselves, but harbor bioluminescent bacteria in specialized light organs. Due to their ease of study, extensive research has been done on bioluminescent bacteria, which has allowed a detailed understanding of the biochemistry and genetics of this process.

Most bioluminescent bacteria are marine in origin and include both free living forms and species that form symbiotic relationships with fish or squid. The light organs are often highly specialized and specifically adapted to harbor essentially pure cultures of the luminescent bacteria (see Figure 1). Bioluminescence in fish is particularly common in the deep sea, where up to 96% of all deep-sea fish are reported to be bioluminescent (Harvey 1952). The exact role of bioluminescence is not clearly known, although numerous advantages to the fish have been proposed, such as warding off predators, attracting prey, or communicating. In return for producing light for the fish, the bacteria are provided with a protected environment and a rich supply of nutrients. Recently, some understanding of the relationship between bioluminescent

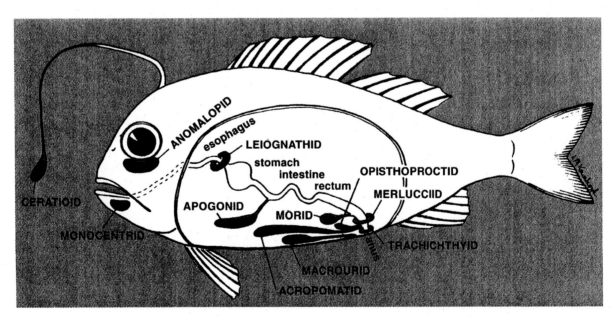

Figure 1
A hypothetical fish illustrating the approximate location and size of light organs (indicated in blue) of the various types of luminescent fishes (modified from Hastings and Nealson 1981, with permission)

marine species and their bacterial symbionts has evolved from the study of the squid *Euprymna scolopes* (see front cover) and its symbiont *Vibrio fischeri* (Boettcher, et. al 1996, McFall-Ngai and Ruby 1991, Ruby and McFall-Ngai 1992), and this system has become an excellent model for studying animal-bacterial symbioses.

Besides the mutualistic relationships, bioluminescent bacteria are common in a variety of other associations with marine animals. Many are saprophytic, found on living or dead fish or shellfish. In fact, one of the easiest ways to isolate bioluminescent bacteria is to allow bacterial growth on the surface of fresh fish and then to examine the fish for bioluminescent patches in a dark room. Luminescent bacteria have also been isolated from stored meat and even open human wounds (Hastings and Nealson 1981). Numerous species are parasites on marine crustaceans, such as sand fleas (Harvey 1952), and other species are commensal in the intestinal tracts of marine animals (Hastings and Nealson 1981).

Bioluminescence is catalyzed by an enzyme known as **luciferase** (Meighen 1988). The bacterial enzyme is a heterodimer with a molecular weight of approximately 80,000 daltons and consists of an α and β subunit with molecular weights of approximately 42,000 and 38,000 daltons, respectively. The active site is on the α subunit, although the β subunit is required for activity. Luciferase is a mixed function oxidase that produces a blue-green light via the simultaneous oxidation of reduced flavin mononucleotide ($FMNH_2$) and a long chain aldehyde (tetradecanal) by O_2:

$$FMNH_2 + O_2 + R\text{-}CHO \longrightarrow FMN + R\text{-}COOH + H_2O + \; light$$

The energy for light production is supplied by the oxidation of the aldehyde and $FMNH_2$. The actual mechanism of light emission is not clearly understood but is thought to result from the formation of a hydroperoxy flavin via the reaction of $FMNH_2$ and O_2. These molecules have been shown to emit light in the presence of aldehydes (Meighen 1988).

Three additional enzymes are necessary to generate the aldehyde required in the reaction. The fatty acids for this fatty acid reductase enzyme complex are removed from the fatty acid biosynthesis pathway via the enzyme **acyl-transferase**. This enzyme reacts with acyl-ACP (*Acyl Carrier Protein*) to release free fatty acids (R-COOH). The fatty acids are then reduced to an aldehyde by a two-enzyme system via the following reaction:

$$R\text{-}COOH + ATP + NADPH \longrightarrow R\text{-}CHO + AMP + PP + NADP^+$$

One enzyme, **acyl-protein synthetase**, activates the fatty acid via the cleavage of ATP to form R-CO-AMP. This serves as the substrate for the final enzyme, **acyl-reductase**, that catalyzes the NADPH-dependent reduction of the activated fatty acid to an aldehyde. The role of each enzyme involved in light production is summarized in Figure 2.

Bacterial bioluminescence is observed only at very high cell densities because of a unique type of regulation known as autoinduction (Nealson and Hastings 1979). The bacteria produce a diffusible compound (an N-acyl homoserine lactone) known as an autoinducer, which induces transcription of the genes encoding the enzymes required for light production. However, light production occurs only when a threshold concen-

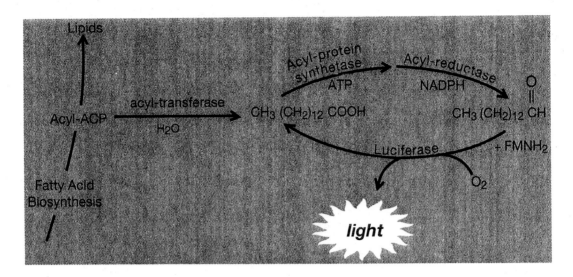

Figure 2
Pathway of aldehyde formation and light production in the bacterial bioluminescence system (modified from Meighen 1988, with permission).

tration of the autoinducer accumulates. Thus, bacteria at high cell density in the light organ of a fish accumulate sufficient autoinducer to bioluminesce, while free living bacteria do not. The requirement for an autoinducer represents a significant ecological adaptation. Because light production requires a tremendous amount of energy and cellular reducing power, free living bacteria in the ocean—which are nutrient limited— will not waste this energy expenditure that probably offers them no benefit.

Interestingly, the autoinduction mechanism has recently been shown to be a general mode of regulation in many gram-negative bacteria. There is considerable interest in this process as a result of the discovery that some plant and animal pathogens, as well as plant symbionts, also produce homoserine lactone autoinducers and have *lux*I and *lux*R analogs to control host colonization. It appears that many bacterial behaviors involved in host colonization (such as bioluminescence) require a large population, or "**quorum**," of bacteria and use the *lux*I and *lux*R system of autoinduction. This has resulted in use of the phrase "**quorum sensing**" (Fuqua, et al. 1994) to describe cell activities that require a threshold cell density.

With the advent of recombinant DNA techniques, it has been possible to clone and determine the genetic organization of the bacterial bioluminescence genes from numerous species (see Meighen 1988, 1991, 1994 for examples). The α and β subunits of luciferase and the three enzymes required for aldehyde formation are encoded in a single operon (the *lux* operon) in all luminescent bacteria examined (see Figure 3). The first two structural genes, *lux*C and *lux*D code for the acyl-reductase and acyl-transferase, respectively. These are followed by *lux*A and *lux*B, which code for the α and β subunits of luciferase, and finally *lux*E, which codes for the acyl-protein synthetase. Two regulatory genes, *lux*I and *lux*R, have also been identified in *Vibrio fischeri* (which has recently been reclassified as *Photobacterium fischeri*). The *lux*I gene is on the same operon as the structural genes, whereas *lux*R is transcribed in the opposite direction. *Lux*R codes for a transcriptional activator that binds the autoinducer synthesized by the *lux*I gene product.

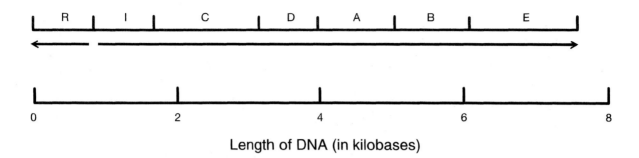

Length of DNA (in kilobases)

Figure 3

The *lux* operon of *Vibrio fischeri*. *lux*C codes for the acyl-reductase; *lux*D codes for the acyl-transferase; *lux*A and *lux*B code for the α and β subunits of luciferase, respectively; *lux*E codes for the acyl-protein synthetase; *lux*R and *lux*I code for regulatory proteins. Arrows under the operon indicate the directions of transcription.

Detailed study of the molecular genetics of bacterial *lux* operons has allowed use of this genetic system in applied and basic research. The *lux* system is now used in toxicity testing (Schiewe, et al. 1985), as a reporter in gene fusions to indicate the level of expression of various operons (Heitzer, et al. 1992, Nealson and Hastings 1991, Selifonova, et al. 1993), in promoter probe vectors (Sohaskey, et al. 1992), and as a method of monitoring the fate of genetically engineered microorganisms in the environment (Shaw, et al. 1992). The genes for bacterial bioluminescence have also been engineered into bacteriophage to allow sensitive testing for bacterial pathogens (Stewart, et al. 1996). Although bioluminescence will always hold its appeal and excitement due to the soothing blue light produced, this unique biological process is likely to find additional practical applications in molecular biology and biotechnology in years to come.

Literature Cited

Boettcher, K. J., E. G. Ruby, and M. J. McFall-Ngai. 1996. Bioluminescence in the symbiotic squid *Euprymna scolopes* is controlled by a daily biological rhythm. *J. Comp. Physiol.* 179:65–73.

Fuqua, W. C., S. Winans, and E. P. Greenberg. 1994. Quorum sensing in bacteria: The LuxR-LuxI family of cell density-responsive transcriptional regulators. *J. Bacteriol.* 176:269–275.

Goode, G. B., and T. H. Bean. 1895. *Oceanic ichthyology: A treatise on the deep-sea and pelagic fishes of the world.* Washington D. C.: Government Printing Office.

Harvey, E. N. 1952. *Bioluminescence.* New York: Academic Press.

Hastings, J. W., and K. H. Nealson. 1981. The symbiotic luminous bacteria. M. P. Starr, H. Stolp, H. G. Truper, A. Ballows, and H. G. Schlegel, ed. 1332–1345. In *The prokaryotes: A handbook on habitats, isolation, and identification of bacteria.* New York: Springer-Verlag.

Heitzer, A., O. F. Webb, J. E. Thonnard, and G. S. Sayler. 1992. Specific and quantitative assessment of naphthalene and salicylate bioavailability by using a bioluminescent catabolic reporter bacterium. *Appl. Environ. Microbiol.* 58:1839–1846.

Meighan, E. A. 1988. Enzymes and genes from the *lux* operons of bioluminescent bacteria. *Ann. Rev. Microbiol.* 42:151–176.

Meighan, E. A. 1991. Molecular biology of bacterial bioluminescence. *Microbiol. Rev.* 55:123–142.

Meighan, E. A. 1994. Genetics of bacterial bioluminescence. *Ann. Rev. Genet.* 42:151–176.

Meighan, E. A., and P. V. Dunlap. 1993. Physiological, biochemical, and genetic control of bacterial bioluminescence. *Adv. Microbial Physiol.* 34:1–67.

McCosker, J. E. 1977. Flashlight fishes. *Scientific American.* 236:106–112.

McFall-Ngai, M. J., and E. G. Ruby. 1991. Symbiont recognition and subsequent morphogenesis as early events in an animal-bacterial mutualism. *Science* 254:1491–1494.

Nealson, K. H., and J. W. Hastings. 1979. Bacterial bioluminescence: Its control and ecological significance. *Microbiol. Rev.* 43:496–518.

Nealson, K. H., and J. W. Hastings. 1991. The luminous bacteria. In *The Prokaryotes.* 2d ed., ed. A. Ballows, H. G. Truper, M. Dworkin, W. Harder, and K. Schleifer. New York: Springer-Verlag.

Robison, B. H. 1995. Light in the ocean's midwaters. *Scientific American.* 273:59–64.

Ruby, E. G., and M. J. McFall-Ngai. 1992. A squid that glows in the night: Development of an animal-bacterial mutualism. *J. Bacteriol.* 174:4865–4870.

Schiewe, M. H., E. G. Hawk, D. I. Actor, and M. M. Krahn. 1985. Use of bacterial bioluminescence assay to assess toxicity of contaminated marine sediments. *Can. J. Fish. Aquat. Sci.* 42:1244–1248.

Selifonova, O., R. Burlage, and T. Barkay. 1993. Bioluminescent sensors for detection of Hg(II) in the environment. *Appl. Environ. Microbiol.* 59:3083–3090.

Shaw, J. J., F. Dane, D. Geiger, and J. W. Kloepper. 1992. Use of bioluminescence for detection of genetically engineered microorganisms released into the environment. *Appl. Environ. Microbiol.* 58:267–273.

Sohaskey, C. D., H. Im, and A. T. Schauer. 1992. Construction and application of plasmid-based and transposon-based promoter-probe vectors for *Streptomyces spp.* that employ a *Vibrio harveyi* luciferase reporter cassette. *J. Bacteriol.* 174:367–376.

Stewart, G. S. A. B., M. J. Loessner, and S. Scherer. 1996. The bacterial *lux* gene bioluminescent biosensor revisited. *ASM News.* 62:297–301.

OVERVIEW OF LABORATORY EXERCISES

The experiments in this manual are designed to provide college-level students with real-world experiences in molecular biology and to provide an appreciation of using the tools of molecular biology to study a biological system. The overall goal of these exercises is to clone and analyze the bioluminescence genes (the *lux* operon) from the bioluminescent bacterium *Vibrio fischeri*. After the identification of luminescent clones, the recombinant plasmids are mapped, subcloned, and sequenced. The laboratory exercises are divided into discreet units, each involving a series of common molecular techniques designed to accomplish one component of the overall project. Collectively, this integrated series of exercises represents the steps in a complete cloning project. The entire series of exercises requires approximately one full semester to complete, with two laboratory periods per week. Smaller portions of the manual are easily adapted to fewer lab periods.

In Part I, **Introductory Techniques**, you will be acquainted with basic equipment and microbiological techniques (Exercise 1). This is important as you will be using fragile and expensive equipment throughout the course, and many of the exercises require cultivating bacteria using sterile techniques. Exercise 2, while optional, provides experience in preparing the media and reagents needed in the course and should give you a sense of the preparation required to make these exercises possible. Exercise 3 is an illuminating exercise that will provide an appreciation for the diversity and habitats of the bioluminescent bacteria! In this exercise you will isolate a variety of bioluminescent strains for use in the PCR exercise (Exercise 23). Exercise 4 introduces the essential skills and concepts associated with the use of restriction enzymes to digest DNA as well as preparing, running, and analyzing agarose gels. These are some of the most fundamental techniques in modern molecular biology and will be used repeatedly throughout the course.

In Part II, **DNA Isolation and Analysis**, you will isolate DNA from the bioluminescent bacterium *Vibrio fischeri* (Exercise 5), which will provide the DNA needed to clone the *lux* genes. You will also isolate the plasmid vectors in Exercise 6 that will be used in the cloning experiments (Parts III and V). In Exercise 7, you will use spectrophotometry to assess the purity and quantity of the DNAs isolated. Further use of the genomic DNA and plasmid vectors requires an accurate measurement of the DNA concentration.

In Part III, **Cloning the *lux* Operon**, you will clone the genes required for bioluminescence. This involves cutting the *Vibrio fischeri* chromosomal DNA and the plasmid vector with the restriction endonuclease *Sal* I (Exercise 8). Following quantification of the insert and vector DNA (Exercise 9), these molecules are ligated together (Exercise 10) to form recombinant DNA molecules. The ligation mix is transformed into a host bacterium, *Escherichia coli* DH5α (Exercises 11 and 12), which results in the formation of a genomic library. The exciting moment of discovering if you cloned the bioluminescence genes comes in Exercise 13, where you will screen thousands of transformants for light production by observing the plates in a dark room. Because the clone of interest has a readily observable phenotype, the often time-consuming process of screening a library is very quick (and fun).

In Part IV, **Restriction Mapping and Southern Blotting**, you will begin the analysis of the *lux* genes from your bioluminescent clones. In Exercises 14 and 15, you will isolate the recombinant plasmids from light-producing clones and prepare restriction maps of the cloned DNA. In Exercise 16 you will map the location of a specific gene *(lux*A) by Southern transfer of DNA from an agarose gel onto a nylon membrane followed by nucleic acid hybridization with a probe for the *lux*A gene.

In order to continue analysis of your cloned DNA, it is necessary to subclone specific genes in the *lux* operon. In Part V, **Subcloning the *lux*A Gene**, you will digest a plasmid vector and plasmid DNA from a light-producing clone (Exercise 17) and purify a DNA fragment containing the *lux*A and *lux*B genes (Exercise 18). Following ligation (Exercise 19) and transformation of the recombinant plasmids into *E. coli* (Exercise 20), you will screen transformants for those containing the genes of interest by colony hybridization (Exercise 21). You will then use a plasmid mini-prep procedure to verify that the selected clones contain the insert DNA of interest (Exercise 22).

In the final exercises of the manual (Part VI, **Advanced Techniques**), you will use a variety of advanced techniques to further characterize your cloned DNA and compare it with bioluminescence genes from other organisms. In Exercise 23, you will use the polymerase chain reaction (PCR) to amplify a portion of the *lux*A gene starting with DNA from a minute fraction of a single colony of a bioluminescent isolate (from Exercise 3). You will then determine the relatedness of the PCR product to the *V. fischeri lux*A gene by Southern hybridization in Exercise 24. The PCR products may also be used as a source of DNA for cloning *lux* genes from other organisms or may be sequenced as a special project for Exercise 28. The subclones generated in Part V will be sequenced in Exercise 25. The DNA sequences generated in this exercise are analyzed by sequence analysis software available on the World Wide Web in Exercise 26. In this exercise, the students are acquainted with the reading of sequencing autoradiographs and the use of computer software to analyze DNA. This is a fun and very useful exercise, reinforcing the relationship between DNA and protein and the basic principles of molecular biology. It also demonstrates one of the most significant uses of computers in the life sciences and opens the door for using resources on the World Wide Web to examine DNA sequence from literally thousands of organisms and tens of thousands of genes.

The final structured exercise (Exercise 27) illustrates the use of pulsed field gel electrophoresis to map a bacterial genome and will allow you to separate and resolve megabase fragments of DNA and even entire chromosomes. Such techniques are commonly used today to analyze genome structure and play a vital role in the Human Genome Project; this exercise will introduce you to one of the most powerful techniques in molecular biology.

After completing all or most of the first 27 exercises, you will have participated in a significant cloning project and be proficient in most of the techniques currently used in molecular biology. However, to many scientists, such a project is often only the beginning of what may lead to a lifelong career studying the intricacies of a given biological system. One limitation of this manual is that you will have largely followed prewritten protocols and the media, reagents, supplies, and equipment will have been provided. Yet, a key component of research in any area of science is designing your own experiments and preparing the needed reagents and supplies. To give you an opportunity to experience the joy and frustration of conducting research of your own design, we have included a final exercise entitled Independent Projects in Molecular

Biology. This exercise will allow you to use your creativity and the skills that you have learned to date in this course to design and conduct a project of your choice.

The laboratory exercises in this manual are followed by 19 appendices that contain a wealth of information on basic molecular biology procedures, the care and handling of enzymes, as well as information on restriction enzymes, gel electrophoresis, various cloning vectors, and suppliers of equipment and supplies.

LABORATORY GUIDELINES AND SAFETY

In this course, you will be working with live bacteria and some potentially hazardous chemical reagents. The majority of the following guidelines are standard procedures for working in microbiology and molecular biology laboratories. The bacterial cultures you will be working with are all nonpathogenic strains of *Escherichia coli* and marine bacteria. However, you should learn and use the bacteriological techniques designed to avoid contamination and self-inoculation in the event that you work with pathogenic bacteria in the future. In addition, bacteriological techniques are similar to the precautions used to avoid nucleases in DNA manipulations. Please familiarize yourself with these guidelines so that you develop safe techniques when working with biochemicals and biological agents.

1. First, **use your head.** Most accidents in the laboratory can be avoided by common sense and by being alert to potential hazards.

2. No drinking, eating, chewing gum, or smoking in the laboratory. Do not put anything in your mouth, such as pencils, fingers, and so on.

3. Wash hands before and after each laboratory. If your hands become contaminated, wash and rinse with disinfectant soap.

4. Wipe the benchtop with disinfectant before and after each laboratory if live cultures are used. If a culture or contaminated sample is spilled, cover with disinfectant and notify your instructor at once.

5. Dispose of all contaminated materials where indicated.

6. Never mouth pipet anything. Dispose of all contaminated or used pipets in the discard jars—do not set contaminated pipets down on your bench.

7. You will be using instruments and glassware that are expensive and often fragile. Always be sure you know the proper operation procedure before using such items. If you are not sure, please ask before proceeding.

8. The adjustable micropipetters that you will be using are expensive and easily damaged. **Never** adjust below or above the volume limits of the pipetter. When not in use, replace them in the rack provided.

9. Dispose of all used micropipet tips in the discard beaker.

10. Turn Bunsen burners off when not in use. Watch that your hair and clothing stays out of the flame. If you have long hair, tie it back prior to lab. Keep all volatile and flammable liquids away from the flame.

11. Some of the exercises require the use of sterile materials and sterile (aseptic) technique, while others do not. Be sure you are clear when aseptic technique and sterile materials are needed and use careful aseptic technique during these exercises.

12. You will be sharing a microcentrifuge with other groups in the laboratory. Before starting long runs (5 to 15 minutes), please check with the other groups to see if they are ready to use the centrifuge.

13. In this laboratory, you will be handling hazardous chemicals that require specific safety procedures. **First**, think of what you are working with and what you are doing with it. Be conscious of dangers to you and to those around you. Wear gloves, a labcoat, and eye protection when handling toxic reagents, and use the fume hoods when doing phenol and chloroform extractions. Hazardous chemicals will be labeled with red tape, and precautions for proper use of toxic chemicals are indicated in the manual. If there is any question concerning proper handling of a chemical, **ask your instructor.**

    Precautions for handling specific toxic chemicals are described in detail in **Appendix XII.**

14. When using radioisotopes, special precautions are required. These are described in detail in **Appendix XIII.**

15. Report any injuries or accidents to the instructor immediately. Do not work with uncovered cuts. Cover them with a sterile adhesive strip.

16. Finally, it bears repeating: **use your head.** Most accidents in the laboratory can be avoided by common sense and by being alert to potential hazards.

Introductory Techniques

EXERCISE 1
Introduction to the Laboratory: Basic Equipment and Bacteriological Techniques

Methods in molecular biology involve extensive use of a variety of specialized but fairly simple equipment. Most reactions in DNA cloning and other manipulations of nucleic acids are performed in small (0.2–1.5 ml) microcentrifuge tubes in volumes as small as 10 µl. Using such small reactions results in considerable savings in amounts of reagents, enzymes, and DNA, but you must be able to dispense volumes as small as 0.5 µl accurately. Micropipetters are capable of dispensing such small volumes accurately, and thus it is essential that you master micropipetting techniques in order to perform the experiments in this manual properly. Various sized micropipetters, including those capable of dispensing volumes up to 1000 µl, will be used throughout the course, and it is important to be able to use these correctly and accurately.

Because recombinant DNA techniques require extensive use of the bacterium *Escherichia coli* as a host, it is also essential for you to become proficient in bacteriological techniques. When working with bacteria, it is necessary to use **aseptic** techniques—a set of procedures designed to prevent contamination of sterile materials (such as media, reagents, pipets, micropipet tips, glassware, etc.) or pure cultures of bacteria. Other important bacteriological techniques include culturing bacteria in solid and liquid media, purifying isolated clones of bacteria, and long-term storage procedures.

In the first part of this exercise, you will practice using micropipetters and microcentrifuges. You should refer to Appendix I (The Metric System and Units of Measure) and Appendix II (Centrifugation) prior to beginning the lab.

In the second part of this exercise, you will learn basic bacteriological techniques. In addition to working with *E. coli,* you will also work with a strain of *E. coli* that contains a plasmid cloning vector (pGEM™-3Zf[+]). The vector contains a gene coding for resistance to the antibiotic ampicillin, and the inclusion of ampicillin in the media is required to ensure that the strain maintains the plasmid. You will also work with two different marine strains of bioluminescent bacteria. One of these, *Vibrio fischeri,* will provide the DNA used to clone the bioluminescence genes. Because the bioluminescent bacteria are from marine habitats, they require special media that contain a high level of salt (see Appendix XV).

PART I. MICROPIPETTING

Reagents

- microcentrifuge tube with 1 ml "10X buffer" (blue)
- microcentrifuge tube with 1 ml "DNA" (green)

- microcentrifuge tube with 1 ml "Reagent A" (yellow)
- microcentrifuge tube with 1 ml H_2O (clear)
- microcentrifuge tube with 1 ml "Enzyme" (clear, 50% glycerol)

Supplies and Equipment

- beaker with sterile 1.5-ml microcentrifuge tubes (assorted colors)
- microcentrifuge tube rack
- microcentrifuge tube opener
- fine-point permanent marker
- micropipetter stand
- 0.5–10 µl adjustable micropipetter

- 10–100 µl adjustable micropipetter
- 100–1000 µl adjustable micropipetter
- sterile 10-µl micropipet tips
- sterile 100-µl micropipet tips
- sterile 1000-µl micropipet tips
- micropipet tip discard beakers
- microcentrifuge

CAUTION: Micropipetters are expensive and can be easily damaged if not handled properly. Always adhere to the following rules when using micropipetters:

- Never adjust the pipetter below the lower volume limit or above the upper volume limit.

- Never leave pipetters on the bench—replace in the stand when not in use.

- Always slowly release the plunger when withdrawing liquids in the pipetter. This is especially important with large volume pipetters (100–1000 µl).

- Always hold a pipetter with the tip pointing down; never horizontal or up. This can allow liquid to run down into the pipetter and contaminate, corrode, or freeze up the piston.

- Be sure you use the proper size tip for each pipetter.

- Always use a new tip for each different reagent.

➡Procedure

1. Remove six 1.5-ml microcentrifuge tubes from the beaker. Note that the tubes have been sterilized by autoclaving. This is done to destroy any nucleases that may be on the tubes. When handling tubes, take care not to contaminate the inside of the tube or lid. With a fine-point permanent marker, label each tube A through F on the frosted lid and on the frosted labeling spot on the side. Place each tube in a microcentrifuge rack.

2. Add volumes of the "reagents" to tubes A and B as indicated in the following table. These volumes and viscosities are similar to those used to set up the reactions in recombinant DNA techniques. Note that the "enzyme" solution contains 50% glycerol. This will have the same viscosity as the enzymes you will be using in this course, which are suspended in a 50% glycerol buffer as a cryoprotectant.

Tube	10X Buffer	DNA	H_2O	Enzyme
A	1 µl	2 µl	6 µl	1 µl
B	1 µl	4 µl	4.5 µl	0.5 µl

3. To add the first volume, remove the 0.5–10 µl pipetter from the stand and set the digital dial to 1.0 µl. Use the following protocol to dispense the volume into the tube (refer to Figure 1.1):

   a. Open the box of sterile 10-µl tips and firmly press the tip of the pipetter into one of the tips. Remove the tip and close the lid of the box to prevent contamination.

   b. Open the tube containing "10X Buffer." A microcentrifuge tube opener is recommended to minimize nuclease contamination from fingers. Press the plunger down to the first stop, hold the tube at eye level, and place the end of the pipet tip into the solution. The end of the pipet tip should be **just** below the surface of the liquid.

   c. **Slowly** release the plunger until it stops. Remove the tip from the liquid, dragging it along the side of the microcentrifuge tube to remove excess liquid on the outside of the tip. Carefully notice how much liquid is in the tip with each volume pipetted. With practice, you will be able to visually recognize if the volume is accurate.

   d. Hold the tube labeled A at eye level and touch the pipetter tip to the inside of the tube (it does not have to be on the bottom of the tube). Push the plunger down to the first stop. This will deliver the volume indicated on the dial. Then press the plunger to the **second** stop as you slowly drag the tip up the side of the tube. This will blow out any remaining liquid in the tip; touching it to the wall creates capillary action that aids in withdrawing the liquid. When dispensing small volumes, you should always **visually** verify that a drop was transferred to the side of the tube.

   e. Hold the pipetter over the tip discard beaker and press the plunger all the way down (the third stop). Doing so will eject the tip.

4. Repeat step 3 to add the additional volumes to tube A. Then, repeat for tube B. Place each new addition (drop) at a new position on the microcentrifuge tube wall. After you make each addition, check off the appropriate volume in the table.

5. After making each addition, snap the lids closed and place the tubes in a balanced configuration in the microcentrifuge. In centrifuges with 18-place rotors, you can balance any combination of tubes except one. Spin the tubes for 2 to 3 seconds

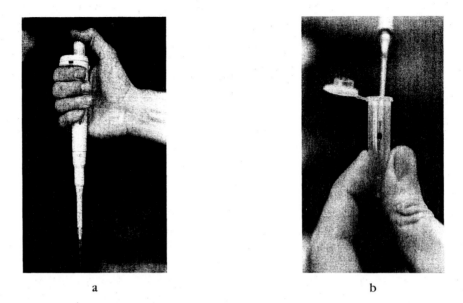

a   b

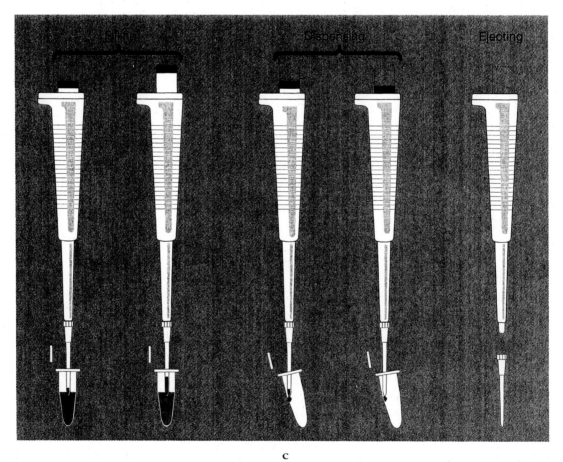

c

Figure 1.1
Proper use of a micropipetter. (a) Holding the micropipetter. (b) Transfering a small volume to the side of a microcentrifuge tube. (c) Withdrawing and transfering sample to a microcentrifuge tube (modified from Brinkman Instruments with permission).

using the pulse button (which allows spins of a few seconds) on the centrifuge. This brief centrifugation will pool and mix each addition in the bottom of the tube.

6. Remove the tubes and place them in your rack. Because you added a total of 10 μl to each tube, set the micropipetter for 10.0 μl and withdraw each tube's contents. If the tube volume exactly fills the micropipet tip, you have pipetted your reagents accurately. If you do not retrieve the 10 μl (as evidenced by air in the tip), or if liquid remains in the tube, repeat steps 2 through 5 until you recover the correct amount.

7. Use the 10–100 μl and the 100–1000 μl pipetters to add the volumes of the colored solutions as indicated in the following table to tubes C, D, E, and F. These pipetters are used in the same manner as the 0.5–10 μl pipetters. As before, become familiar with what the given volumes look like in the pipet tip. Be particularly careful to release the plunger very slowly when using the 100–1000 μl pipetter, as rapid release can squirt liquid up into the body of the pipetter. You will be setting up reactions similar to these when digesting large amounts of DNA or when doing bacterial transformations or plasmid mini-preps.

Tube	10X Buffer	DNA	Reagent A	H$_2$O	Enzyme
C	—	25 μl	50 μl	15 μl	10 μl
D	10 μl	60 μl	—	20 μl	10 μl
E	100 μl	250 μl	500 μl	150 μl	—
F	100 μl	200 μl	600 μl	100 μl	—

8. Mix each tube briefly on a vortex mixer, and pulse for 2 to 3 seconds in a microcentrifuge. Because a total of 100 μl was added to tubes C and D, set the 10–100 μl pipetter to 100 μl and remove the contents of each tube to check your pipetting accuracy. Repeat the procedure if you do not recover the correct amount.

9. Set the 100–1000 μl pipetter to 1000 μl and remove the contents of tubes E and F to check your pipetting accuracy. Remember to release the plunger very slowly. Repeat the procedure if you do not recover the correct amount.

Practice with the micropipetters until you are comfortable with them. We will be using them extensively throughout the course and it is essential that you be proficient at using all sizes of micropipetters.

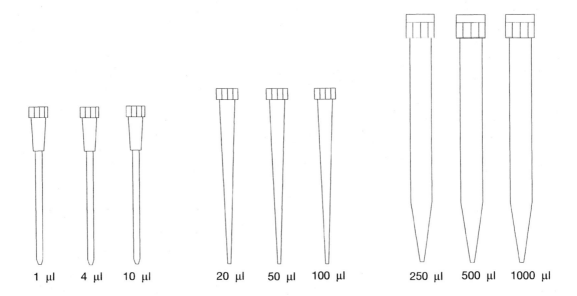

1 µl   4 µl   10 µl   20 µl   50 µl   100 µl   250 µl   500 µl   1000 µl

Results and Discussion

1. Were you able to use the micropipetters accurately and gain familiarity with what different volumes look like in a tip? Draw a line to indicate the liquid height in the tips above.

2. What errors could result in pipetting volumes greater than the desired volume?

3. What errors could result in pipetting volumes smaller than the desired volume?

4. Micropipetters are not 100% accurate; accuracy and precision data are generally provided by the supplier. For example, a typical 0.5-10 µl micropipetter may be accurate to ±0.1 µl. What is the percent error if you are pipetting 10.0 µl? What is the percent error if you are pipetting 0.5 µl?

    % error at 10 µl: _____   % error at 0.5 µl: _____

5. Knowing the required amount of accuracy in various procedures is important in virtually all types of lab work. In some cases, high accuracy (<1% margin of error) may be required, while in other cases a 10 or 20% margin of error may be acceptable. What degree of accuracy do you think is required for setting up reactions where DNA molecules, enzymes, and buffers are added together? For example, will a DNA (substrate) or enzyme concentration off by 2, 10, or 50% affect the outcome of the reaction? If the buffer concentration or pH is off by 2, 10, or 50%, will it affect the reaction? Your instructor will provide some guidance in answering this question.

PART II. BACTERIOLOGICAL TECHNIQUES

Cultures

- *Photobacterium phosphorium* broth culture
- *Vibrio fischeri* MJ1 broth culture
- *E. coli* DH5α broth culture
- *E. coli* DH5α (pGEM™-3Zf[+]) broth culture
- photobacterium plate streaked with *Photobacterium phosphorium*
- GVM plate streaked with *Vibrio fischeri* MJ1
- LB plate streaked with *E. coli* DH5α
- LB + 50 μg/ml ampicillin plate streaked with *E. coli* DH5α (pGEM™-3Zf[+])

Media

- sterile photobacterium broth in 25 × 150-mm tubes (1 per group)
- sterile GVM broth in 25 × 150-mm tubes (1 per group)
- sterile LB broth in 25 × 150-mm tubes (1 per group)
- sterile LB broth + 50 μg/ml ampicillin in 25 × 150-mm tubes (1 per group)
- sterile photobacterium agar plates (1 per group)
- sterile GVM agar plates (1 per group)
- sterile LB agar plates (5 per group)
- sterile LB agar plates + 50 μg/ml ampicillin (LB/Ap^{50}; 2 per group)
- sterile agarose stab vials (0.1X LB in 0.3% agarose) (4 per group)
- sterile agarose stab vials (1X GVM in 0.3% agarose) (2 per group)
- sterile agarose stab vials (1X photobacterium broth in 0.3% agarose) (2 per group)

Supplies and Equipment

- adhesive dot labels
- inoculating loop
- inoculating needle
- sterile toothpicks
- Bunsen burner
- sterile 9.9-ml dilution blanks
- sterile 9.0-ml dilution blanks
- sterile 1.0-ml pipets
- sterile 10-ml pipets
- vortex mixer
- spreading triangle
- alcohol jar
- spreading turntable
- 37°C incubator
- 37°C shaking water bath

➡ Procedure

Note: Disinfect your bench before and after working with bacteria.

A. Streaking Agar Plates

1. Label one of the LB plates and one of the LB/Ap^{50} plates "DH5α." Label a second set of plates "DH5α (pGEM™-3Zf[+])."

2. Label the GVM plate *"V. fischeri"* and the photobacterium plate *"P. phosphorium."*

3. Sterilize your inoculating loop in the Bunsen burner flame until red hot (see Figure 1.2). Adjust the Bunsen burner until it has an internal blue cone and hold the loop just above the apex of the cone—this is the hottest spot in the flame. Allow to cool 10 seconds. Do not set the sterile loop down!

Figure 1.2
Flame-sterilization of an inoculating loop in a Bunsen burner. Note that the loop is **above** the apex of the internal blue cone, which is the hottest part of the flame.

4. Lift the lid of the LB plate streaked with *E. coli* DH5α and touch the loop to an area of the plate without any growth to be sure the loop is cool. Then select a well-isolated colony and touch the colony with the loop [Figure 1.3(a)]. Replace the lid.

5. Raise the lid of the sterile LB plate labeled "DH5α" and streak the loop across approximately one-quarter of the plate as demonstrated [see Figure 1.3(b)]. Hold the loop loosely in your hand like a pencil and lightly flick the loop across the surface of the agar. You should streak rapidly rather than drawing on the plate. This prevents gouging the agar and allows streak lines to be closer together. Hold the lid just above the plate while streaking to minimize airborne contaminants.

6. Replace the lid on the Petri dish and resterilize your loop in the Bunsen burner. At an angle to the original streaks, streak the sterile loop five or six times, across the original streaks into a second quarter of the plate [see Figure 1.3(c)]. Continue to streak the second quarter **without** passing the loop into the previous streaks.

7. Repeat step 6 with a third and fourth quarter of the plate and replace the lid. Your streaks should cover virtually all the plate's area.

8. In a similar manner, streak *E. coli* DH5α onto the LB/Ap[50] plate and *E. coli* DH5α (pGEM™-3Zf[+]) onto an LB plate and an LB/Ap[50] plate. Streak *P. phosphorium* on the photobacterium plate and *V. fischeri* on the GVM plate. If you wish, you may streak from the broth cultures rather than the plate [Figure 1.4(a)]. Incubate the *E. coli* strains at 37°C and the luminescent strains at room temperature in your drawer. All plates should be incubated in an inverted position.

B. Inoculation of Broth Cultures

1. Flame-sterilize your inoculating loop and allow it to cool. Pick up the broth culture of *E. coli* DH5α and remove the cap using the hand that is holding the loop

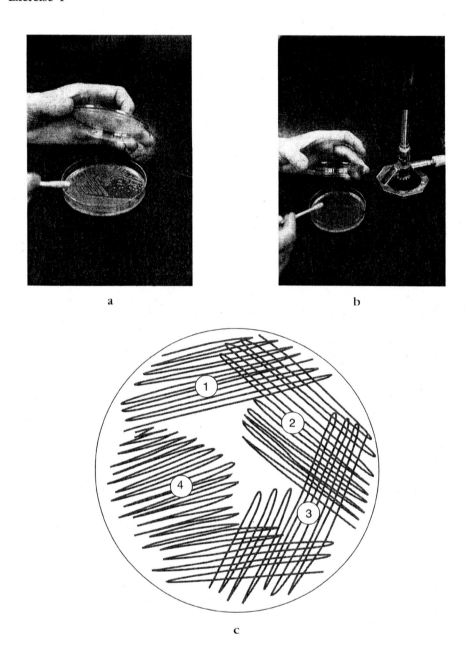

Figure 1.3
Procedure for streaking plates to obtain isolated colonies. (a) Picking an isolated colony with a sterile inoculating loop. (b) Streaking a sterile agar plate to spread out the organisms.
(c) Pattern of a properly streaked plate.

[see Figure 1.4(a)]. Put the loop into the broth, remove, and then replace the cap on the tube.

2. Pick up a tube of sterile LB broth, remove the lid with the hand holding the loop, and dip the loop with the cells into the sterile broth [see Figure 1.4(b)]. Remove the loop and resterilize it in the flame. When resterilizing, heat the wire closest to the handle first and slowly heat toward the loop. This eliminates "spattering" of live culture and the possibility of contaminating your working area.

a

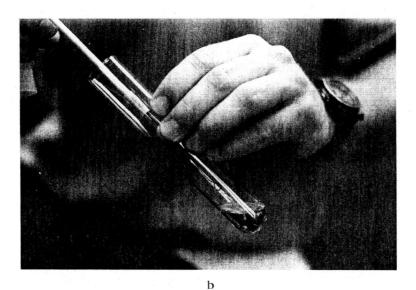

b

Figure 1.4
Inoculation of a broth culture. (a) Removing an inoculum from a broth culture. (b) Transfering a loopful to a sterile broth.

3. In a similar manner, inoculate the LB + ampicillin broth with *E. coli* DH5α (pGEM™-3Zf[+]), GVM broth with *V. fischeri,* and photobacterium broth with *P. phosphorium.* You may use either isolated colonies from a plate or a broth culture to inoculate the sterile broths. Incubate the *E. coli* strains at 37°C and the luminescent strains at room temperature. All cultures should be shaken to ensure aerobic growth.

C. Inoculation of Stab Cultures

1. Each group will inoculate a stab or slant culture with each of the strains used in this exercise. Many bacteria remain viable for long periods of time in low-nutrient soft agar stabs, and you may keep the stabs throughout the course. Be sure the *E. coli* strains are placed in the 0.1X LB stabs, *V. fischeri* go in the GVM stabs, and *P. phosphorium* go in the photobacterium stabs.

2. To inoculate a stab, remove a sterile toothpick from the beaker and touch the tip of the toothpick into a well-isolated colony on one of the plates. Open the appropriate stab vial and stab the soft agarose 4 to 5 times with the toothpick and replace the lid loosely (see Figure 1.5). You may also inoculate a stab of photobacterium agar with *P. phosphorium*. Be sure you **verify the bioluminescent phenotype** of the luminous strains you pick (i.e., be sure that they glow in the dark; nonluminous mutants are very common).

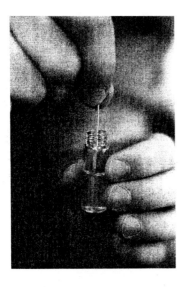

Figure 1.5
Inoculation of a stab vial with a sterile toothpick.

3. Write the strain name and date on an adhesive dot label and place it on the lid. You can write additional information on the glass vial with a fine-point marker. Place a piece of see-through tape around the vial to prevent the markings from coming off.

4. Incubate the *E. coli* strains at 37°C and the luminescent strains at room temperature. After growth is observed, tighten the lid. For long-term storage, it is also useful to stretch Parafilm™ around the lids to prevent the media from drying.

D. Bacterial Enumerations and Spread Plating

1. Using the 9.0-ml and 9.9-ml dilution blanks, prepare serial $1/10$ and $1/100$ dilutions of the *E. coli* DH5α culture, as demonstrated by your instructor. Dilute until you have 10^{-5}, 10^{-6}, and 10^{-7} dilutions (see Figure 1.6). The objective is to dilute the culture so that when an aliquot is spread on a plate, you will be able to count individual colonies. A plate with 30 to 300 colonies will allow a statistically accurate enumer-

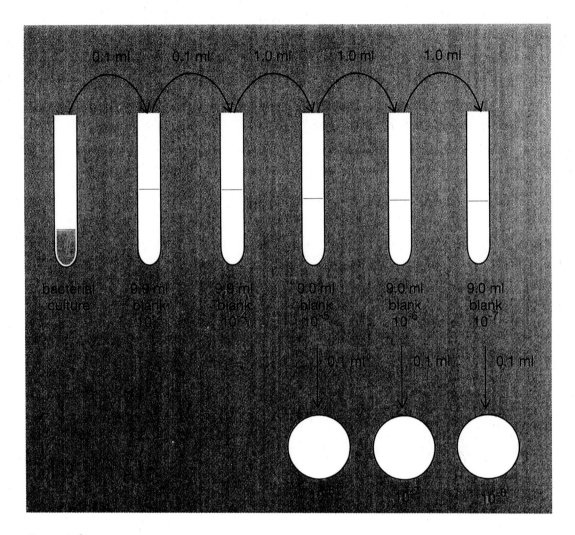

Figure 1.6
Serial dilutions of *E. coli* DH5α for plate-counting.

ation of the bacteria in the culture—such plates are referred to as "countable." This number also provides a reasonable density of colonies for selecting bacterial clones containing recombinant DNA.

Follow steps a through e below to prepare the dilutions—your instructor will demonstrate. Be sure you have labeled all tubes and plates before starting. This procedure takes dexterity and practice to do alone, but works well with two people if one person holds the tubes while the other does the pipetting.

a. Remove the lid from a can of sterile 1.0-ml pipets (or peel back the wrapper of a prepackaged sterile pipet). Touching just the mouth end of the pipet, remove it from the can while taking care not to drag it across the ends of the other pipets in the can. Do not touch the lower three-quarters of the pipet.

b. Holding the pipet 2 to 4 cm from the mouth end, insert it firmly into a pipet aid.

CAUTION: Never mouth pipet anything in the laboratory. Many of the solutions contain potential biohazards (bacteria) or toxic chemicals. Always use a pipet aid.

c. Aseptically remove the cap from the bacterial culture, insert the sterile pipet, and remove 0.1 ml. Replace the cap and set the culture tube in your rack.

d. Next, remove the cap from a sterile 9.9-ml dilution blank and transfer the 0.1 ml of culture to it. Replace the cap and set the dilution in your rack. Immediately place the contaminated pipet into a pipet discard jar containing disinfectant. **Never set a contaminated pipet on your lab bench!**

e. Vortex the dilution blank thoroughly (your instructor will demonstrate), and prepare the remaining dilution blanks in a similar manner. Be sure you use aseptic technique and that you thoroughly vortex each dilution blank before making another transfer. Use a separate sterile pipet for each dilution and place the pipet immediately in the pipet discard jar (containing disinfectant) when finished with it.

2. Label three LB plates 10^{-6}, 10^{-7}, and 10^{-8}. Using a sterile pipet, transfer 0.1 ml of each of the last three dilutions to the plates. You may do this with one pipet if you start with the most dilute tube (10^{-8}). Note that the dilution of the plate is the dilution of the blank multiplied by the volume plated.

3. To spread the diluted culture evenly on the surface of the agar plate, set the plate on the turntable. Dip the spreading triangle into the jar of alcohol, drain off any excess alcohol, and pass the triangle through the Bunsen burner to ignite it. Do not hold the triangle in the flame; just let the alcohol burn out. Repeat the flaming two times, and allow the triangle to cool for 10 to 15 seconds.

CAUTION: Alcohol is highly flammable. Use extreme caution and be sure that you are not working over your notebook or other flammable materials.

Remove the lid from the plate with one hand and give the turntable a spin with the hand holding the triangle (see Figure 1.7). Touch the triangle to the surface of the agar and move it back and forth across the surface of the plate as it rotates. When the turntable stops, replace the lid and resterilize the triangle in the alcohol. Repeat with the other plates.

Figure 1.7
The spread plate technique to obtain isolated colonies evenly distributed on the surface of an agar plate.

4. Allow the plates to set upright for 5 to10 minutes to absorb the liquid, then invert and incubate at 37°C. Be sure that you label your plates on the bottom with your group, organism, and dilution.

Results and Discussion

1. Examine all of the cultures the following day. Check your streaked plates to see if you obtained isolated colonies. You may take your plates streaked with the bioluminescent bacteria into a dark room to check for bioluminescence. Circle several bioluminescent colonies.

   Wrap Parafilm™ around the plate of *E. coli* DH5α and store it in your drawer. You will need this to prepare the competent cells in Exercise 11.

2. Indicate in the following table which of the *E. coli* strains were able to grow on the media with and without ampicillin. Explain your results.

Bacterial Strain	Growth on LB	Growth on LB + Amp
E. coli DH5α		
E. coli DH5α (pGEM™-3Zf[+])		

3. Can you see visible growth in the stabs? If so, screw the lids down tight and wrap Parafilm™ around the caps to ensure that they do not dry out. Store the stabs in your drawer.

4. Look at the three plates that you spread with the diluted *E. coli* culture. If you prepared the dilutions correctly, you should have roughly tenfold fewer colonies on each plate as you go to the higher dilutions. Is this the case with your plates? If not, suggest an explanation.

5. Do you have a countable plate (30–300 colonies)? Count all the colonies on this plate and calculate the number of bacteria per ml of culture from the following formula:

$$\text{Bacteria/ml} = \frac{\text{number of colonies per plate}}{\text{dilution of plate counted}} = \underline{\hspace{2cm}}$$

Is the number of organisms present per ml surprising? What are some advantages of working with an organism which can produce such large numbers during an overnight incubation?

EXERCISE 2
Preparation of Media and Reagents Used in Molecular Biology

Methods in molecular biology involve the use of living organisms (generally bacteria) and a wide variety of reagents, enzymes, and DNA molecules. The bacteria used in recombinant DNA techniques must be cultivated on media that allows them to grow optimally. Most bacteriological media contain rich sources of nutrients such as tryptones or peptones (which are rich sources of amino acids), yeast extract (which is a rich source of vitamins and other nutrients), and various salts. Because many of the bacterial strains used in molecular biology harbor plasmids that confer resistance to one or more antibiotics, antibiotics are commonly added to media. Similarly, other additives are often added to media so that various phenotypic changes in the organisms can be visualized by color reactions. For example, when a gene is cloned into an appropriate vector (for example, pGEM™), bacteria that harbor a fragment of foreign DNA can be distinguished from those that do not by adding a chromogenic substrate (X-gal) to the media. This results in the formation of blue colonies by bacteria lacking foreign DNA, while those with foreign DNA form white colonies.

Bacteriological media are generally prepared in either a liquid form (broth) or in a solid form (agars). Agar is a product of a marine alga that is a nongrowth substrate for most bacteria, does not melt until it reaches 100°C, and, once molten, does not solidify until about 44°C. Agar plates allow the formation of isolated colonies of bacteria (see Exercise 1), while liquid media allow the growth of large amounts of cells for procedures such as DNA isolations. Prior to use, all media are sterilized, generally by autoclaving. An autoclave is essentially a large pressure cooker that heats the media to a temperature of 121°C by injecting steam under pressure into the autoclave chamber. The pressure prevents the media from boiling at the elevated temperature. An alternate method of sterilizing media and other liquids is to filter it through a 0.2-μm membrane filter. This is tedious, but some media ingredients (such as antibiotics) will break down under autoclave temperatures and are added to media from filter-sterilized stocks after autoclaving.

Molecular biology procedures also require a wide variety of chemical reagents. These contain a myriad of components, but common features of all molecular biology reagents are that they are (a) made with highly purified water (deionized or double-distilled) and (b) generally autoclaved to destroy any DNases present. High-purity water is important as metal ions can interfere with many of the enzymatic reactions involving DNA. The need to eliminate DNase activity in molecular biology reagents is of obvious importance for any work with DNA.

Throughout most of this course, you will be provided with required media and reagents. However, to give you some experience in preparing these materials, you will make both solid and liquid media, and some of the reagents used in these exercises.

Media

☐ LB broth powder (or separate bottles of NaCl, Tryptone, and yeast extract)

☐ Bacto™ agar

Reagents

☐ ampicillin (50 mg/ml; filter sterilized)
☐ X-gal (5-bromo-4-chloro-3-indolyl-β-D-galactoside; 40 mg/ml in N-N-dimethylformamide)
☐ 4 N NaOH

☐ molecular biology–grade chemicals for making reagents
☐ deionized or glass-distilled water

Supplies and Equipment

☐ top loading balance
☐ pH paper
☐ 100- and 1000-ml graduated cylinders
☐ 0.5–10 µl micropipetter
☐ 100–1000 µl micropipetter
☐ sterile 10-µl micropipet tips
☐ sterile 1000-µl micropipet tips
☐ 1-liter beaker
☐ 1-liter Erlenmeyer flasks

☐ 250-ml shake flasks with caps (Bellco™ or equivalent)
☐ 25 × 150-mm culture tubes with caps
☐ 2-inch stir bars
☐ aluminum foil
☐ autoclave tape
☐ stir plates
☐ sterile disposable 100 × 15-mm Petri plates (20/group)

▶ Procedure (work in groups of two)

A. Media Preparation

Each group will make 600 ml of LB (Luria-Bertani) broth, a common rich medium used for growing *Escherichia coli* strains used in molecular biology. You will add agar to 500 ml of your LB broth for preparation of agar plates and use the remainder for broths. Antibiotics will be added to some of the media for use with strains containing antibiotic resistance genes and X-gal will be added to the plates used for screening clones. Refer to Appendix XV for hints on media preparation.

1. LB broth is made by adding 25 g of LB powder per liter (refer to Appendix XV if you are making LB from individual ingredients). Calculate the amount of LB needed to prepare 600 ml of broth. Check your calculation with your instructor before proceeding.

_____ g LB/600 ml

2. Measure 600 ml of deionized water and pour it into a 1-liter beaker containing a stir bar. Place on a stir plate.

3. Slowly add the weighed amount of LB powder to the beaker with slow stirring. Stir until completely dissolved. If lumps form, these may be squashed with a stirring rod.

4. Adjust the pH to 7.4 (±0.1) with 4 N NaOH (about 1 ml per liter). Check the pH with a pH meter or pH paper.

5. Dispense 50 ml of the broth into a 250-ml shake flask and place the cap on top. Place a piece of autoclave tape and a piece of labeling tape on the flask. Label it "LB broth" and your group number. You will use this to grow *E. coli* in Exercise 11 to make competent cells.

6. Dispense 5 ml of broth into each of six 25 × 150-mm tubes and cover each with a cap. Put your tubes in a rack and place a piece of autoclave tape and a piece of labeling tape on the rack. Label the rack with "LB/Ap^{50} broth" and with your group number. The Ap^{50} refers to the concentration of ampicillin in μg/ml. **Do not add the antibiotic yet!** Place a second colored piece of tape on the rack labeled "Needs ampicillin." You will add the antibiotic to these tubes after autoclaving and use this medium to grow bacteria for small-scale plasmid DNA isolations.

7. Pour 500 ml of the LB broth into a 1-liter Erlenmeyer flask to make the agar medium. Place a stir bar in the flask. Agar is added to media at a concentration of 1.5% (w/v). Calculate the amount of agar needed to make 500 ml. Check your calculation with your instructor before proceeding.

_____ g agar/500 ml

8. Weigh the required amount of agar and add it to the flask of LB broth. The agar will sink to the bottom and will not dissolve until heated in the autoclave. Cover the flask with aluminum foil so that the foil folds down at least 1 inch over the edge of the flask. Place a piece of autoclave tape and labeling tape on the flask and label it "LB/Ap^{50}/X-gal" and add your group number. Place a second colored piece of tape on the flask labeled "Needs ampicillin and X-gal." Double-check to see that your agar flask has a stir bar in it.

9. When all groups have completed their media, place your flasks and tubes in an autoclave tray. Your instructor will demonstrate the use of the autoclave and sterilize your media by autoclaving for 15 minutes at 121°C. Note that a slow exhaust cycle is used when autoclaving liquid media. This allows the pressure to be slowly released from the chamber and results in a slow decrease in the media temperature. Why is this necessary? What would happen to liquid media at 121°C if the pressure was released rapidly?

10. After autoclaving, remove the broths and set them at your bench.

CAUTION: Media coming out of an autoclave is extremely hot and may be super-heated. Wear insulated gloves and eye protection and use caution when removing hot materials and swirling flasks.

Gently swirl the flask of LB agar to evenly mix the agar. Because the agar was a layer of crystals on the bottom of the flask when you placed it in the autoclave, the molten agar will still be layered at the bottom. Place the flasks of agar in a 55°C water bath for at least 30 minutes. This will prevent the media from solidifying but will cool it sufficiently to allow the addition of the antibiotics and X-gal without thermal denaturation.

11. Ampicillin will be added to the broth tubes to a final concentration of 50 µg/ml. Note the concentration of the ampicillin stock and calculate the volume of ampicillin to add to each tube containing 5 ml of medium. Check your calculation with your instructor before proceeding.

_____ µl ampicillin/5 ml

Aseptically add the required amount of ampicillin to each tube, with a micropipetter and a sterile tip. Place the antibiotic on the side of the tube, being careful not to touch the tube with the pipetter. Vortex the tube to mix the antibiotic into the medium.

12. Ampicillin will be added to the LB agar to give a final concentration of 50 µg/ml, and the X-gal will be added to give a final concentration of 40 µg/ml. Note the concentration of the ampicillin and X-gal stocks and calculate the volume of each required for 500 ml of LB agar. Check your calculations with your instructor before proceeding.

_____ µl ampicillin/500 ml   _____ µl X-gal/500 ml

13. While you are waiting for your flask of agar to cool, cut open the end of a bag of sterile Petri plates and remove the plates. Save the bag—you will store your plates in this after you have poured media into them. Be careful not to remove the lids from the plates. Label the plates on the bottom (along the edge with a fine-point permanent marker) with your group number and "LB/Ap⁵⁰/X-gal."

14. When your flask of LB agar has cooled to 55°C, remove it from the water bath and place it on the stir plate with slow stirring (avoid the formation of bubbles). Aseptically add the required amounts of ampicillin and X-gal with a micropipetter and sterile tip. Allow the additions to run down the side of the warm flask so they warm up prior to reaching the medium. Allow the medium to stir for one minute to ensure that the additions are completely mixed.

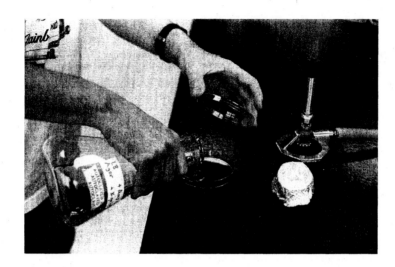

Figure 2.1
Pouring agar into Petri plates. Note holding the lid of the sterile Petri plate over the bottom to minimize airborne contaminants.

15. Pour the molten agar into the Petri dishes as demonstrated by the instructor (see Figure 2.1). Each plate should contain 20-25 ml of medium. Allow the plates to solidify (for about 15 minutes), and then invert.

16. Remove one plate from your batch and dry for 10 minutes in a 60-80°C drying oven. You will use this plate to be sure the antibiotic and X-gal were added properly. Divide the plate into thirds, and streak each portion with one of the following strains:

☐ *E. coli* DH5α   ampicillin sensitive
☐ *E. coli* DH5α (pGEM™-3Zf[+])   ampicillin resistant (blue)
☐ *E. coli* DH5α (pUWL501)   ampicillin resistant (white)

17. Dry the remaining plates overnight at room temperature. The following day, check your test plate and place the remaining plates in the bag (inverted). Tape the bag shut and label the bag with your group number, today's date, and "LB/Ap[50]/X-gal." Store at 4°C until needed.

B. Reagent Preparation

Your instructor will assign your group a specific stock reagent to make. Follow the recipes in Appendix XV. These will be used to prepare the various buffers and reagents used in the course.

EXERCISE 3
Isolation of Luminescent Bacteria from Natural Sources

Biological luminescence is relatively common in nature and is carried out by a few bacteria, fungi, protozoa, and higher animals. With one exception, luminescent bacteria are all marine organisms. They may be found free living in seawater, on the surface of marine animals, or living as symbionts in specialized organs in marine fish or squid. In addition to the luminescent marine bacteria, one luminescent soil bacterium, *Photorhabdus luminescens,* has recently been described.

The marine luminescent bacteria are found in the genera *Photobacterium, Shewanella,* and *Vibrio.* They are all gram-negative, motile rods that possess a fermentative metabolism. Because they are widespread in seawater and associated with marine animals, they are relatively easy to isolate. In this course, we will be working with the bacterium *Vibrio fischeri* MJ1, which was isolated from a marine fish (*Monocentris japonicus*) harboring the bacterium as a symbiont in a light organ (see figure in Acknowledgments). However, many other species of luminescent bacteria exist and can readily be isolated from marine samples (Nealson and Hastings 1992).

In this exercise, we will attempt to isolate diverse luminescent species. We will attempt to isolate these illuminating bacteria from the surface of fresh fish, squid, or shrimp. In addition, we will also attempt to isolate luminescent bacteria directly from seawater. There is no selective enrichment for these organisms. However, because of their high numbers in the ocean and on marine animals, and because of their easy identification when colonies are formed, they are relatively easy to isolate if fresh seafood is available.

Media and Reagents

- ☐ fresh shrimp, fish, or squid
- ☐ fresh seawater samples
- ☐ 3% NaCl solution
- ☐ photobacterium agar plates

- ☐ GVM agar plates
- ☐ photobacterium agar stab vials
- ☐ GVM agar stab vials

Supplies and Equipment

- ☐ Petri dishes
- ☐ sterile toothpicks

- ☐ dark room
- ☐ nose plugs

▶Procedure (work individually)

A. Isolation from Fresh Seafood

1. Place a shrimp, squid, or piece of fish into each half of a Petri dish. Flood the dish with 3% NaCl until the seafood samples are about two-thirds covered. Be sure that some of the specimen is above the level of the saltwater.

2. Incubate one dish at room temperature (less than 28°C) and one at 15 to 18°C. Many of the bioluminescent bacteria are not viable or will not glow at temperatures exceeding 28°C.

3. Examine your samples in a dark room several times a day for 1 to 2 days. They will not smell particularly pleasant. Allow your eyes to dark adapt for 3 to 5 minutes and check carefully for luminescent patches in your Petri dish. If you do not observe luminescent patches within two days, discard your plate; further incubation is futile (and will cause serious deterioration of lab air quality!).

4. If you observe luminescence, take a sterile toothpick and hold it over the luminescent spot. Have your partner open a door so that enough light enters to see the shrimp, fish, or squid, and touch the luminescent spot. Turn on the light and rub the toothpick on the surface of a photobacterium plate and then on the surface of a GVM plate. Use a sterile inoculating loop to streak each plate for isolation as described in Exercise 1.

5. Incubate your plates at room temperature for 20 to 24 hours and examine for luminescent colonies. Reexamine several times for up to 2 days if you are not initially successful. Restreak any bioluminescent colonies to be sure that you have a pure culture. You may have more than one bioluminescent colony morphology, indicating the presence of different bioluminescent species or strains.

6. Give each of your isolates a strain designation and record the colony morphology (size, shape, color, transparency, and so on) of each. Bacterial strains are usually named with a combination of letters and numbers. The letters may be your initials or initials of anyone or anything of interest (for example your spouse, pet, college or university, and so on). Use numbers to distinguish the different strains you isolate.

7. When you have each strain in pure culture, inoculate them into a stab vial of the appropriate medium for long-term storage (see Exercise 1). Incubate at room temperature with the lids loosened for 1 to 2 days, then tighten the lids. These strains may be used to amplify the *lux*A gene by the polymerase chain reaction (PCR; see Exercise 23).

Note: Viability of bioluminescent bacteria in stabs is often poor. It is a good idea to restreak and transfer brightly glowing colonies to fresh media every 4 to 6 weeks. Alternately, you may freeze the cultures at −70°C in 10% glycerol (see Exercise 11).

B. Isolation from Seawater

1. Spread 0.1 ml of fresh seawater directly on two plates of photobacterium agar as described in Exercise 1. Repeat with two plates of GVM agar.

2. Incubate one plate of each medium type at room temperature and one of each at 15 to 18°C.

3. Examine your plates in the dark (allow your eyes to dark adapt for 3 to 5 minutes) several times a day for 1 to 2 days. Check carefully for luminescent colonies.

4. If you observe luminescent colonies, pick them with a toothpick as described in Part A and streak onto the appropriate media.

5. Incubate your plates at room temperature for 1 to 2 days and examine them for luminescent colonies. Restreak these to be sure that you have a pure culture. Give each isolate a strain designation, describe colony morphology, and inoculate into stab vials.

Results and Discussion

1. Describe each of your bioluminescent isolates in the following table (indicate light intensity as + for faint, ++ for moderate brightness, and +++ for very bright):

Strain	Source	Colony morphology/comments	Light intensity

2. Was your shrimp or seawater procedure more successful in producing luminescent bacteria? What is the role of the bioluminescent bacteria in each habitat?

Literature Cited

Nealson, K. H., and J. W. Hastings. 1992. The luminous bacteria. In *The prokaryotes.* A. Balows, H. G. Truper, M. Dworkin, W. Harder, and K. H. Schleifer, 625–639. New York: Springer-Verlag.

EXERCISE 4
Restriction Digestion and Agarose Gel Electrophoresis of DNA

Restriction endonucleases (or restriction enzymes) are some of the most powerful tools in modern molecular biology. These enzymes, produced by various strains of bacteria, recognize specific sequences in DNA molecules and then cleave the phosphodiester bonds between the nucleotides. Type II restriction enzymes are most widely used because they recognize and cut within specific DNA sequences. This results in the generation of DNA fragments (**restriction fragments**) from a DNA molecule having these sequences. Thus, digestion of a specific DNA molecule with a given enzyme will always result in identical restriction fragments. Digestion of different DNA molecules with the same enzyme will yield a different set of restriction fragments. Such **restriction analyses** can, therefore, be used as a **fingerprint** to characterize any DNA molecule.

Type II restriction enzymes usually recognize palindromic or symmetric sequences in DNA (see Figure 4.1). For example, the widely used restriction enzyme *Eco*R I recognizes the sequence GAATTC. This is considered a palindrome because the complementary strand will have the identical sequence in the opposite direction. Restriction enzymes may cut anywhere within this sequence, but a given enzyme will always cut between the same two nucleotides. Some (such as *Hinc* II or *Eco*R V) cut in the center of the recognition sequence, which results in the formation of blunt ends in DNA after both strands are cut. Most type II restriction enzymes (such as *Eco*R I) make cuts one or two bases away from the axis of symmetry, which results in the formation of single-stranded ends two or four base pairs long on the end of each fragment. Because these ends are complementary, they are referred to as "cohesive" or "sticky" ends. Because a given restriction enzyme produces identical ends in any DNA molecule, restriction fragments from one source can be recombined with DNA from other sources if they are both cut with the same enzyme. This characteristic allows the formation of **recombinant DNA molecules** and is the basis of most DNA cloning protocols. Some restriction enzymes, such as *Eco*R V, result in blunt ends. These ends do not stick together but are useful in cloning as they can be attached to other blunt ends.

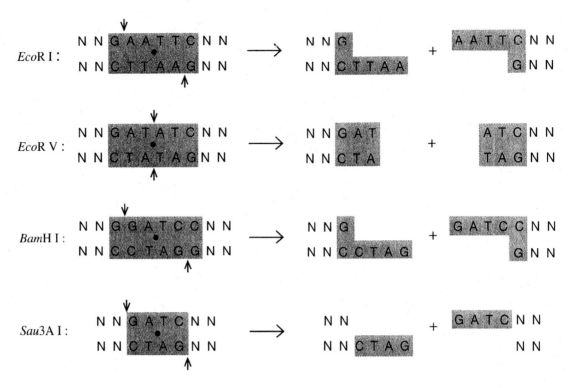

Figure 4.1
Various activities catalyzed by type II restriction endonucleases. The recognition sequence is indicated in color, and arrows indicate the site of cleavage. The • indicates the axis of symmetry. Although most enzymes result in sticky ends, some, such as *Eco*R V, cut at the axis of symmetry, resulting in blunt ends. Note that *Bam*H I and *Sau*3A I recognize different sequences (6 and 4 bases, respectively) but that the resultant fragments have compatible ends and will stick to each other. There are numerous combinations of enzymes that produce compatible ends and these are useful in cloning.

To date, more than 2500 restriction enzymes have been isolated and more than 200 of these are commercially available, offering a wide variety of options for cutting DNA molecules at different sites. Restriction digestions are done by adding the DNA to be digested and an enzyme to a buffer that is optimal for that particular enzyme. Commercial enzyme manufacturers supply the appropriate buffer with each enzyme. Optimal buffer conditions vary widely for different enzymes. Sometimes it is necessary to use two or more enzymes in the same digestion. This can usually be accomplished with a compromise buffer that allows sufficient activity from each enzyme. An excellent discussion of restriction endonucleases and their applications is given by Sambrook, et al. (1989).

In order to characterize DNA restriction fragments, verify that restriction digestions have worked, or to purify restriction fragments, it is necessary to separate and identify the various sized fragments. Originally, this could only be done by ultracentrifugation, which was time-consuming and could only determine relative sizes. In the early 1970s, DNA fragments were successfully separated by polyacrylamide gel electrophoresis. Because the phosphate residues in DNA confer a net negative charge to the molecule, DNA will migrate toward the positive electrode in an electric field. Although more effective than centrifugation, this procedure was still time-consuming, involved toxic reagents, and could not resolve large fragments. Several years later, agarose gels were used to separate DNA restriction fragments. This offered several advantages over previous techniques. Agarose gels are simple to prepare and run and can separate DNA molecules ranging in size from several hundred base pairs to more than 50,000 base pairs. In addition, it is easy to visualize the DNA by a brief staining with the fluorescent dye, ethidium bromide. Ethidium bromide is a planar molecule that intercalates into DNA molecules. Thus, it is concentrated in the DNA fragments, which will fluoresce when illuminated with UV light. The binding of ethidium bromide is proportional to the mass of the DNA fragment. Therefore, large restriction fragments will fluoresce more brightly than smaller fragments resulting from the same digest. This allows one to quantify DNA in ethidium bromide gels.

Agarose is a highly purified form of agar, a polymer extracted from brown algae. As a DNA molecule passes through an agarose gel in an electrical field, its migration is inversely proportional to the log of its size. Thus, small fragments will migrate faster through the gel than larger fragments. By loading appropriate DNA size standards in the gel, it is possible to accurately determine the size of sample DNA restriction fragments after staining and photographing the gel. The concentration of agarose in the gel may be varied to allow resolution of differing sizes of DNA fragments (see Table 4.1). Agarose gel electrophoresis has become one of the most widely used techniques in molecular biology. It is essential to cloning procedures, purifying restriction fragments, restriction mapping, Southern blotting, DNA fingerprinting, and restriction fragment length polymorphism (RFLP) analysis.

TABLE 4.1 Effect of agarose concentration on the size range of linear DNA molecules efficiently separated in agarose gels.

Percent agarose in gel (w/v)	Size range of DNA molecules efficiently separated (kb)
0.6	1.0–20
0.8	0.6–12
1.0	0.4–8
1.2	0.3–6
1.5	0.2–4

In this exercise, you will digest plasmid and bacteriophage lambda DNA with restriction enzymes and separate the resulting restriction fragments by agarose gel electrophoresis. After staining and photographing the gel, you will examine the specificity of different restriction enzymes and calculate the size of the restriction fragments formed. You will also prepare DNA standards for future gels that you will run. Both restriction digestions and agarose gel electrophoresis are common techniques in molecular biology and will be used repeatedly throughout the course. Further information on these techniques may be found in Appendix IV (Agarose and Polyacrylamide Gel Electrophoresis), Appendix VIII (Care and Handling of Enzymes), and Appendix IX (Restriction Endonucleases).

Reagents

- lambda DNA, 0.2 μg/μl (on ice)
- pBR322 DNA, 0.2 μg/μl (on ice)
- *Hin*d III (on ice; 8 μl/group)
- *Eco*R I (on ice; 3 μl/group)
- *Pst* I (on ice; 3 μl/group)
- 10X restriction buffer (on ice)

- sterile deionized water (on ice)
- loading dye (room temperature)
- 1X TAE buffer
- LE agarose (low EEO)
- ethidium bromide stain (1.0 μg/ml)

Supplies and Equipment

- mini ice buckets (1/group)
- 1.5-ml microcentrifuge tubes
- freezer storage box (for microcentrifuge tubes)
- microcentrifuge tube opener (1/group)
- microcentrifuge tube rack (1/group)
- microcentrifuge tube float (1/group)
- 0.5–10 μl micropipetter
- 10–100 μl micropipetter
- sterile 10-μl micropipet tips
- sterile 100-μl micropipet tips
- micropipet tip discard beaker
- 250-ml Erlenmeyer flask (1 per 2 groups)
- 25-ml Erlenmeyer flask (1 per 2 groups)

- microwave oven
- autoclave gloves
- top loading balance
- 37°C water bath
- 60°C water bath
- microcentrifuge
- horizontal mini-gel electrophoresis chamber
- gel casting tray
- 6-tooth comb
- gel staining tray
- power source
- transilluminator
- Polaroid camera with type 667 film
- 3-cycle semi-log paper

➡Procedure

A. Restriction Endonuclease Digestion

1. Fill your mini ice bucket with crushed ice and place each of the required cold enzymes, buffers, and DNAs in it. Each group will be provided its own set of reagents and enzymes containing slightly more than the required volume.

2. Label six sterile 1.5-ml microcentrifuge tubes A through F with a permanent marker and place them in your microcentrifuge tube rack.

3. Using micropipetters and sterile pipet tips, add the following reagents to each of the tubes. Be sure to read the Reminders below the table before proceeding.

Tube	10X Buffer	Lambda	pBR322	H₂0	*Hin*d III	*Eco*R I	*Pst* I
A	2 µl	1 µl	—	17 µl	—	—	—
B	10 µl	50 µl	—	32 µl	8 µl	—	—
C	2 µl	2 µl	—	15 µl	—	1 µl	—
D	2 µl	—	1 µl	17 µl	—	—	—
E	2 µl	—	1 µl	16 µl	—	—	1 µl
F	2 µl	—	2 µl	14 µl	—	1 µl	1 µl

Reminders:

- Add reagents in the order listed in the table headings. Always add buffer, water, and DNA to tubes prior to adding the enzymes.

- Carefully read Appendix VIII on care and handling of enzymes prior to dispensing enzymes. The total volume of enzyme added should never exceed one-tenth the reaction volume or the enzyme may be inhibited by glycerol in the storage buffer.

- Touch the tip to the side of the tube when adding each reagent. Be sure you **see** the liquid transferred to the side of the tube, especially with volumes <2 µl.

- You may use the same tip to add the same reagent to different tubes, but be sure to use a new sterile tip for each reagent.

- Check off each addition in the table as you add it. Omissions will prevent the restriction digestion from working.

Tubes A and D have no added enzyme and serve as controls for any nonspecific nuclease activity. Tube B contains a large amount of lambda DNA digested with *Hin*d III as this results in a set of restriction fragments that are excellent size standards for determining sizes of unknown restriction fragments. You will save and use this digest in future gels as a size standard.

4. After adding all ingredients, cap the tubes and pool the reagents in the bottom by spinning the tubes in the microcentrifuge for 2 to 3 seconds. Be sure that the tubes are in a balanced configuration. Tap or vortex each tube to remix, and then recentrifuge to pool the digestion mix in the bottom of the tube.

5. Push each tube into the foam microcentrifuge tube holder and float in a 37°C water bath. Incubate for 20 to 30 minutes. While the restriction digestions are proceeding, prepare and cast an agarose gel (see Part B).

Note: Digests can be frozen after digestion and analyzed at a later date.

B. Agarose Gel Electrophoresis

1. Two groups will prepare one batch of agarose to pour their two gels. Label a 250-ml Erlenmeyer flask "0.8% agarose," and add 80 ml 1X TAE (Tris Acetate EDTA; see Appendix XV) buffer to it.

   This is the same buffer that the electrophoresis will be done in. Thus, the gel will have the same ionic strength as the electrophoresis buffer. Failure to prepare the agarose in the electrophoresis buffer will result in erratic separation of restriction fragments.

2. Calculate the amount of LE agarose to give a 0.8% (w/v) gel. Check your calculations with your instructor.

   _____ g agarose/80 ml = 0.8%

   Add LE agarose to the TAE buffer and swirl. It is best to let the agarose set in the buffer for 5 to 10 minutes and swirl occasionally to allow the agarose to hydrate prior to heating. This speeds up the melting of the agarose and decreases the tendency of the agarose to boil over.

CAUTION: In the next step you will be heating agarose to boiling. Agarose can become superheated and boil over in the microwave or when swirled. Wear gloves and eye protection when removing the agarose and swirling heated flasks.

3. Place an inverted 25-ml Erlenmeyer flask in the mouth of the 250-ml flask and heat in a microwave on medium heat until all the agarose has completely dissolved (agarose melts at 96°C). It is best to stop the microwave several times and swirl the flask to ensure thorough mixing.

   Heat just enough to melt the agarose, but avoid overheating. Traces of unmelted agarose appear as small lenses in solution. If not melted, they will distort the movement of DNA fragments in the gel. The small Erlenmeyer flask allows pressure to escape, but prevents significant liquid loss from the flask.

4. Allow the agarose to cool to about 60°C or place in the 60°C water bath until ready to use.

5. Seal the ends of the gel casting tray with tape or with the end gates. Press the tape tightly against the tray with your fingernail. If trays with end gates are used, do not overtighten the set screws and seal the ends of the tray with a piece of tape. Set the tray on a level portion of the bench.

   Benches that are slightly off-level will have minimal effect on mini gels, but it is important to pour larger gels on a level benchtop to prevent significant variations in the thickness of the gel. This will distort the rate that DNA molecules migrate through the gel.

6. Pour the agarose into the casting tray to a depth of 4 to 5 mm. Note that the meniscus of the agarose is about 1 to 2 mm above the level of agarose in the tray. Place a **6-tooth comb** in the appropriate slots in the casting tray. These are set so that the bottom of the comb teeth will be about 0.5 mm above the bottom of the tray, resulting in a thin agarose bottom on each well when the comb is removed.

7. Allow the gel to solidify on your bench (for about 10 minutes). The gel will become translucent when solidified.

8. Pour 1X TAE into the horizontal electrophoresis chamber to a level of about 0.5 cm above the center tray support. If the 1X TAE buffer is left in the box from a previous run, you may reuse it, but mix the buffer by rocking the chamber back and forth to equilibrate the ions that carry the current during electrophoresis.

9. Pour a small amount of 1X TAE around the comb and carefully remove the comb. Remove the tape from the ends of the casting tray (or lower the gates), exposing the ends of the gel. If using trays with gates, tighten the screws gently to hold the gates down—the electrical current will have a very difficult time trying to go through the plexiglass if they slide up!

10. Place the gel tray in the electrophoresis chamber with the wells toward the negative (black) electrode. Check the level of the electrophoresis buffer. The buffer should just cover the gel so that the dimples from the wells disappear. If there is a large amount of buffer over the gel, less current will pass through the gel, resulting in longer electrophoresis time.

11. After the restriction digestion has incubated for at least 20 to 30 minutes, remove the tubes from the water bath and place them in your microcentrifuge tube rack. Transfer 5 µl of the *Hind* III digest of lambda (tube B) to a clean microcentrifuge tube labeled B' and add 5 µl of water. Place the remainder of the lambda digest (tube B) **back** in the 37°C water bath and digest for an additional 30 to 60 minutes. You will use this as a size standard for future experiments.

12. Add 2 µl of loading dye to tubes A, B', C, D, E, and F. **Do not** add loading dye to tube B in the water bath. Pulse each tube for 2 to 3 seconds in the microcentrifuge to mix the dye with the restriction digestion.

    The loading dye contains two tracking dyes (bromphenol blue and xylene cyanol) that migrate at different rates and a high concentration of sucrose, which makes the sample sink to the bottom of the well when loaded into the gel (see Appendix XV for the complete recipe of the loading dye).

13. For the tubes containing lambda DNA, float the microcentrifuge tubes in a 60 to 65°C water bath for 2 to 3 minutes.

    Lambda DNA is a linear molecule 48.5 kb long that has complementary 12-base single-stranded ends (referred to as the *cos* sites). These can anneal with each other to form a circular molecule. Thus, the two restriction fragments from each end of a linear lambda DNA molecule may anneal to form an additional DNA fragment with a size equal to the sum of the two restriction fragments. This will cause a smaller amount of the two individual fragments to be seen. You may wish to omit this step and compare your gel with another group that heated their lambda digests.

14. Use the micropipetter to load 10 µl of the digestion into the wells (see Figure 4.2). You may add 20 µl if your wells are large enough. By convention, gels are viewed with the wells on top and read left to right. Therefore, you should load sample A in the well closest to you, and so on.

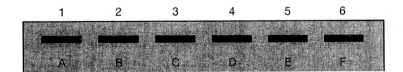

The wells will be clearly visible over a black lab bench top. If you have lightly colored bench tops, set the gel box over a piece of black paper. When loading wells, rest one hand on the gel box and steady the pipetter with both hands. Lower the pipet tip just below the top of a well and slowly dispense the sample. The dense dye solution will make the sample sink to the bottom of the well. Be careful not to lower the pipet tip too far as you might puncture the bottom of the well and your sample will leak out of the gel.

Figure 4.2
Loading samples into an agarose gel.
Note the use of two hands to steady
the pipetter.

15. Place the lid on the electrophoresis chamber and connect the leads to the appropriate jack in the power supply. Check to be sure that the wells are next to the negative (black) electrode so that the DNA will migrate toward the positive (red) electrode.

16. Make sure that the current knob is turned down, and then turn on the power supply. Set the voltage to 100 V and electrophorese until the bromphenol blue (the leading, darker dye) has migrated about three-quarters of the way to the end of the gel. Do not allow the bromphenol blue to run off the end. This dye corresponds to a DNA fragment approximately 300–400 base pairs (bp) long.

17. When the electrophoresis is complete, turn the voltage down to the lowest level and switch the power source off. Remove the lid containing the power leads so that the gel can be safely removed.

C. Staining and Photodocumentation of Agarose Gel

1. Label a staining tray with your group number. Remove the tray supporting the gel from the electrophoresis box and slide the gel into the staining tray [see Figure 4.3(a)]. Be careful as the gels are slippery and can easily slide off the tray.

CAUTION: In the next step, you will be working with ethidium bromide, which is a mutagen and suspected carcinogen. Wear gloves, eye protection, and a lab coat, and work over absorbent plastic-backed paper at all times when handling it. Carefully read precautions for handling and disposing of ethidium bromide (Appendix XII) before using ethidium bromide stain.

2. At the staining station, flood the gel with the ethidium bromide solution (1 μg/ml) and allow to stain for 5 to 10 minutes [see Figure 4.3 (b)]. It is a good policy for one partner in each group to wear gloves for staining with ethidium bromide, while the other does not wear gloves and operates the camera, opens doors, and so on.

3. Pour the ethidium bromide solution back into the storage bottle using the funnel provided [see Figure 4.3(c)]. The staining solution can be reused 10 to 12 times. Rinse the gel several times in tap or distilled water. Gels are often destained for 5 to 10 minutes in distilled water to reduce background fluorescence, but this is usually unnecessary unless very low amounts of DNA need to be visualized.

4. Use a plastic kitchen spatula to carefully lift the gel from the tray and slide it onto the transilluminator at the photodocumentation station [Figure 4.3(d)]. Be careful to avoid trapping bubbles under the gel. The transilluminator's filter glass is very expensive and can be damaged if scratched. This may be minimized by covering the glass with a sheet of plastic wrap and laying the gel on the plastic. Some gel trays are UV transparent and may be placed directly on the transilluminator with the gel in them.

CAUTION: UV light is hazardous and can damage your eyes. Never look at an unshielded UV light. Always view through a UV blocking shield or wear UV blocking glasses. Some transilluminators have a switch on the lid so that the UV light switches off when the lid is opened, but many transilluminators do not have this safety feature.

5. Lower the UV blocking lid on the transilluminator and briefly observe the DNA fragments in the gel. Did you obtain restriction fragments with the digested DNA? How does the digested DNA compare with the undigested DNA?

Figure 4.3
Staining an agarose gel with ethidium bromide. Note the use of gloves, plastic-backed absorbent paper, and eye protection. (a) Transfer of the gel into a gel staining tray. (b) Flooding the gel with ethidium bromide. (c) Returning ethidium bromide stain to the storage bottle. (d) Set up of photodocumentation station.

6. To photograph the gel, set the handheld hooded Polaroid camera on the transilluminator. The exposure time should be set at 1 second and the F-stop between 5.6 and 8. Steady the camera with one hand, hold your breath, and depress the trigger. For some gels with faint bands, it may be necessary to do a two- or three-second exposure (by repeatedly depressing the trigger without moving the camera).

7. Steady the top of the camera and pull the white tab protruding from the camera pack straight out with a firm motion. This will expose a second tab (with arrows). Pull this tab straight out with a firm, steady motion.

8. Allow the photograph to develop for approximately 45 seconds. Peel the print away from the back of the negative by pulling up on one corner of the print. Be careful not to get any of the caustic developing jelly on your hands. Immediately place the negative containing the caustic jelly in the trash.

Tape your gel photograph in your notebook below:

D. Preparation of DNA Size Standards

1. After the *Hin*d III digest of lambda has digested at least one hour, remove it from the water bath and set it in a microcentrifuge tube rack. You should have 95 μl remaining that contains 9.5 μg of DNA (0.1 μg/μl). Add 25 μl of loading dye and enough deionized water to make the total volume 190 μl:

restriction digest:	95 μl
loading dye:	25 μl
deionized water:	_____ μl
Total volume:	190 μl

This will result in a final concentration of 0.05 μg/μl. As your gel photograph indicates, a *Hin*d III digest of lambda (λ) DNA results in a series of well-separated bands on an agarose gel. For this reason, this digest is often used for DNA size standards. Because you know the concentration of the standards you have just prepared, they can also be used to quantify DNA (see Exercise 9).

2. Vortex the tube thoroughly for 20 to 30 seconds and transfer the entire contents to a clean tube labeled "λ/*Hin*d III, 0.05 µg/µl."

3. Cap the tube and place it in your group's storage box in the freezer. If you will be using the standard frequently, it can also be stored in a refrigerator at 4°C. To use as a standard in future experiments, load 5 to 10 µl of this stock into a well.

Results and Discussion

1. Examine the photograph of your gel. Can you explain the results of each digestion? Note the difference in mobility between the uncut lambda DNA and the uncut plasmid DNA. What accounts for the difference?

2. *Pst* I cuts pBR322 only one time. How many restriction fragments does it generate? If an enzyme cuts lambda only once, how many restriction fragments would it generate?

3. Based on your results, how many times does *Eco*R I cut pBR322?

4. The size of the restriction fragments generated by digesting lambda DNA with *Hin*d III are given in the following table. Measure the distance migrated by each of these bands and record your results (in mm) in the table. Measure from the front of each well to the leading edge of each band.

Restriction Digestion of Lambda DNA

Fragment #	*Hin*d III		*Eco*R I	
	Size (kb)	Distance (mm)	Size (kb)	Distance (mm)
1	23.1			
2	9.42			
3	6.56			
4	4.36			
5	2.32			
6	2.03			
7	0.56			

5. Linear DNA migrates in an agarose gel at a rate that is inversely proportional to the log of its molecular weight. When dealing with nucleic acids, size in base pairs is often substituted for molecular weight. Therefore, plotting the distance migrated versus the log of the fragment size will yield a straight line that can be used as a standard curve to determine the size of other DNA molecules.

Plot the distance migrated by the *Hin*d III fragments on the x-axis of the 3-cycle semi-log paper following this exercise and the size (in kb) on the log scale (y-axis). Connect the data with a straight line. (NOTE: the relation between distance and log bp will not be linear over the entire range).

6. Measure the distance migrated by the *Eco*R I restriction fragments and record the results in the table in step 4. Using this data and the standard curve, calculate the size of each *Eco*R I fragment and enter it in the table. Do the sizes add up to the total length of a lambda DNA molecule? If not, what might account for the discrepancy? Refer to the restriction map of lambda DNA (Appendix XI) to determine if you calculated the correct sizes.

7. Measure the distance migrated by each of the pBR322 restriction fragments and record the values in the following table. Use the standard curve made from the *Hin*d III digest of lambda to calculate the size of each fragment.

Restriction Digestion of pBR322

Fragment #	*Pst* I		*Eco*R I + *Pst* I	
	Size (kb)	Distance (mm)	Size (kb)	Distance (mm)
1				
2				

8. Based on your measurements, what is the size of pBR322? What does your data tell you about the physical structure of the plasmid (i.e., is it a linear or circular molecule)? In the circle below, draw a line at the 12:00 position and label it with a 0 and the calculated size of the plasmid (i.e., 7.5/0 for a 7.5 kb plasmid). Arbitrarily designate this as the site where *Eco*R I cuts. Indicate the location of the *Pst* I site(s). Verify your calculations by examining the map of the plasmid in Appendix XI.

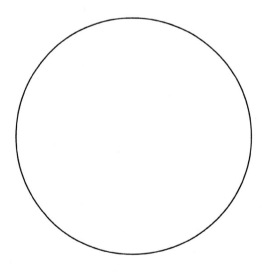

Literature Cited

Sambrook, J., E. F. Fritsch, and T. Maniatis. 1989. *Molecular cloning: A laboratory manual.* 2d ed. New York: Cold Spring Harbor Laboratory Press.

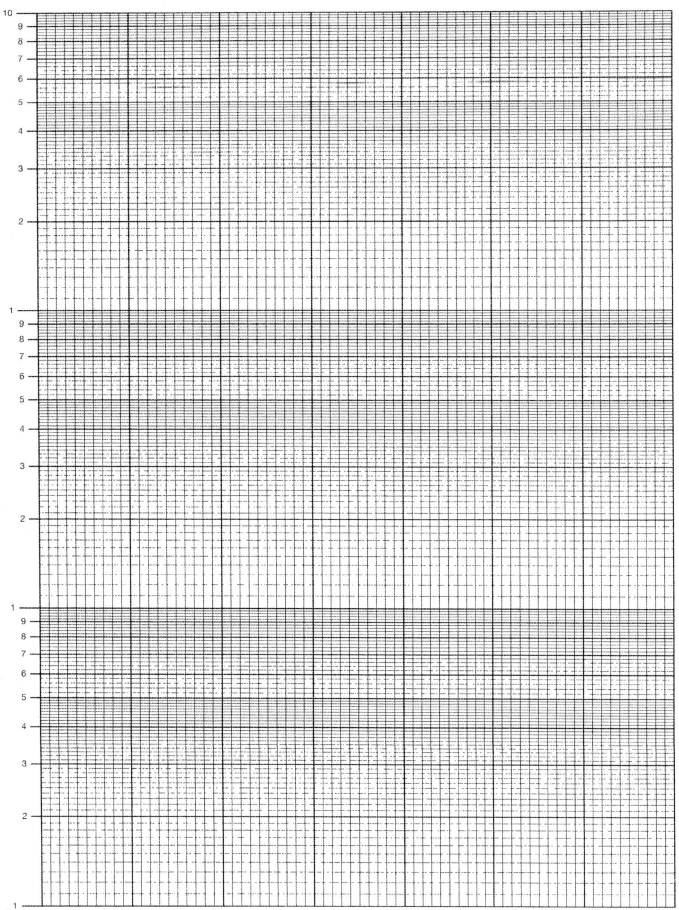

Semi Log 3 x 10

DNA Isolation and Analysis

EXERCISE 5
Isolation of Chromosomal DNA from *Vibrio fischeri*

The isolation of DNA is one of the more commonly used procedures in many areas of bacterial physiology, genetics, molecular biology, and biochemistry. Purified DNA is required for many applications, such as studying DNA structure and chemistry, examining DNA-protein interactions, carrying out DNA hybridization, sequencing or PCR, performing various genetic studies, or gene cloning. The isolation of DNA from bacteria is a relatively simple process. The organism to be used should be grown in a favorable medium at an optimum temperature and should be harvested in late log to early stationary phase for maximum yield. The cells can then be lysed and the DNA isolated by one of several methods. The method of choice depends in part on the organism of interest and the intended use of the DNA after purification. Following lysis, other cellular constituents are selectively removed. Once this is accomplished, DNA can be precipitated from solution with alcohol and dissolved in an appropriate buffer.

Lysis of the bacteria is initiated by resuspending a bacterial pellet in a buffer containing lysozyme (an enzyme that degrades peptidoglycan) and EDTA (a chelator). In addition to inhibiting DNases, the EDTA disrupts the outer membrane of the gram-negative envelope by removing the Mg^{2+} from the lipopolysaccharide (LPS) layer. This allows the lysozyme access to the peptidoglycan. After partial enzymatic disruption of the peptidoglycan, a detergent such as sodium dodecyl sulfate (SDS) is added to lyse the cells. Most gram-negative cells will lyse after this treatment, and many can even be lysed without lysozyme. Once the cells are lysed, the solution should be treated gently to prevent excessive shearing of the DNA strands.

Subsequent steps involve the separation of the DNA from other macromolecules in the lysate. Both phenol (that has been equilibrated with Tris buffer to decrease the acidity) and chloroform (with isoamyl alcohol as a defoaming agent) are commonly used to dissociate protein from nucleic acids. These reagents also remove lipids and some polysaccharides. Proteolytic enzymes such as pronase or proteinase K are often added to further remove protein. Proteinase K is a particularly useful enzyme in that it is not denatured by SDS and, in fact, works more effectively in the presence of SDS. The nucleic acids (including RNA) may then be precipitated in ethanol if the ionic strength of the solution is high. This is followed by RNase treatment to degrade the RNA. The DNA in solution may then be reprecipitated with ethanol. In this second precipitation, the ribonucleotides from RNase treatment will remain in solution, leaving purified DNA in the pellet. The pellet can then be dissolved in an appropriate buffer.

Alcohol precipitations of DNA and RNA are widely used in molecular biology and are valuable because they allow the nucleic acids to be concentrated by removing them from solution as an insoluble pellet. If concentrations of DNA are relatively high (>1 μg/ml), DNA can be effectively precipitated in 10 to 15 minutes by shielding the negative charge with monovalent cations (0.3 M sodium or 2.5 M ammonium ions are commonly used) followed by the addition of two volumes of ethanol. Numerous factors affect the effectiveness of alcohol precipitations and are reviewed in Appendix VII.

A major consideration in any DNA isolation procedure is the inhibition or inactivation of DNases, which can hydrolyze DNA. The buffer in which the cells are suspended should have a slightly basic pH (8.0 or greater), which is above the optimum of most DNases. EDTA is also included in the resuspension buffer to chelate divalent cations (such as Mg^{2+}), which are required by DNases. The SDS also reduces DNase activity by denaturing these enzymes. DNase activity is further controlled by keeping cells and reagents cold, using proteolytic enzymes such as pronase or proteinase K, and a heating step that will thermally denature DNase (but must not denature the DNA).

In this exercise, we will use a modification of the lysozyme/SDS lysis procedure (Marmur 1961) to isolate DNA from the bioluminescent bacterium *Vibrio fischeri*. This procedure is useful for isolating DNA from a large variety of gram-negative bacteria. It yields partially purified DNA of sufficient quality for techniques such as restriction digestion, ligation, and cloning. The DNA from this exercise will be used to prepare a genomic library that will be screened to obtain a clone containing the bioluminescence genes (the *lux* operon) of *V. fischeri*.

Cultures

❑ 100 ml overnight culture of *V. fischeri* MJ1 grown in GVM broth

Reagents

❑ sterile TES buffer (10 mM Tris [pH 8.0], 5 mM EDTA, 1.5% NaCl) (on ice)
❑ lysozyme (10 mg/ml in 10 mM Tris buffer, pH 8.0) (on ice)
❑ 10 mg/ml proteinase K (on ice)
❑ 10 mg/ml RNase A (in 10 mM Tris [pH 7.5], 15 mM NaCl) (on ice)
❑ 20% sodium dodecyl sulfate (SDS)
❑ "phenol" ("phenol" is equilibrated with Tris buffer [pH 8.0] and contains 0.1% 8-hydroxyquinoline and 0.1% β-mercaptoethanol)

❑ "phenol:chloroform" (1:1 v/v; "chloroform" is chloroform:isoamylalcohol [24:1])
❑ chloroform:isoamyl alcohol (24:1)
❑ sterile 3.0 M sodium acetate (pH 5.2)
❑ 95% ethanol (in freezer)
❑ sterile TE buffer (10 mM Tris [pH 8.0], 1 mM EDTA) (on ice)

Supplies and Equipment

❑ ice chests
❑ 250-ml centrifuge bottles (1/group)
❑ sterile 50-ml polyallomer Oak Ridge centrifuge tube (3/group)
❑ sterile 50-ml disposable centrifuge tube (2/group)
❑ sterile 1.5-ml microcentrifuge tubes (2/group)
❑ 10–100-µl micropipetter

❑ 100–1000-µl micropipetter
❑ sterile 100-µl micropipet tips
❑ sterile 1000-µl micropipet tips
❑ sterile 5- and 10-ml pipets
❑ sterile 8-inch glass rods (1 cm diameter; 2/group)
❑ vortex mixer
❑ refrigerated high-speed centrifuge
❑ shaking water bath

➡Procedure

1. Take your overnight culture of *Vibrio fischeri* MJ1 into a dark room to verify that it is bioluminescent (swirl the flask to saturate the medium with oxygen). This is the first step in the cloning of the *lux* genes!

2. Transfer the culture (100 ml) to a 250-ml centrifuge bottle. Balance your bottle with another group and centrifuge at 6000 × *g* (6000 rpm in a GSA rotor) for 10 minutes at 4°C. Decant the supernatant into a discard beaker containing disinfectant. Shake the bottle to remove the last traces of the supernatant.

3. Slurry the cells on a vortex mixer (this helps break up the cell pellet) and resuspend the cells in 10 ml of ice-cold TES buffer, making certain that there are no clumps of cells. This may be done by repeatedly drawing the buffer up into the pipet and forcibly squirting it into the pellet.

4. Transfer the suspension to a 50-ml polyallomer Oak Ridge centrifuge tube and rinse the centrifuge bottle with an additional 2 ml of ice-cold TES buffer. Combine both suspensions in the 50-ml tube.

5. Add 1.0 ml of a freshly made lysozyme solution (10 mg/ml) to the cell suspension and mix. Place on ice for 10 to 15 minutes.

6. Add proteinase K to a final concentration of 50 µg/ml and mix. Incubate in a 55°C shaking water bath for 10 minutes with gentle shaking. Although proteinase K will act in the presence of SDS (used in the next step to lyse the cells), this incubation is done to degrade periplasmic nucleases that are released upon disruption of the outer membrane by the EDTA. This minimizes nuclease activity prior to lysing the cells in the next step.

CAUTION: The next step involves the use of phenol. Phenol is toxic and can cause severe burns if it contacts your skin. Wear gloves, eye protection, and lab coats when using. Perform all work with phenol in the fume hood. If phenol contacts your skin, rinse immediately with cold water and notify your instructor.

7. Lyse cells by adding 20% SDS to a final concentration of 1%. What results in the lysis? Mix the suspension thoroughly by gently rocking the tube back and forth. Notice any change in the appearance of the cell suspension. To what is this due?

8. Place the tube in a 55°C shaking water bath for 15 to 20 minutes with gentle shaking. This incubation may be extended up to 60 minutes, which may result in better lysis. The solution should become extremely viscous—this is the DNA in solution.

9. Add an equal volume of "phenol" and mix by **gently** rocking the tube back and forth until you get an emulsion. It is important that the two phases are mixed to allow the solvent to precipitate the protein.

10. Continue mixing for 5 minutes once an emulsion is formed. This may be done in a 45°C shaking water bath with gentle shaking. Cool to room temperature.

11. Centrifuge the "phenol" extracted lysate at room temperature for 10 minutes at 17,000 × *g* (12,000 rpm in an SS34 rotor). You should see a layer of white precipitate at the interface between the "phenol" layer and the aqueous layer—this is precipitated protein.

Figure 5.1
Withdrawing DNA solution with the "mouth end" of a pipet.

12. Transfer the upper aqueous layer to a sterile 50-ml disposable centrifuge tube by aspirating the liquid off with the "mouth end" of a sterile 10-ml pipet. This reduces shearing of the DNA, which may occur by passing the solution through the small orifice of the pipet. The DNA is present in this layer. Be careful not to disturb or aspirate any of the material at the interface, which is primarily denatured protein. Place the tube containing the DNA solution on ice. Place the Oak Ridge tube containing the "phenol" in a fume hood.

13. Add two volumes of cold 95% ethanol to the tube with the DNA solution. Watch as you do this—the DNA will start to precipitate out at the interface. Refer to Appendix VII for a review of the principles of nucleic acid precipitations. Note that the cations required to shield the nucleic acid charges were present in the TES buffer added in step 3.

14. Invert the tube gently several times to mix and incubate on ice for 5 to 10 minutes. The large molecular weight DNA should float to the top. The precipitated DNA may be stored at $-20°C$ at this point if desired.

15. Carefully wind the DNA onto a sterile glass rod as demonstrated (see Figure 5.2). You may need more than one rod if you have a large yield of DNA. Gently press the DNA against the side of the tube to remove excess liquid.

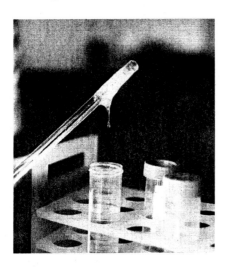

Figure 5.2
Spooling DNA on a glass rod.

16. Place the glass rod in an upright position and allow it to drain for at least 10 to 15 minutes. It is important for the ethanol to evaporate, as alcohol prevents DNA from going into solution and may inhibit subsequent enzymatic reactions. Placing the rod in a warm (37°C) incubator will speed up drying.

17. Transfer the DNA to a sterile 50-ml Oak Ridge centrifuge tube containing 15.0 ml of sterile TE. Gently swirl the glass rod back and forth, but don't try to force the DNA into solution, as this may cause shearing. Once the DNA is off the glass rod, the solution can be refrigerated or frozen and the purification completed at a later date.

18. When the DNA is in solution, add RNase to a final concentration of 100 μg/ml and cap the centrifuge tube. The RNase solution has been treated by boiling for

10 minutes to inactivate any DNase. Incubate at 45°C for 15 to 30 minutes to degrade the RNA. This will also assist in getting any remaining DNA into solution. The DNA should be completely dissolved before proceeding.

 CAUTION: The next step involves the use of chloroform as well as "phenol." Chloroform is toxic and is dangerous to inhale. Wear gloves, eye protection, and lab coats when using it. Perform all work with chloroform in the fume hood.

19. Add proteinase K to a final concentration of 50 μg/ml. Incubate for an additional 15 to 30 minutes at 45°C. This will inactivate the RNase and help remove remaining proteins.

20. Add an equal volume of "phenol:chloroform" (1:1) to the DNA solution. Shake gently for 5 minutes. You should get an emulsion.

21. Centrifuge for 5 minutes at 17,000 × *g* (12,000 rpm in an SS34 rotor) at room temperature to separate the phases. Transfer the top aqueous layer (containing the DNA) to a sterile 50-ml Oak Ridge centrifuge tube. Use the mouth end of a sterile 10-ml pipet (see Figure 5.1).

22. Add an equal volume of chloroform:isoamyl alcohol and mix gently by rocking back and forth for 5 minutes. **Note previous Caution!** You should get an emulsion.

23. Centrifuge for 5 minutes at 17,000 × *g* (12,000 rpm in an SS34 rotor) at room temperature to separate the phases. Transfer the top aqueous layer (containing the DNA) to a sterile 50-ml disposable centrifuge tube. Use the mouth end of a sterile 10-ml pipet (see Figure 5.1).

24. Add one-tenth volume of 3.0 M sodium acetate (pH 5.2) and reprecipitate the DNA with two volumes of 95% ethanol as described in steps 13-14. Wind on a sterile glass rod as done in step 15.

25. Remove excess ethanol by inverting the glass rod and allowing it to drain at least 10 to 15 minutes. Again, it is important for all the ethanol to evaporate. Dissolve the DNA completely in 2.0 ml of TE buffer. Use more buffer if your DNA won't dissolve.

 Note: If the DNA does not wind, it is either at too low a concentration or has been sheared during isolation. If this is the case, **gently** mix the alcohol and aqueous phase, and place the tube in the freezer overnight. The DNA will precipitate and can be recovered by centrifugation.

The DNA is now sufficiently pure for cloning the *lux* genes. We will determine the concentration and purity of the DNA (Exercise 7) so that we will be able to use a known amount of DNA to clone the bioluminescence genes (Exercises 8-13).

Results and Discussion

1. Why are sterile materials used when isolating DNA? Is it more important to use careful sterile technique at the beginning or at the end of the procedure?

2. Why are cations required in DNA solutions in order to precipitate DNA? What would result if they were absent when the ethanol was added?

3. What, if any, modifications do you think would be necessary to isolate DNA from plant cells? From animal cells? Why?

Literature Cited

Marmur, J. 1961. A procedure for the isolation of deoxyribonucleic acid from micro-organisms. *J. Mol. Biol.* 3:208–218.

EXERCISE 6
Large-Scale Purification of Plasmid DNA

A variety of methods have been developed to isolate plasmid DNA. Different procedures vary primarily in the method used to lyse the cells and the strategy used to selectively recover plasmid DNA. For plasmids that are less than 10 kilobases (kb) in length, the cells may be lysed by relatively harsh treatments, such as by boiling or by alkali. However, for large plasmids (>10 kb) gentler methods are used, such as lysis by detergents such as SDS or Triton X100. In any procedure, the objective after lysing the cells is to separate the plasmid DNA not only from other cellular components but also from chromosomal DNA. To accomplish the latter, plasmid purification procedures exploit two major differences between bacterial chromosomal DNA and plasmid DNA:

1. Bacterial chromosomes are much larger than plasmids. For example, most plasmids used as cloning vectors are less than 10 kb, whereas the *E. coli* chromosome is about 4000 kb.

2. Because DNA is a rather fragile molecule, the extremely long, thin molecule of chromosomal DNA breaks into many pieces (generally about 100 kb long) when the cell lysate is mixed or pipetted. In contrast, the much smaller plasmids are not sheared by manipulation and remain in a supercoiled, covalently closed, circular form.

Most protocols, therefore, involve a differential precipitation step to remove chromosomal DNA. The long strands of chromosomal DNA become entangled with precipitated cell debris (protein, lipids, and wall polymers) and are removed by centrifugation, leaving the smaller plasmid DNA in solution. The procedures also take advantage of the properties of closed, circular DNA. The two strands of plasmid DNA will remain interlocked under denaturing conditions, such as high temperature (boiling) or alkali (up to pH 12.5). Thus, when cooled or returned to a neutral pH, plasmids will instantly return to their original circular, supercoiled state because the complementary strands are physically associated with each other. The complementary strands of chromosomal DNA, however, are not physically associated and become trapped with the precipitate of cellular debris.

After the cellular debris and most of the chromosomal DNA is removed by centrifugation, the plasmids will remain in solution with RNA molecules. The plasmid DNA and RNA can then be precipitated with alcohol or polyethylene glycol (PEG; see Appendix VII), and the RNA may be removed by treatment with RNase. This results in a partially purified preparation of the plasmid, although it still contains some protein and small fragments of chromosomal DNA. The partially purified plasmid DNA, however, is suitable for some applications, such as transformation or restriction analysis. Many of the rapid, small-scale plasmid isolation procedures (or "mini-preps") yield partially purified DNA such as this (see Exercise 14).

Following an initial purification, the plasmid may be further treated to yield highly purified DNA free of protein, RNA, or chromosomal DNA contaminants. This has been done traditionally by ultracentrifugation through a cesium chloride gradient. The purification takes advantage of plasmid DNA's supercoiled, covalently closed circular structure. Supercoiled plasmid DNA does not bind intercalating dyes, such as ethidium bromide, as readily as does linear DNA or nicked circular DNA. The extensive binding of intercalating dyes by the chromosomal DNA decreases its density, and plasmid DNA will band at a higher density (a lower position) in a cesium chloride density gradient (see Appendix II for the theory and principles of density gradient centrifugation). The ethidium bromide also allows the DNA bands to be seen via fluorescence under a mid-range UV light (short range UV lights should not be used as they cause photo nicking of DNA), and bands of plasmid DNA may be removed with a syringe. The ethidium may then be extracted with alcohol and the CsCl removed by dialysis. The resulting plasmid DNA from density gradient centrifugation is suitable for applications such as cloning where highly purified DNA is desired.

Recently, several companies have developed small chromatography columns that yield plasmid DNA equivalent in purity to CsCl purified DNA. Most of these protocols take advantage of DNA's ability to bind tightly to powdered glass under high-salt conditions. The glass can be washed to remove contaminants, and the pure plasmid is then eluted with a low ionic strength buffer or water. These procedures are less time-consuming than ultracentrifugation, use nonhazardous materials, and provide a useful option if an ultracentrifuge is unavailable.

Many plasmids exist in the cell at a relatively low copy number, and it is often useful to increase, or amplify, the copy number prior to isolation. Adding the antibiotic chloramphenicol to mid- to late-log phase cells blocks the replication of chromosomal DNA (and hence cell division) but not replication of plasmids. Continued incubation for several more hours will result in greatly increased numbers of plasmids per cell. Amplified plasmids often contain significant amounts of ribonucleotides because the cell runs out of deoxyribonucleotides for the increased plasmid replication. Many of the small plasmids used for cloning vectors (such as the pUC plasmids and their derivatives) have a naturally high copy number, and amplification is not necessary.

In this exercise, we will use a modification of the alkaline lysis procedure of Birnboim and Doly (1979) to purify plasmid cloning vectors and *lux* clones to be used in future experiments. The plasmid harboring cells (grown in media containing an antibiotic that selects for the plasmid) are lysed with lysozyme and SDS under alkaline conditions. The sodium hydroxide denatures both the plasmid and chromosomal DNA. Upon neutralization with an acidic potassium acetate solution, most of the chromosomal DNA is precipitated out with the cell debris, while the covalently closed circular plasmids reanneal and remain in solution. The plasmids are then alcohol precipitated, washed, and dissolved in TE buffer (10 mM Tris [pH 8.0], 1 mM EDTA). The crude plasmid prep will then be further purified by a MaxiPreps™ column marketed by Promega Corporation. Similar purification kits are sold by a variety of suppliers. Alternately, the crude plasmid prep may also be purified on a cesium chloride gradient.

Cultures

☐ 200 ml of an overnight culture of a plasmid-bearing strain of *E. coli*

Reagents

☐ cell resuspension solution (50 mM Tris [pH 7.5], 10 mM EDTA, 100 µg/ml RNase A)
☐ cell lysis solution (0.2 M NaOH, 1% SDS)
☐ neutralization solution (2.55 M potassium acetate [pH 4.8])
☐ Wizard™ MaxiPrep DNA purification resin
☐ Wizard™ column wash solution (85 mM NaCl, 8.5 mM Tris [pH 7.5], 2 mM EDTA, 55% ethanol) (Note: the user must add ethanol to this reagent prior to use.)

☐ Wizard™ Maxicolumns with reservoirs
☐ isopropanol
☐ 80% ethanol
☐ TE buffer (10 mM Tris [pH 8.0], 1 mM EDTA) (on ice)
☐ TE buffer (70°C)

Supplies and Equipment

☐ sterile 5- and 10-ml pipets
☐ 100–1000 µl micropipetters
☐ sterile 1000-µl micropipet tips
☐ sterile 4-inch funnel
☐ sterile cheesecloth
☐ sterile 50-ml graduate cylinder
☐ sterile 1½-inch, 18-gauge hypodermic needle
☐ vacuum manifold or flask
☐ vacuum pump or aspirator

☐ 250-ml centrifuge bottle (1/group)
☐ sterile 50-ml polyallomer Oak Ridge centrifuge tube (3/group)
☐ sterile 50-ml disposable centrifuge tube (1/group)
☐ sterile 1.5-ml microcentrifuge tubes
☐ refrigerated high-speed centrifuge with rotors for 50-ml tubes and 250-ml bottles
☐ clinical centrifuge with rotor for 50-ml tubes

➡️Procedure

A. Preparative Column Purification

Note: Each group will receive 200 ml of an overnight culture of an *E. coli* strain containing a plasmid. The plasmids used in cloning all have a gene encoding resistance to an antibiotic (e.g., ampicillin). To ensure that all the bacteria in the culture contain the plasmid of interest, the inoculum originally came from a colony on a plate containing the antibiotic. The appropriate antibiotic has also been added to the medium. Thus, only bacteria containing the plasmid can grow in the flask.

1. Transfer the 200-ml overnight culture to a 250-ml centrifuge bottle and balance your bottle (±0.5 g) with another group. Centrifuge at 6000 × *g* (6000 rpm in a GSA rotor) for 10 minutes at 4°C.

2. Discard the supernatant into a discard beaker containing disinfectant and resuspend the pellet in 12 ml of the cell resuspension solution. Run the buffer up and down in the pipet until the pellet is completely homogenized. Transfer the cell suspension to a sterile 50-ml Oak Ridge centrifuge tube.

3. Add 12 ml of the lysis solution (SDS/NaOH), cap the tube, and mix by gently inverting several times. Incubate at room temperature for 10 minutes and mix occasionally. Lysis of the cells is indicated by the suspension becoming clear and very viscous. Verify that lysis has occurred before proceeding.

4. Add 12 ml of neutralization solution, cap the tube, and invert the tube sharply several times. This will neutralize the NaOH and cause the SDS and protein to form a flocculent precipitate. Verify that a precipitate forms before proceeding.

5. Balance your tube with a tube from another group and centrifuge at 17,000 × *g* (12,000 rpm in an SS34 rotor) for 15 minutes at 4°C.

6. The cellular debris and SDS will form a large pellet at the bottom and side of the tube, and the plasmid DNA will be in the supernatant. Place the sterile cheesecloth in a sterile funnel, and pour the supernatant through the cheesecloth into a sterile 50-ml graduate cylinder. Note the volume in the graduate and pipet equal volumes (approximately 18 ml) to each of two sterile 50-ml polyallomer Oak Ridge centrifuge tubes.

7. Add 0.6 volumes (approximately 10 ml) of isopropanol to each Oak Ridge tube. Mix by rapidly inverting the tubes. What results from this addition? Can you explain what you observe?

8. Balance the tubes (if necessary) with isopropanol and centrifuge **immediately** at 17,000 × *g* (12,000 rpm in an SS34 rotor) for 10 minutes at room temperature. Prolonged incubation with isopropanol will result in precipitation of protein.

9. Carefully decant the supernatant into a discard beaker. Remove as much liquid as possible with a sterile pipet and use a sterile cotton swab to remove the last traces.

10. Dissolve each pellet in 1 ml of sterile TE buffer and combine into one of the Oak Ridge tubes.

11. Add 10 ml of Wizard™ MaxiPrep DNA purification resin to the DNA solution and swirl to mix.

CAUTION: In the next step, you will be using a very sharp sterile hypodermic needle. Use extreme care when handling the needle and discard in a Sharps container when done.

12. Insert a sterile 18-gauge needle vertically into the stopper in a vacuum flask or manifold. Attach the Maxicolumn to the needle. Be careful not to touch the lure tip of the needle or the tip of the Maxicolumn as this may contaminate them with nucleases. Transfer the DNA-resin solution to the Maxicolumn with a sterile 5-ml pipet. Apply a vacuum until the resin mixture is packed in the bottom of the column.

13. Add 12 ml of the column wash solution to the Oak Ridge tube that contained the DNA-resin mixture. Swirl and immediately pour the wash into the Maxicolumn. Apply a vacuum until the wash solution has been completely pulled through the column.

14. Repeat step 13 with a second 12 ml of the column wash solution.

15. Add 5 ml of 80% ethanol to the Maxicolumn and turn on the vacuum to draw the ethanol through the column. Leave the vacuum on for 10 minutes to dry the column resin.

16. Remove the Maxicolumn from the vacuum flask and place it into the sterile 50-ml disposable centrifuge tube. Add 1.5 ml of sterile TE buffer (preheated to 70°C) to the column and wait **exactly** one minute. **Be sure that all groups sharing a clinical centrifuge add the hot TE at the same time.**

17. **Immediately** place the tube containing the Maxicolumn in a swinging bucket rotor in a clinical centrifuge. Be sure that your tube is balanced with one from another group. Elute the DNA from the column by centrifuging for 5 minutes at 1300 × g (full speed on the clinical centrifuge).

18. Remove and discard the Maxicolumn. Transfer the eluate (use a micropipetter with a sterile tip) to a sterile microcentrifuge tube and centrifuge at 12–14,000 × g (high speed in a microcentrifuge) for 5 minutes. This will pellet resin fines that may be present in the eluate.

19. Label a second sterile microcentrifuge tube with the name of the plasmid purified and transfer the supernatant into it. Place the purified DNA on ice and quantify as described in Exercise 7. The plasmid DNA may also be quantified using fluorescent techniques described in Exercise 9. Record the data in the table on the following page and write the concentration of your plasmid on the microcentrifuge tube.

Plasmid	Dil.	A_{260}	A_{280}	A_{234}	A_{260}/A_{280}	A_{234}/A_{260}	Concentration µg/ml

B. Verification of Plasmid by Restriction Digestion (Optional)

1. It is also a good idea to do a restriction digest and run the uncut and digested plasmid on an agarose gel to verify that you have the correct plasmid. Refer to Exercise 4 for restriction digestion protocols and refer to a map of your plasmid to choose appropriate enzymes to use. Maps of cloning vectors are provided in Appendix XI, and your instructor will provide you with maps of other plasmids. It is best to choose one enzyme that will cut the plasmid once (i.e., linearize it) so that you may accurately check its size and another enzyme or combination of enzymes that will give you two or more restriction fragments that are characteristic of that plasmid. If the plasmid contains cloned genes, it is often useful to choose an enzyme(s) that will excise the cloned DNA from the vector.

2. Digest approximately 0.2 µg of plasmid (generally 1 to 2 µl) with one or more enzymes that will generate characteristic restriction fragments of your plasmid. You may wish to use a larger amount of DNA for enzymes that result in small fragments (<1 kb) that you wish to see.

3. Run an aliquot of your digestion and 0.1 µg of undigested plasmid on an agarose gel. Choose an agarose concentration that will effectively resolve the DNA fragments you are interested in (see Exercise 4 and Appendix IV). Place the photograph of your gel in your notebook below:

4. Place the tube in your group's box in the freezer. If your plasmid isolation was successful, these plasmids will be used in subsequent cloning and mapping experiments.

Literature Cited

Birnboim, H. C., and J. Doly. 1979. A rapid alkaline extraction procedure for screening recombinant plasmid DNA. *Nucleic Acids Res.* 7:1513.

Promega Corporation. 1994. Wizard™ MaxiPreps DNA Purification System. Technical Bulletin 139. Madison, WI.

Sambrook, J., E. F. Fritsch, and T. Maniatis. 1989. *Molecular cloning: A laboratory manual.* 2d ed. New York: Cold Spring Harbor Laboratory Press.

EXERCISE 7
Spectrophotometric Analysis of DNA

Ultraviolet (UV) spectrophotometry can be used to determine the quantity of DNA present in a sample and estimate its purity. It works because nucleic acids and proteins have different absorbance spectra. DNA and RNA both strongly absorb UV light with a maximum absorbance at 260 nm. This allows the quantification of these molecules by UV absorbance at concentrations as low as 2 μg/ml. A solution of pure double-stranded DNA with a concentration of 50 μg/ml has an A_{260} of 1.0. This value varies slightly depending on the mole% G + C of the DNA, but this variation can generally be ignored.

Measurement of the absorbance of a nucleic acid solution at wavelengths other than 260 nm is useful for characterizing purity. To assess the purity of a nucleic acid preparation, the relevant spectrum lies between 320 nm and 220 nm. The absorbance at 320 nm should be less than 5% of the A_{260}. An A_{320} greater than this indicates the presence of particulates in the solution or dirty cuvettes (Schleif and Wensink 1981) and will result in erroneous readings.

Proteins absorb UV light maximally at 280 nm, and the A_{260}:A_{280} ratios of a DNA solution are often used as indicators of protein or RNA contamination. The A_{260}:A_{280} ratio from a preparation of pure double-stranded DNA should be between 1.8 and 1.9. Higher ratios are often due to RNA contamination; lower ratios can indicate the presence of protein.

Absorption of light at 280 nm by proteins is due to the presence of aromatic amino acids; the absorbance per mass can vary because it is dependent upon the presence of tyrosine, phenylalanine, and tryptophan, which account for a small and variable portion of the amino acids in proteins. Additionally, the extinction coefficient for protein is much lower than the extinction coefficient for nucleic acid. Therefore, the A_{260}:A_{280} is not a very sensitive indicator of protein contamination. The A_{234}:A_{260} ratio is a better indicator. Nucleic acids have an absorbance minimum at 234 nm, and protein or phenol contamination causes an increase in this ratio. An A_{234}:A_{260} ratio greater than 0.50 is indicative of protein contamination (Johnson 1981).

In this exercise, you will use a UV-visible spectrophotometer to quantify and assess the purity of the DNA that was isolated in the previous exercises. This information will be required to calculate the amount of DNA to be used to initiate the cloning of the *lux* genes in Exercise 8. Further information on the principles and use of spectrophotometers is found in Appendix III.

Reagents

- *Vibrio fischeri* DNA stock solution from Exercise 5
- plasmid DNA stock solution from Exercise 6
- sterile TE buffer
- deionized water (in squirt bottle)

Equipment and Supplies

- vortex mixer
- matched set of semi-micro quartz cuvettes
- plastic cuvette racks
- lens paper
- Kimwipes™
- UV-visible spectrophotometer
- discard beaker

- 13 × 100-mm test tubes
- racks for 13 × 100-mm test tubes
- 10–100 μl micropipetters
- 100–1000 μl micropipetters
- sterile 100-μl micropipet tips
- sterile 1000-μl micropipet tips

Procedure

Note: To measure wavelengths in the ultraviolet (UV) range, a special deuterium bulb is required because the visible (tungsten) lamp does not produce sufficient energy in this region of the electromagnetic spectrum to allow accurate measurements. In addition, the short wavelengths required to measure absorbance of nucleic acids are absorbed by glass or plastic cuvettes. Therefore, quartz cuvettes are required for these measurements. Quartz cuvettes are supplied in optically matched sets and are very expensive. Handle them and the spectrophotometers with care. **Never** set a glass or quartz cuvette on the bench where it may be accidentally knocked over—keep it in a cuvette rack at all times when not in the spectrophotometer. If in doubt as to the proper usage of the spectrophotometers, ask your instructor.

1. Turn on the spectrophotometer and the UV (deuterium) lamp 20 minutes prior to use. This allows the output of the electronics and the light to stabilize. The visible lamp should be turned off. Familiarize yourself with the operating instructions of the spectrophotometer before proceeding.

2. Set the spectrophotometer to read absorbance and set the wavelength to 260 nm.

3. Dilute an aliquot of your DNA stock solution with the sterile TE buffer provided. The final volume of your dilution should be 500 μl. A ½₀ dilution is usually a reasonable starting dilution.

4. Remove the matched quartz cuvettes from their box. Handle them only by the frosted sides, and do not touch the optical surfaces. Rinse each carefully with

deionized water, then carefully shake the water droplets out. Wipe the outside of the optical surfaces of each cuvette with lens paper to remove dirt and finger-prints. **Remember to keep the cuvettes in the rack at all times when not in use!**

5. Zero the spectrophotometer before you read the absorbance of your DNA. To do this, place 1 ml of TE buffer in the cuvette and wipe the optical surfaces with a Kimwipe™. This will be the background cuvette. Open the chamber door of the spectrophotometer and place the cuvette inside the holder. Make sure that you orient the cuvette with the optical surfaces in the path of the light or it will not transmit light. Your instructor will demonstrate the proper orientation.

6. Adjust the spectrophotometer to 0 absorbance (100%T).

7. Remove the background cuvette and place it in the cuvette rack (do not discard the TE buffer). Place a small piece of tape labeled B (for blank) next to it.

8. Transfer your diluted DNA sample into the second cuvette and wipe the optical surfaces with a Kimwipe™. Place the cuvette in the chamber and close the door. The display will show the absorbance of your sample. An absorbance between 0.1 and 1.0 is generally within the accurate reading range of most spectropho-tometers, and many are accurate to absorbances of 2 or 3. If the value is outside of the valid range, redilute your sample so that it falls within this range. Record your reading and the dilution of the sample in the table at the end of this exer-cise. Remove the cuvette containing the sample and place it in the cuvette rack.

9. To determine the A_{280} of the diluted DNA sample, change the wavelength of the spectrophotometer to 280 nm.

10. Rezero the spectrophotometer using the cuvette containing the TE buffer.

11. Place the cuvette containing the sample in the spectrophotometer, close the door and record the A_{280} of the diluted DNA sample in your data table.

12. Determine the A_{234} and A_{320} of the diluted DNA sample following the same proce-dures outlined in steps 9–11.

13. If a scanning UV spectrophotometer is available, scan your DNA sample from 220 to 320 nm. If you do not have a scanning spectrophotometer, you may manually construct an absorbance spectrum by taking absorbance readings every 10 nm from 220 to 320 nm. Follow the procedure outlined in steps 9–11 for each mea-surement. Plot your data on graph paper. Take absorbance readings at additional wavelengths if needed to clarify the absorbance peaks and minimums.

Results and Discussion

1. What is the absorbance at 320 nm? What does this tell you about your sample?

2. Correct your measured absorbance for the dilution by the following formula:

$$\text{Corrected } A_{260} = \frac{A_{260}}{\text{dilution}}$$

   Given that an A_{260} of 1.0 is equivalent to a DNA concentration of 50 µg/ml, calculate the concentration of DNA in your sample. Record your result in the data table.

3. Calculate the $A_{260}{:}A_{280}$ ratio. If the ratio is greater than 1.9, there is RNA contamination in the DNA preparation. The greater the ratio, the greater the contamination. This method does not allow you to calculate the quantity of RNA contamination present, but it does provide a relative estimate of DNA purity. If a preparation free of RNA is necessary, the solution can be further treated with RNase. Following the RNase treatment, the solution should be extracted with phenol and/or chloroform followed by an ethanol precipitation (see Appendix VII).

4. Calculate the $A_{234}{:}A_{260}$ ratio. If the ratio is greater than 0.5, there is protein and/or phenol contamination in the DNA preparation. The greater the contamination, the greater the ratio. This method does not allow you to quantify the contamination, but it does provide a relative estimate. Proteins can be removed with phenol and chloroform extractions or by treatments with proteolytic enzymes (such as proteinase K or pronase) followed by organic extractions and ethanol precipitation.

DNA sample	Dil.	A_{260}	A_{280}	A_{234}	A_{320}	A_{260}/A_{280}	A_{234}/A_{260}	Final concentration (µg/ml)

Literature Cited

Johnson, J. L. 1981. Genetic characterization. In *Manual of methods for general bacteriology*, 451–475. ed. P. Gerhardt, Washington, D.C.: American Society for Microbiology.

Schleif, R. F., and P. C. Wensink. 1981. *Practical methods in molecular biology.* New York: Springer-Verlag.

Cloning the *lux* Operon

III

EXERCISE 8
Restriction Digestion of *Vibrio fischeri* Genomic DNA and Plasmid Vector

In order to clone a gene or operon from the genome of an organism, a genomic library is constructed. A library is a large collection of clones (bacteria that each contain a fragment of the target organism's genomic DNA) that represents all regions of the genome. Most genomic libraries are constructed using a method called shotgun cloning. As the name implies, the technique is not very precise. To construct such a library, one digests the genomic DNA into fragments. The restriction enzyme used in these digestions must leave the entire gene or operon of interest on a single restriction fragment. A suitable cloning vector (plasmid or bacteriophage) is then cut with the same enzyme (or an enzyme that generates compatible ends) so that the chromosomal DNA can be ligated into the vector. These recombinant DNA molecules are then transferred into a suitable host organism. If the library is complete (i.e., all portions of the genome are present), one must now begin the often tedious task of screening the library to find the clone of interest. Because genomes are large, libraries often consist of thousands (for bacterial genomes) or even millions (for mammalian genomes) of clones.

There are several potential problems that can arise in preparing genomic libraries. Most of the widely used restriction enzymes recognize a six base pair (bp) sequence of DNA. Statistically, any given six base pair sequence of DNA would appear at an average of every 4^6 (4096) bp, which would result in an average restriction fragment length of about 4 kilobases (kb). This can vary widely, however, depending on the mole% G + C in the organism's genome and the G + C content in the enzyme's recognition sequence. For example, *V. fischeri* has a mole% G + C of 40% while *Sal* I cuts at sites with a high G + C content (GTCGAC). Therefore, *Sal* I would be expected to cut less frequently in *V. fischeri*. In contrast *Hin*d III (AAGCTT) would likely cut much more frequently. With these "six-base cutters," however, it is possible that an enzyme may cut within the sequence you wish to clone, or that the restriction fragment containing the gene of interest may be too large to clone successfully. This problem may be avoided by individually trying several different enzymes. Alternately, restriction maps of the region of interest may be available and allow selection of an enzyme that will yield a restriction fragment suitable for cloning.

An alternate approach is to perform a partial digest with a restriction enzyme that recognizes a four base pair sequence (such as *Sau*3A I). Any given four base pair sequence in DNA will appear at an average of every 4^4 (256) bp. Thus, if genomic DNA were completely digested with a restriction enzyme that has a four base pair recognition sequence, most of the pieces would be too small to contain an entire gene or operon. However, if only a fraction of the restriction sites recognized by such "four-base cutters" were cleaved, a random set of larger, overlapping fragments that are suitable in length for cloning would be obtained. Some of these would likely contain the gene of interest. For example, if you were interested in cloning a region that was 2 kb long, by cutting approximately every tenth site, you would generate a set of random fragments that should contain the intact sequence of interest. Such partial digestions can be done by either diluting the enzyme or decreasing the time of digestion. Generally, several different digestions are performed, each to a different degree of completeness. By pooling these digestions and purifying a fragment size suitable for cloning, one can be reasonably assured of obtaining a fragment with the gene(s) of interest.

Partial digestions with restriction enzymes possessing four base pair recognition sequences are commonly used when it is unknown whether the gene or genes of interest are present on a single restriction fragment. This procedure is also mandatory when bacteriophage lambda or cosmid vectors are used, which require that an insert be within a specific size range. With these vectors, the digested DNA must be of a uniform size fraction prior to cloning.

Vectors used in DNA cloning are the molecular vehicles that carry the foreign DNA so that it can be maintained in a host organism. An essential component of all vectors is an origin of replication that is recognized by the host organism. There are five different types of vectors that are commonly used for cloning DNA: (i) plasmids; (ii) bacteriophages; (iii) cosmids; (iv) phagemids; and (v) artificial chromosomes. Of these five, plasmids are most commonly used when creating a library from prokaryotic DNA.

Most modern plasmid vectors are derivitives of the pUC plasmids designed and constructed by Joachim Messing (Vieria and Messing 1982). They have a number of features that make them very useful and versatile in cloning. They are small, high copy-number plasmids containing the β-lactamase gene (from the widely used vector pBR322), which confers resistance to the antibiotic ampicillin. This allows transformants to be readily selected on media containing ampicillin. The pUC plasmids also contain a gene for the β-galactosidase α-peptide (*lacZα*) that allows for the visual screening of clones containing inserts by α-complementation of a mutant β-galactosidase gene from the host. pUC18 and pUC19 also have a multiple cloning site that contains unique recognition sequences for 13 different restriction enzymes (see Appendix XI). This allows the vector to be cut by many enzymes commonly used in gene cloning.

In this exercise, you will be using a derivative of a pUC plasmid vector, pGEM™-3Zf(+) (see Figure 8.1). The pGEM™ plasmids (designed and constructed by Promega Corporation) have all the features found in pUC18/19 plasmids plus several additional ones. pGEM™-3Zf(+) includes two bacteriophage promoters (T7 and SP6) that can be used to produce single-stranded RNA molecules that are useful as nucleic acid probes. These twin promoters are the basis of the naming of pGEM™ vectors—the "Gemini" twins. This vector also contains the origin of replication for the single-stranded bacteriophage f1, which allows single-stranded phage to be generated *in vivo* directly from the plasmid. This eliminates the need to subclone the inserted DNA into an M13 vector in order to produce single-stranded DNA, which is often used for DNA sequencing.

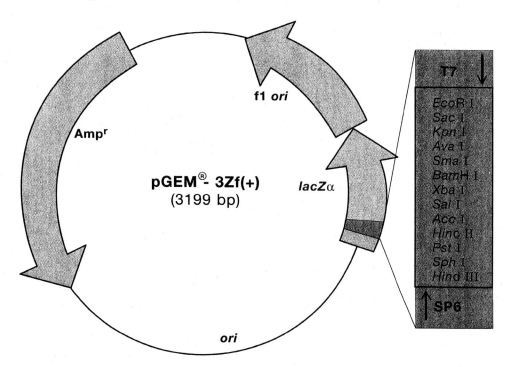

Figure 8.1
Map of the cloning vector pGEM™-3Zf(+). Modified from Promega Corporation 1996, with permission.

In this exercise, we will construct a *Vibrio fischeri* MJ1 genomic library in a plasmid vector by shotgun cloning the DNA isolated in Exercise 5. The *lux* operon in *V. fischeri* is known to be located on a 9 kb *Sal* I restriction fragment (Engebrecht, et al. 1983) and on an 18 kb *Bam*H I restriction fragment (Boylan, et al. 1985). Because it is easier to clone smaller restriction fragments, we will attempt to clone the *Sal* I fragment. Both the genomic DNA and the pGEM™-3Zf(+) vector will be digested with the restriction enzyme *Sal* I. Using the same enzyme to prepare the vector and the insert ensures that the ends of these molecules will be compatible and can be ligated together by the enzyme DNA ligase (see Exercise 10).

When the vector is cut with only one enzyme, it is possible for the ends of the vector to be ligated to each other, reforming the original plasmid. Therefore, many of the transformants may contain vectors without an insert. If desired, this can be avoided by phosphatasing the vector with enzymes such as calf intestine alkaline phosphatase (CIAP) following the restriction enzyme digestion. This removes the 5' phosphate from the ends of the cut vector and prevents the vector from ligating back to itself (see Exercise 10). It can, however, still be ligated to a DNA insert that possesses a 5' phosphate. Phosphatase treatment of the vector results in a high frequency of transformants containing foreign DNA inserts, but lowers the total number of transformants.

Because the size of the restriction fragment is known, one could further enhance the chance of obtaining the correct clone by size fractionating the genomic digest. Sucrose gradient fractionation or purifying fragments from agarose gels may be used to isolate restriction fragments of the desired size. By ligating fragments that are similar in size to the *lux* fragment, you would need to screen fewer clones to obtain the clone of interest. It is, however, relatively straightforward to clone the *lux* operon directly from the total genomic digest, because bacterial genomes are relatively small and the method of screening is very easy. Following transformation of the host bacterium with the ligation mix, the library can be screened visually on the transformation plates. Because the entire *lux* operon will be cloned, *Escherichia coli* transformants containing the *lux* operon will be bioluminescent, making the often time-consuming task of screening the library easy.

Reagents

- ❑ *V. fischeri* DNA from Exercise 5 (on ice)
  Concn. = _____ μg/μl
- ❑ pGEM™-3Zf(+), 0.2 μg/μl (on ice)
- ❑ lambda DNA, 0.2 μg/μl (on ice)
- ❑ *Sal* I (on ice)
  Activity = _____ units/μl
- ❑ 10X *Sal* I restriction buffer (on ice)

- ❑ sterile deionized water (on ice)
- ❑ sterile TE buffer (on ice)
- ❑ 3 M sodium acetate (pH 5.2) (on ice)
- ❑ loading dye (room temperature)
- ❑ 1X TAE buffer
- ❑ LE agarose (low EEO)
- ❑ ethidium bromide stain (1.0 μg/ml)

Supplies and Equipment

- ❑ mini ice buckets (1/group)
- ❑ sterile 1.5-ml microcentrifuge tubes
- ❑ microcentrifuge tube opener (1/group)
- ❑ microcentrifuge tube rack (1/group)
- ❑ microcentrifuge tube float (1/group)
- ❑ 0.5–10 μl micropipetter
- ❑ 10–100 μl micropipetter
- ❑ 10-μl micropipet tips
- ❑ 100-μl micropipet tips
- ❑ micropipet tip discard beaker
- ❑ 250-ml Erlenmeyer flask (1 per 2 groups)
- ❑ 25-ml Erlenmeyer flask (1 per 2 groups)

- ❑ microwave oven
- ❑ top loading balance
- ❑ 37°C water bath
- ❑ 60°C water bath
- ❑ microcentrifuge
- ❑ horizontal mini-gel electrophoresis chamber
- ❑ gel casting tray
- ❑ 12-tooth comb
- ❑ gel staining tray
- ❑ power source
- ❑ transilluminator
- ❑ Polaroid camera and type 667 film

▶ Procedure

A. Restriction Enzyme Digestion

1. Fill your mini ice bucket with crushed ice and place the *V. fischeri* and pGEM™ DNA, *Sal* I, and reagents in it. Each group will be provided with its own set of reagents and enzyme containing slightly more than the required volume.

2. Label seven sterile 1.5-ml microcentrifuge tubes A through G with a permanent marker and place them in your microcentrifuge tube rack.

3. Use the table on the following page to set up your digestions. Use the micropipetters to add reagents to each of the tubes. You will need to use the concentration of the *V. fischeri* DNA determined in Exercise 7 to calculate the necessary volumes of chromosomal DNA. Then calculate the volumes of water or TE buffer to give the total volume indicated. Be sure to read the Reminders below the table before proceeding.

Tube	10X Buffer	*Vibrio fischeri*	pGEM™	Lambda	H₂O	TE	*Sal* I	Total vol.
Chromosomal DNA digestion								
A	5 μl	____ μl (10 μg)	—	—	____ μl	—	____ μl (50U)	50 μl
B	2 μl	____ μl (1 μg)	—	—	____ μl	—	—	20 μl
C	—	____ μl (1 μg)	—	—	—	____ μl	—	20 μl
Vector digestion								
D	2 μl	—	5 μl (1 μg)	—	____ μl	—	2 μl	20 μl
E	2 μl	—	1 μl	—	____ μl	—	—	20 μl
Lambda controls								
F	2 μl	—	—	2 μl	____ μl	—	1 μl	20 μl
G	2 μl	—	—	1 μl	____ μl	—	—	20 μl

Reminders:

- Always add buffer, water, and DNA to tubes prior to adding enzymes.

- Touch the tip to the side of the tube when adding each reagent. Be sure you **see** the liquid transferred to the side of the tube, especially with volumes <2 μl.

- You may use the same tip to add the same reagent to different tubes, but be sure to use a **new** sterile tip for each reagent.

- Check off each addition in the table as you add it. Omissions or incorrect volumes will usually prevent the restriction digestion from working.

4. The rationale for each tube is as follows:

Tube A contains the *V. fischeri* chromosomal DNA to be digested by *Sal* I for the cloning. This digest will be ligated with the vector to construct the library.

Tube B contains chromosomal DNA and restriction buffer, but no enzyme. This serves as a control for any nuclease activity in the restriction buffer, H₂O, or the DNA.

Tube C contains only chromosomal DNA in TE buffer. This tube will be incubated on ice to serve as a control for tube B to show what the undigested DNA looks like when the digests are analyzed on a gel. In addition, this tube may be used to quantify the concentration of genomic DNA (see Exercise 9).

Tube D contains the vector to be digested by *Sal* I. The digested vector will be used to create recombinant plasmids by ligating to the genomic *Sal* I restriction fragments prepared in tube A.

Tube E contains vector without any enzyme. This serves as an uncut vector control to compare with the digested vector (tube D) on a gel.

Tube F contains lambda DNA, buffer, and *Sal* I. This serves as a control to verify that the restriction enzyme is cutting effectively. Refer to Appendix XI to see where *Sal* I cuts lambda DNA.

Tube G contains lambda DNA without any enzyme. This serves as a control to compare digested lambda (tube F) with undigested lambda on a gel.

5. After adding all ingredients, cap the tubes and tap gently to mix. Pool the reagents in the bottom of the microcentrifuge tube by spinning the tubes in the microcentrifuge for 2 to 3 seconds. Be sure that the tubes are in a balanced configuration. Tap each tube again to mix, and then recentrifuge to pool the digestion mix in the bottom of the tube.

6. Push all the tubes except tube C into the foam microcentrifuge tube holder and float in a 37°C water bath. Be sure your group number is labeled on the holder. Place tube C in your ice bath. Incubate tubes for at least 60 minutes. At about 30 minute intervals, remove all tubes from the water bath and vortex. This ensures complete mixing of the enzyme and DNA fragments. Pulse the tubes in a microcentrifuge and return to the water bath. While the restriction digestions are proceeding, prepare and cast an agarose gel. This will be used to check to see that the DNAs are being properly digested.

 Note: Digests can be frozen at this point and analyzed at a later date.

B. Agarose Gel Electrophoresis

1. Two groups will prepare one batch of agarose to pour their two gels. Label a 250-ml Erlenmeyer flask "0.8% agarose" and add 80 ml 1X TAE (<u>T</u>ris <u>A</u>cetate <u>E</u>DTA; see Appendix XV) buffer to it.

   This is the same buffer that the electrophoresis will be done in. Thus, the gel will have the same ionic strength as the electrophoresis buffer. Failure to prepare the agarose in the electrophoresis buffer will result in erratic separation of restriction fragments.

2. Calculate the amount of LE agarose to give a 0.8% gel. Check your calculations with your instructor.

   _____ g agarose/80 ml = 0.8%

   Add LE agarose to the TAE buffer and swirl. It is best to let the agarose set in the buffer for 5 to 10 minutes and swirl occasionally to allow the agarose to hydrate prior to heating. This speeds up the melting of the agarose and decreases the tendency of the agarose to boil over.

CAUTION: In the next step, you will be heating agarose to boiling. Agarose can become superheated and boil over in the microwave or when swirled. Wear gloves and eye protection when removing the agarose and swirling heated flasks.

3. Place an inverted 25-ml Erlenmeyer flask in the mouth of the 250-ml flask and heat in a microwave on medium heat until all the agarose has completely dissolved. It is best to stop the microwave several times and swirl the flask to ensure thorough mixing.

4. Allow the agarose to cool to about 60°C or place in the 60°C water bath until ready to use.

5. Seal the ends of the gel casting tray with tape or with the end gates. Press the tape tightly against the tray with your fingernail. If trays with end gates are used, do not overtighten the set screws and seal the ends of the tray with a piece of tape. Set the tray on a level portion of the bench.

6. Pour the agarose into the casting tray to a depth of 4 to 5 mm. Place a **12-tooth comb** in the appropriate slots in the casting tray.

7. Allow the gel to solidify on your bench (for about 10 minutes). The gel will become translucent when solidified.

8. Add 1X TAE buffer to the electrophoresis chamber to a level of about 0.5 cm above the center tray support. If 1X TAE buffer is left in the box from a previous run, you may reuse it, but mix the buffer by rocking the chamber back and forth to equilibrate the ions that carry the current during electrophoresis.

9. Pour a small amount of 1X TAE around the comb and carefully remove the comb. Remove the tape from the ends of the casting tray (or lower the gates), exposing the ends of the gel.

10. Place the gel tray in the electrophoresis chamber with the wells toward the negative (black) electrode. Adjust the level of the electrophoresis buffer so that the buffer just covers the gel and the dimples from the wells disappear.

11. After the restriction digestions have incubated for **50 to 60 minutes,** remove the tubes from the water bath and place them in your microcentrifuge tube rack.

12. Label two sterile 1.5-ml microcentrifuge tubes A' and D'. Add 10 μl (about 2 μg) of the digested *V. fischeri* DNA (tube A) to the tube labeled A'. Add 2 μl (0.1 μg) of the digested vector (tube D) and 8 μl water to the tube labeled D'. **Return the chromosomal and vector digestions (tubes A and D) to the water bath to continue digesting.**

13. Add 2 μl of loading dye to tubes A' and D'. Add 5 μl of loading dye to tubes B, C, E, F, and G. **DO NOT** add loading dye to the chromosomal and vector digestions (tube A and D). Pulse each tube for 2 to 3 seconds in the microcentrifuge to mix the dye with the restriction digestion.

14. Retrieve your tube of the *Hin*d III digest of lambda DNA prepared in Exercise 4 (from your group's storage box in the freezer). Float the tube in the 60°C water bath for 2 to 3 minutes to melt the *cos* sites in lambda.

15. Use the micropipetter to load 10 μl of the lambda standard to wells 2 and 11. It is best not to use the outer lanes if possible because the current runs slightly different near the outer edges of the gel. Load the seven reaction samples in wells 3 through 10 as follows:

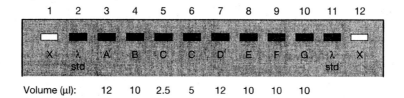

Volume (μl):   12   10   2.5   5   12   10   10   10

16. Place the lid on the electrophoresis chamber and connect the leads to the power supply. Check to be sure that the wells are next to the negative (black) electrode so that the DNA will migrate toward the positive (red) electrode.

17. Set the voltage at 120 V and electrophorese until the bromphenol blue (the leading, darker dye) has migrated to about three-quarters of the way to the end of the gel.

18. When the electrophoresis is complete, turn the voltage down to the lowest level and switch the power source off. Remove the lid of the gel box so that the gel can be removed.

C. Staining and Photodocumentation of Agarose Gel

1. Label a staining tray with your group number. Remove the tray supporting the gel from the electrophoresis box and slide the gel into the staining tray. Be careful as the gels are slippery and can easily slide off the tray.

 CAUTION: In the next step, you will be working with ethidium bromide, which is a mutagen and suspected carcinogen. Wear gloves, eye protection, and a lab coat, and work over absorbent plastic-backed paper at all times when handling it. Carefully read precautions for handling and disposing of ethidium bromide (Appendix XII) before using ethidium bromide stain.

2. At the staining station, flood the gel with ethidium bromide (1 μg/ml) and allow to stain for 5 to 10 minutes. Be sure to wear gloves when handling ethidium bromide. It is a good policy for one partner in each group to wear gloves for staining with ethidium bromide, while the other does not wear gloves and operates the camera, opens doors, and so on.

3. Pour the ethidium bromide solution back into the storage bottle using the funnel provided. The staining solution can be reused 10 to 12 times. Rinse the gel several times in tap or distilled water. Gels are often destained for 5 to 10 minutes in distilled water to reduce background fluorescence, but this is usually unnecessary unless very low amounts of DNA need to be visualized.

4. Use a plastic kitchen spatula to carefully lift the gel from the tray and slide it onto the transilluminator. Be careful to avoid trapping bubbles under the gel. The transilluminator's filter glass is very expensive and can be damaged if scratched. Some gel trays are UV transparent and may be placed directly on the transilluminator with the gel in them.

CAUTION: UV light is hazardous and can damage your eyes. *Never* look at an unshielded UV light. Always view through a UV blocking shield or wear UV blocking glasses. Some transilluminators have a switch on the lid so that the UV light switches off when the lid is opened, but most transilluminators do not have this feature.

5. Lower the UV blocking lid and briefly observe the DNA fragments produced.

6. To photograph the gel, set the handheld Polaroid camera on the transilluminator. The exposure time should be set at 1 second and the F-stop between 5.6 and 8. Steady the camera with one hand, hold your breath, and depress the trigger. For some gels, it may be necessary to do a two- or three-second exposure.

7. Steady the top of the camera and pull the white tab protruding from the camera pack straight out with a firm motion. This will expose a second tab (with arrows). Pull this tab straight out with a firm, steady motion.

8. Allow the photograph to develop for approximately 45 seconds. Peel the print away from the back of the negative by pulling up on one corner of the print. Be careful not to get any of the caustic developing jelly on your hands. Immediately place the negative with the caustic jelly in the trash.

Tape your gel photograph in your notebook below:

D. Final Preparation of Vector and Chromosomal DNA for Cloning

1. Examine the photograph of your gel. Completely digested plasmid vector will appear as a single band while any undigested plasmid will appear as several bands. If the digestion of the vector is complete, retrieve tube D from the 37°C water bath and label it "pGEM™/Sal." Place this tube in your freezer box.

   It is essential that the vector be completely digested. Any undigested vector will be in the supercoiled form, which transforms much more effectively than relaxed DNA. This will result in a high percentage of transformants that lack inserts.

2. Examine the digested chromosomal DNA. If the chromosomal DNA was successfully digested, hundreds of fragments will be formed, which will appear as a nearly continuous smear on the gel. If this is observed, and most of the high molecular weight DNA runs farther than the uncut control, you can assume that the digestion was complete.

   If the digestion of the chromosomal DNA appears complete, retrieve tube A from the 37°C water bath, label it "V.f./Sal," and place it in your freezer box.

Results and Discussion

1. What range of restriction fragment sizes are generated from the *Sal* I digestion of *V. fischeri* DNA? Because *Sal* I recognizes a six base pair sequence, is the size distribution what you would expect? Why or why not? What characteristics of an enzyme or the genome would result in a smaller average size? A larger average size?

2. Based on your results, what sizes of recombinant plasmid molecules would be formed when the digest is ligated to the vector?

3. What range of restriction fragment sizes would you expect if you digested *V. fischeri* DNA with *Hin*d III (refer to Appendix IX for the recognition sequence of *Hin*d III)? Would you expect it to be the same as your *Sal* I digest? Why or why not?

4. What range of restriction fragment sizes would you expect if you digested *V. fischeri* DNA with *Sau*3A I (refer to Appendix IX for its recognition sequence)? Would you expect it to be the same as your *Sal* I digest? Why or why not?

5. Suppose you wish to clone a large gene (2 kb) from a bacterium that has a mole% G + C of 70%. Suggest a possible restriction enzyme (or enzymes) that you would use to digest the chromosomal DNA such that you could recover your gene intact. Justify your answer.

Literature Cited

Boylan, M., A. G. Graham, and E. A. Meighen. 1985. Functional identification of the fatty acid reductase components encoded in the luminescence operon of *Vibrio fischeri*. *J. Bacteriol.* 163:1186-1190.

Engebrecht, J., K. Nealson, and M. Silverman. 1983. Bacterial bioluminescence: Isolation and genetic analysis of functions from *Vibrio fischeri*. *Cell* 32:773-781.

Vieira, J., and J. Messing. 1982. The pUC plasmids, an M13mp7-derived system for insertion mutagenesis and sequencing with synthetic universal primers. *Gene* 19:259.

EXERCISE 9

Quantification of Genomic DNA by Fluorometry and Agarose Plate Fluorescence

There are a number of important steps in the production of recombinant DNA molecules. None of them are more essential than the ligation of the vector DNA to the insert DNA; this is the step where the recombinant molecules are formed. In order to set up these reactions, the concentration of the vector and the insert DNA must be known. By adjusting the concentration of the insert and vector DNA in the ligation reaction, the chances of producing the desired recombinant DNA molecule can be increased.

In Exercise 7, you used ultraviolet light spectrophotometry to quantify DNA and assess its purity. This is a fairly sensitive quantification technique, but it requires substantial volumes of the DNA-containing solutions. This becomes a constraining factor when the quantities of DNA are limited, as is usually the case when cloning. However, DNA can be detected at low levels (<0.1 µg/ml) in small volumes (2-5 µl) through the use of ethidium bromide or Hoechst 33258 enhanced fluorescence of the DNA sample. When these compounds associate with the DNA, there is a marked increase in their fluorescence and the amount of fluorescence is proportional to the mass of the DNA present. With both of these fluorochromes, a standard curve is prepared from dilutions of a standard DNA solution. By comparing the fluorescence of the sample DNA to that of the DNA standards, the quantity of DNA present in the sample can be estimated.

Ethidium bromide is a planar, polycyclic molecule that intercalates between the bases of double-stranded DNA and RNA. When the ethidium is intercalated, there is a large increase in the fluorescence. Ethidium bromide absorbs ultraviolet light at 302 nm and produces an orange fluorescence (506 nm). The amount of the fluorescence is proportional to the mass of the DNA as long as the ethidium bromide is not limiting. RNA will also fluoresce in the presence of ethidium bromide, but its fluorescence is less than that of an equal mass of DNA. This is because ethidium bromide intercalates into double-stranded nucleic acids and RNA is largely single-stranded. Various contaminants present in DNA preparations, such as SDS, can reduce ethidium bromide fluorescence. When working with DNA samples that contain low molecular weight, fluorescence-quenching contaminants, the ethidium bromide method is especially useful because the contaminants can diffuse out of the DNA sample and into the agarose, resulting in a more accurate estimation of DNA quantity. The ethidium bromide method involves adding the dye to agarose plates and spotting samples and standards on the plate that is then viewed on a transilluminator. Alternately, DNA concentrations can be estimated from restriction fragments on a gel if the concentration of DNA standards (such as a *Hin*d III digest of lambda DNA) is accurately known.

The second method of DNA quantification by fluorescence uses the dye Hoechst 33258 (bis-benzimidazole). This dye shows a large fluorescent enhancement when it binds to DNA. It preferentially binds A-T rich areas, but does not appear to intercalate like ethidium bromide. Hoechst 33258 binds twice as well to double-stranded DNA as it does to single-stranded DNA. The fluorescent enhancement from RNA is less than 1% of that produced by an equal mass of DNA, so this procedure can be used to measure DNA in crude preparations that contain significant quantities of RNA. While this method is very quick and more sensitive than the ethidium bromide–agarose plate method, it requires the use of a fluorometer.

When the Hoechst dye–DNA complex is excited with 365 nm light, it fluoresces at 460 nm. The fluorometer used in this exercise (Hoefer TKO 100 Mini-Fluorometer™) is dedicated to these two wavelengths (although any fluorometer may be used). The fluorescence of Hoechst 33258 is quenched by the presence of SDS or ethidium bromide. Consequently, it is essential to completely remove the ethidium bromide from DNA isolated by CsCl-ethidium bromide ultracentrifugation or from agarose gels before quantifying DNA by this method.

In this exercise, you will use both of these fluorescence methods to determine the concentration of your *Sal* I-digested chromosomal and vector DNA. This will allow you to accurately set up appropriate insert-to-vector ratios when performing ligations in subsequent exercises.

Reagents

- digested vector DNA from Exercise 8
- digested insert DNA from Exercise 8
- DNA stock solution (1000 µg/ml calf thymus DNA)
- TE buffer

- ethidium bromide-agarose plates (1/group)
- 1× TNE buffer (100 mM Tris [pH 7.4], 10 mM EDTA, 1.0 M NaCl) containing 0.1 µg/ml Hoechst 33258 dye (in repipet bottle)

Supplies and Equipment

- 0.5–10 µl micropipetter
- 10–100 µl micropipetter
- sterile 10-µl micropipet tips
- sterile 100-µl micropipet tips
- microcentrifuge tubes
- microcentrifuge tube opener
- microcentrifuge tube rack

- transilluminator
- Polaroid camera with type 667 film
- Hoefer TKO 100 mini-fluorometer (or equivalent)
- fluorometer cuvette
- 13 × 100-mm test tubes
- test tube rack

➡Procedure

A. Quantification of DNA by Ethidium Bromide Fluorescence in Agarose Gels

Note: This method works well for estimating the concentration of restriction fragments and is advantageous in that you can use the same gel photograph that you used to check your digestions of vector and insert DNA. It is, however, difficult to estimate the concentration of chromosomal DNA or chromosomal digests with this technique, because they are a smear of DNA on a gel rather than a single fragment.

1. Refer to the photograph taken of the gel you ran in Exercise 8. You loaded 0.5 µg of a *Hin*d III digest of lambda DNA as standards. A molecule of lambda DNA is 48.5 kb long and is digested by *Hin*d III into fragments with the sizes indicated in the following table. Complete the table to determine the amount of DNA (in ng) in each fragment visualized on your gel.

Fragment #	Size (kb)	% of lambda	ng of DNA/fragment
1	23.1		
2	9.42		
3	6.56		
4	4.36		
5	2.32		
6	2.03		
7	0.56		

Thus, the lambda restriction fragments can be used as standards with each fragment representing a known amount of DNA.

2. (Optional) Compare the intensity of the **uncut** chromosomal DNA (lanes 5 and 6) with the lambda standards, and estimate the amount of DNA in each lane (note that it is not possible to estimate the concentration of the digested chromosomal DNA as it is a long smear rather than a distinct fragment).

DNA sample	Lane	ng DNA per lane	Volume loaded	% loaded	µl DNA per lane	DNA concn.
Chromosome (uncut)	5		2.5 µl			
Chromosome (uncut)	6		5 µl			

Based on the amount of *V. fischeri* DNA added to tube C (Exercise 8) and the amount of the tube loaded onto the gel, calculate the concentration (in ng/µl) of the chromosomal DNA. To do this you will need to refer to Exercise 8 and record the volume of DNA placed into tube C and the total volume after the loading dye was loaded to the tube:

Volume of *V. fischeri* DNA in tube C: _____

Total volume in tube C (including loading dye): _____

Based on the volume loaded per lane and the total volume of the tube, calculate the actual volume of *V. fischeri* DNA added per lane. From this you can calculate the concentration. Note that this is the concentration of the uncut chromosomal DNA and not the *Sal* I digest. You will have to use this to calculate the concentration of digested DNA. If this procedure gets too arduous, don't worry; there are easier ways to determine the concentration (see Parts B and C).

3. Repeat this estimate with the digested vector (lane 7) and calculate the concentration (in ng/µl) of the digested vector. Because you loaded the entire contents of the tube (Tube D'), you can directly calculate the concentration of DNA in the digest from the volume of vector removed for the gel.

DNA sample	Lane	ng DNA per lane	µl DNA per lane	DNA concn.
Vector digest	7			

B. Quantification of DNA by Ethidium Bromide–Agarose Plate Fluorescence

1. Label nine microcentrifuge tubes 1 through 9. Add sterile TE buffer to each tube as indicated in the table on the following page:

Tube	TE Buffer	1000 μg/ml DNA stock	100 μg/ml DNA stock	10 μg/ml DNA stock	Final concn.
1	5 μl	5 μl	—	—	500 μg/ml
2	8 μl	2 μl	—	—	200 μg/ml
3	18 μl	2 μl	—	—	100 μg/ml
4	5 μl	—	5 μl	—	50 μg/ml
5	8 μl	—	2 μl	—	20 μg/ml
6	18 μl	—	2 μl	—	10 μg/ml
7	5 μl	—	—	5 μl	5 μg/ml
8	8 μl	—	—	2 μl	2 μg/ml
9	18 μl	—	—	2 μl	1 μg/ml

2. Add the DNA standard solution (1000 μg/ml) to tubes 1 through 3 as indicated in the table. Mix thoroughly by repeatedly drawing the solution up and down with your micropipetter set at 10 μl. Pool the liquid in the bottom of the tube by pulsing for 2 to 3 seconds in a microcentrifuge.

3. Repeat step 2 with the 100 μg/ml standard (note that tube 3 is the 100 μg/ml standard) to prepare tubes 4 through 6. Be sure you have thoroughly mixed tube 3 before proceeding. In a similar manner, use tube 6 as the 10 μg/ml standard to prepare tubes 7 through 9.

4. Label two microcentrifuge tubes "I ⅕" and "I ½₅" and add 8 μl of TE buffer to each tube. Place 2 μl of your insert DNA solution in the tube labeled "⅕" and mix thoroughly to make a ⅕ dilution. Transfer 2 μl of this tube to the tube labeled "½₅" and mix to make the ½₅ dilution.

5. Prepare similar dilutions of the digested vector DNA in tubes labeled "V ⅕" and "V ½₅."

6. Write "STD" on the back of an ethidium bromide–agarose plate near the edge. To the right of this write "V" and to the left write "I" (see Figure 9.1). Spot 5 μl of each dilution of the DNA standard onto the ethidium bromide–agarose plate in a column down the center of the plate under the STD mark. Using a fresh tip for each transfer, place the drops in the same order the dilutions were made (1 through 9). Try to make all of the drops the same size and be careful to place discrete drops to prevent spreading.

7. Place 5 μl from each of the vector dilutions on one side of the standards. Place 5 μl of the insert dilutions on the other side of the concentration standards.

8. Incubate the plate undisturbed until the drops have been absorbed. Place the plates on the transilluminator and photograph. The fluorescence will increase if the plates are allowed to incubate for 1 to 2 hours before observing.

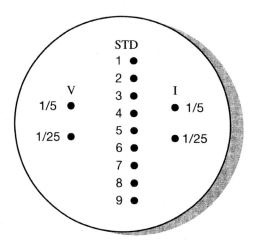

Figure 9.1
Spotting DNA standards and samples on ethidium bromide–agarose plates.

9. Determine the concentration of the DNA present in the sample by comparing the brightness of the vector and insert drops with that of the standard dilutions. Choose a standard with similar intensity as your spot and correct for the dilution of your sample as follows:

$$\text{Concentration of sample (µg/ml)} = \frac{\text{concentration of std.}}{\text{dilution}}$$

Check the concentration of both of the dilutions of each sample to see if they give you the same value. Both dilutions should indicate approximately the same concentration for the samples as long as the intensity of spots fall within the range of the DNA standards.

C. Quantification of DNA by Fluorometry

1. Turn on the fluorometer 15 minutes prior to use. This allows the electronics and light output to stabilize.

2. Use the 1000 µg/ml DNA stock solution to prepare 250 µg/ml, 100 µg/ml, and 25 µg/ml DNA standard solutions. Make each dilution in a total volume of 20 µl. Check your calculations with your instructor before proceeding.

3. Use the repipet to place 2.0 ml of the 1X TNE-dye solution into the cuvette. Carefully wipe all four sides of the cuvette with a Kimwipe™ to remove all moisture, dust, and fingerprints. Insert the cuvette (with the mark on the cuvette forward) into the fluorometer and close the lid. The cuvette should always be placed in the fluorometer in the same orientation.

4. Set the SCALE knob at the most sensitive setting (fully clockwise). Adjust the ZERO knob until the digital display reads 000. Do not touch the SCALE knob again.

5. Remove the cuvette. Add 2 µl of the 100 µg/ml DNA stock dilution. Cover the top of the cuvette with Parafilm™ and mix thoroughly by inversion. Wipe the sides of the cuvette, then place it in the fluorometer (with the mark forward). Adjust the SCALE knob until the readout shows 100, indicating 100 µg/ml (the concentration of the diluted standard solution).

6. Remove the cuvette and pour out the contents. Rinse thoroughly with the 1X TNE-dye solution, then drain. Add 2.0 ml of the 1X TNE-dye, then check the zero reading. Adjust the zero if necessary but do not change the scale setting.

7. Remove the cuvette and add 2 µl of the 25 µg/ml DNA dilution. Mix as in step 5 and read in the fluorometer. Repeat this procedure with the 250 µg/ml standard. These last steps are to check the linearity of the response. The fluorometer readings should be within five units of the standard's concentration. This completes the calibration of the fluorometer.

8. To measure the DNA concentration of your samples, label two 13 × 100-mm tubes I and V. Place 2.0 ml of TNE-dye solution into each tube. Add 2 µl of the insert DNA to **tube I** and 2 µl of the vector DNA to **tube V**. Cover the tubes with Parafilm™ and mix thoroughly by inversion.

9. Rinse the cuvette twice with 1X TNE-dye solution. Fill the cuvette with the 2.0 ml of the TNE-dye solution containing one of your samples. Close the cover and record the value displayed on the digital readout. The value shown is the concentration of DNA in the sample in µg/ml.

10. Repeat step 9 with the other sample and record the concentrations in the following table.

Concentrations of Insert and Vector DNA

Sample	DNA concentration (ng/µl)		
	Agarose gel	Ethidium bromide plate	Fluorometer
Vector			
Insert			

Results and Discussion

1. This exercise involves numerous quantitative dilutions and different units of measure. Such dilutions and interconversion of units are common in molecular biology. Answering the following questions will enhance your ability to use dilutions and to convert various units of measure:

a. If you make a ⅕ dilution of a 1.0 mg/ml solution of DNA,

What is its concentration in µg/ml?
What is its concentration in µg/µl?
What is its concentration in ng/µl?

b. What is the relationship between mg/ml and µg/µl?

c. What is the relationship between µg/ml and ng/µl?

d. You add 5 µl of loading dye to 20 µl of a linearized vector sample, and load 5 µl of this on an agarose gel. You also load 10 µl of a 0.05 µg/µl of a *Hind* III digest of lambda DNA in an adjacent lane. After staining the gel, the fluorescence of the unknown DNA fragment is the same as the fluorescence of the 2.03 kb lambda fragment. What is the concentration of your linearized vector DNA? If you added 4.0 µl of a vector stock to the 20 µl digestion, what was the concentration of the stock?

2. Compare the concentrations determined by the ethidium bromide fluorescence with those determined by the fluorescence of Hoechst 33258. Are the two values comparable? If they are not, which value would you be more likely to believe? Why?

3. What is the lowest amount of DNA (in ng) that you can detect in ethidium bromide–stained agarose gels or on ethidium bromide plates?

4. Why is it important to melt the *cos* sites on the lambda digest prior to loading it onto a gel if you wish to use the lambda fragments to quantify DNA?

5. What are the sources of error in these procedures? How accurate do you think your measurements are (i.e., how many significant figures are justified for your values)? What degree of accuracy do you think is required for DNA concentrations used in gene cloning?

6. When the samples were placed in the fluorometer, a $\frac{1}{1000}$ dilution was made. Based on the concentration of DNA in your DNA samples, what concentration was the fluorometer able to detect? How does the fluorometer sensitivity compare with ethidium bromide fluorescence?

Literature Cited

Brunk, C. F., K. C. Jones, and T. W. James. 1979. Assay for nanogram quantities of DNA in cellular homogenates. *Anal. Biochem.* 92:497–500.

Cesarone, C. F., C. Bolognesi, and L. Santi. 1979. Improved microfluorometric DNA determination in biological material using 33258 Hoechst. *Anal. Biochem.* 100:188–197.

Labarca, C., and K. Paigen. 1980. A simple, rapid, and sensitive DNA assay procedure. *Anal. Biochem.* 102:344–352.

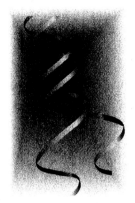

EXERCISE 10
Ligation of Restriction Fragments of *Vibrio fischeri* DNA to a Plasmid Vector

D NA ligation is the reaction that forms recombinant DNA molecules *in vitro* by covalently bonding together two restriction fragments with compatible ends. This reaction is one of the most important steps in any cloning project. The ligation reaction synthesizes an ester linkage between a 3'-OH and a 5'-phosphate (5'-P) of two different DNA molecules (or two ends of the same molecule). This reaction is catalyzed by the enzyme **DNA ligase**—the same enzyme that closes nicks during DNA synthesis. The DNA ligase commonly used in molecular cloning is T4 DNA ligase (see Appendix X), which is purified from an *Escherichia coli* culture that is infected by the bacteriophage T4 (a bacteriophage is a virus that infects bacteria).

In order to carry out the ligation reaction, the DNA ligase requires a 3'-OH from one end of a restriction fragment directly adjacent to a 5'-P of another end. With blunt-ended DNA molecules, this is a rather rare occurrence (although T4 ligase can ligate blunt-ended molecules under certain conditions). However, when DNA is digested with a restriction enzyme that generates complementary cohesive ends, the proper substrate for ligase is formed when the two complementary ends hydrogen bond together. The few hydrogen bonds will not form a stable duplex, and cohesive ends are constantly annealing and separating. However, the weak bonding does hold the cohesive ends together long enough for ligase to catalyze its reaction. In addition to an adjacent 5'-P and a 3'-OH, ligase also requires an energy source (ATP), a Mg^{2+} cofactor, and proper incubation conditions (a reducing buffer at 4–37°C).

When ligating restriction fragments into a cloning vector, the objective is to form recombinant molecules where a restriction fragment from the DNA being cloned (the "insert") is ligated into the vector. Assuming that the DNA preparations are free from any contaminants that might affect the ligase, the formation of the recombinant molecules is a matter of chance. If the ends of all molecules in the reaction (insert and vector) have the same complementary ends, there are a number of possible products. First, the two ends of the vector could ligate to produce the intact vector. Similarly, the insert molecules could ligate to themselves. One of the remaining possibilities, which is the product desired, is that a single vector ligates to both ends of an insert to form a covalently closed circular recombinant molecule. The actual number of possible ligation products is almost limitless. For example, ligations can yield multiple inserts or multiple vectors recombining, multiple inserts within a vector, and so on.

The rate of ligation reactions is essentially a function of the total concentration of complementary ends of DNA. In addition, the success of ligation reactions in forming the desired recombinant DNA molecules containing single inserts is also a function of the relative concentrations of vector and insert. If the total DNA concentration is low, it is likely that two complementary termini that pair will belong to the **same** molecule. This results in circularization of linear molecules. As the concentration of DNA termini is increased, it becomes more likely that a given end will pair with the end of **another** molecule, resulting in the formation of recombinant molecules. If the concentration is increased too much, however, recombinant molecules containing numerous fragments (concatamers) will form. The formation of concatamers should be avoided when using plasmid vectors but is required when cloning in lambda and cosmid vectors. Thus, by carefully adjusting the concentration of the insert and vector DNA in the ligation mix, you can increase your chances of producing the desired recombinant molecules.

Several proposed theories assist in the determination of the concentration and insert-to-vector ratio of DNA in ligation reactions (Dugaiczyk, et al. 1975; Sambrook, et al. 1989). In general, ligations should have a one- to threefold molar excess of insert DNA over vector DNA. From a practical point of view, it is best to set up several ligation reactions and vary the insert-to-vector ratio while keeping the total DNA concentration constant. After transforming these reactions into the host bacteria, it will be readily apparent which mixture has the optimal ratio because it will yield the most transformants containing vectors with inserts. The frequency of transformants containing inserts can also be increased by phosphatasing the vector (see Exercise 8). This significantly reduces the recircularization of the vector and results in a high frequency of vectors with inserts.

In this exercise, you will set up ligation reactions with the *Sal* I digested vector and *Sal* I restriction fragments of *Vibrio fischeri* (Exercise 8). Based on the quantification of the digested insert and vector DNA determined in Exercise 9, you will be able to accurately set up specific vector-to-insert ratios to optimize the cloning. The ligation will result in a large number of recombinant DNA molecules that do not contain the *lux* genes and hopefully a few that do contain the *lux* genes. Because we know that the entire *lux* operon is on a single *Sal* I fragment and that the genes are expressed in *E. coli,* we should be able to readily identify the clones of interest.

PART I. SETTING UP LIGATION REACTIONS

Reagents

- *Sal* I digest of *V. fischeri* DNA (on ice)
  concn. = _____ µg/µl
- *Sal* I digest of pGEM™-3Zf(+) (0.05 µg/µl;
  on ice)   concn. = _____ µg/µl
- T4 DNA ligase (on ice)
- 10X ligation buffer with ATP (on ice)

- sterile deionized water (on ice)
- sterile 3.0 M sodium acetate (pH 5.2) (on
  ice)
- 95% ethanol (ice cold)
- 70% ethanol (ice cold)
- sterile TE buffer (on ice)

Supplies and Equipment

- mini ice buckets (1/group)
- sterile 1.5-ml microcentrifuge tubes
- microcentrifuge tube opener (1/group)
- microcentrifuge tube rack (1/group)
- 0.5–10 µl micropipetter
- 10–100 µl micropipetter
- sterile 10-µl pipet tips

- sterile 100-µl pipet tips
- micropipet tip discard beaker
- 37°C water bath
- 65°C water bath
- microcentrifuge
- low-temperature incubator or chiller
  (12–16°C)

➡Procedure

1. Fill your mini ice bucket with crushed ice and place each of the required enzymes, buffers, and DNAs in it. Each group will be provided with its own set of enzymes and reagents containing slightly more than the required volume.

2. Heat the *Sal* I digests of the vector (the tube labeled "pGEM™/Sal") and the digested *V. fischeri* DNA (the tube labeled "V.f./Sal") at 65°C for 15 minutes. This inactivates the restriction enzyme. Some enzymes are not heat-inactivatable and must be destroyed by extractions with phenol and chloroform (see Sambrook et al. 1989).

3. Label four sterile microcentrifuge tubes L1, L2, L3, and L4. You will use these to set up ligation reactions with different insert-to-vector ratios. Note that these are **molar** ratios. For example, because the *Sal* I fragment containing the *lux* genes is about 9 kb, an equimolar amount of vector (about 3 kb) would be one-third as much DNA (in µg). We will assume that the average size of the digested *V. fischeri* digest is about 9 kb (the approximate average size of restriction fragments generated by an enzyme with a six base pair recognition site is about 4 kb, but because *V. fischeri* has a low G + C content and the recognition sequence for *Sal* I has a high G + C content, the average size of fragments should be larger).

4. Using micropipetters and sterile tips, combine the digested vector and the insert DNA as indicated in the table on the following page. Add sterile deionized water and 10X ligase buffer (+ATP) to give a final volume of 30 µl. See steps 10–16 in this exercise if your DNA concentrations are too low to combine in a 30 µl volume. After all additions have been made, return remaining digested vector and genomic DNA to your freezer box.

Tube	I:V ratio	Digested vector	Genomic digest	H₂O	10X buffer	Total volume
L1	1:1	2 μl (0.1 μg)	_____ μl (0.3 μg)	_____ μl	3 μl	30 μl
L2	2:1	2 μl (0.1 μg)	_____ μl (0.6 μg)	_____ μl	3 μl	30 μl
L3	3:1	2 μl (0.1 μg)	_____ μl (0.9 μg)	_____ μl	3 μl	30 μl
L4	4:1	2 μl (0.1 μg)	_____ μl (1.2 μg)	_____ μl	3 μl	30 μl

5. After adding all ingredients, cap the tubes and vortex to mix. Pool the reagents in the bottom of the microcentrifuge tube by spinning the tubes in the microcentrifuge for 2 to 3 seconds. Be sure that the tubes are in a balanced configuration.

6. Label four sterile microcentrifuge tubes L1/T_0, L2/T_0, L3/T_0, and L4/T_0. Add 5 μl of the corresponding ligation mix to each tube. Spin the tubes in a microcentrifuge for 2 to 3 seconds to pool the mixtures in the bottom of the tubes. Place the tubes in your freezer box. These will be used in the next period to analyze the ligation mixes before and after ligation (by agarose gel electrophoresis).

7. Add 1 μl of T4 DNA ligase to each ligation reaction (tubes L_1 through L_4). Be sure you actually **see** a small drop of ligase transferred to the tube. Pulse the tubes in the microcentrifuge to pool the ligation mix in the bottom. Vortex the tubes, and pulse again in the microcentrifuge to pool the liquid.

8. Incubate your ligation reactions at 10-12°C overnight. Ligations may also be incubated at 16°C for 1 to 4 hours.

9. Clearly label the tubes and place them in your freezer storage box.

Note: If your DNA samples are so dilute that they cannot be combined with the ligase buffer in less than 30 μl, you will have to combine the vector, insert DNA, and sterile H₂O in a volume of 50 μl and concentrate the DNA by ethanol precipitation as described below.

10. Add 0.1 volumes of 3 M sodium acetate to each tube and vortex. Add 2 volumes of ice-cold 95% ethanol and mix. Incubate the tubes in your ice bucket for 10 minutes to precipitate the DNA.

11. Centrifuge at high speed in a microcentrifuge (12-14,000 × g) for 10 minutes to pellet the DNA. Be sure that the hinges on the centrifuge tubes are pointed toward the edge of the rotor so that you will know where in the tube the pellet is located (it will be invisible).

12. Remove the microcentrifuge tubes and carefully remove the supernatant with a 100-µl micropipetter. Slide the tip of the pipetter down the side of the tube **opposite** the pellet. You probably won't be able to see the pellet, but it will be along the side below the microcentrifuge tube tab. Add 350 µl of 70% ethanol to the tube and gently rock to wash the pellet. This removes any traces of salt that may inhibit enzymatic reactions.

13. Centrifuge for 5 minutes at high speed in the microcentrifuge (12-14,000 $\times$ g). Be sure that the hinges on the microcentrifuge tubes are again pointing toward the edge of the rotor.

14. Carefully pour off the ethanol wash and spin the tube in the microcentrifuge for 2 to 3 seconds to pool any remaining liquid. Remove traces of liquid with a micropipetter, being careful not to disturb the pellet.

15. Dry the pellet by setting the open tube on your bench until you can no longer smell any traces of ethanol. Alternately, you may evaporate the ethanol by briefly blowing a stream of warm air from a hair dryer across the top of the tube.

16. Add 3 µl of the 10X ligase buffer (ATP is present in the buffer) and 27 µl of sterile water to each tube. Resuspend the precipitated restriction fragments in the buffer. Use the micropipetter to run the buffer up and down the entire side of the tube where the pellet is located. Do this repeatedly to ensure that all of the DNA is dissolved. Now complete steps 5-9 above.

PART II. CHECKING THE SUCCESS OF THE LIGATION REACTIONS

Reagents

☐ loading dye (room temperature)

☐ 1X TAE buffer

☐ LE agarose (low EEO)

☐ ethidium bromide stain (1.0 µg/ml)

Supplies and Equipment

☐ mini ice buckets (1/group)

☐ 1.5-ml microcentrifuge tubes

☐ microcentrifuge tube opener (1/group)

☐ microcentrifuge tube rack (1/group)

☐ 0.5–10 µl micropipetter

☐ sterile 10-µl pipet tips

☐ micropipet tip discard beaker

☐ 250-ml Erlenmeyer flask (1 per 2 groups)

☐ 25-ml Erlenmeyer flask (1 per 2 groups)

☐ microwave oven

☐ autoclave gloves

☐ top loading balance

☐ spatula

☐ 60°C water bath

☐ microcentrifuge

☐ horizontal mini gel electrophoresis chamber

☐ gel casting tray

☐ 12-tooth comb

☐ gel staining tray

☐ power source

☐ transilluminator

Prior to transforming the competent *E. coli* (which you will prepare in Exercise 11) with the recombinant DNA, we will run an agarose gel to analyze the success of the ligation reactions. This will provide an indication of the number and sizes of ligation products. See Exercise 4 for details on preparing, running, staining, and visualizing agarose gels.

➡️Procedure

A. Agarose Gel Electrophoresis

1. Prepare one batch of **0.8% agarose** in 1X TAE buffer per two groups.

 CAUTION: In the next step, you will be heating agarose to boiling. Agarose can become superheated and boil over in the microwave or when swirled. Wear gloves and eye protection when removing the agarose and swirling heated flasks.

2. Melt the agarose in a microwave oven and cool to about 60°C or place in the 60°C water bath until ready to use. Cast a gel with a **12-tooth** comb. When solidified, place the gel in the electrophoresis chamber.

3. While the gel is solidifying, label four microcentrifuge tubes L1/T$_{end}$, L2/T$_{end}$, L3/T$_{end}$, and L4/T$_{end}$. Remove the ligation tubes from the incubator (or your freezer box) and add 5 µl of each ligation mixture to the corresponding tube that you just labeled. Spin in a microcentrifuge for 2 to 3 seconds to pool the contents at the bottom of the tube. Return the ligation reactions to the freezer.

4. Remove and thaw the four aliquots of the ligation mixes you froze prior to adding the DNA ligase (L1/T_0 through L4/T_0). With these samples, you will be able to compare the unligated and ligated DNA for each ligation reaction.

5. Add 1 μl of loading dye to each T_0 and T_{end} tube. Pulse the tubes for 2 to 3 seconds in the microcentrifuge to mix the dye with the ligation mixture.

6. Remove the *Hin*d III digest of lambda DNA from your freezer box and heat the standards in the 65°C water bath for 2 to 3 minutes to melt the *cos* sites in lambda.

7. Load 5 μl of the *Hin*d III digest of lambda into wells 2 and 11. Do not load any sample in the first and last wells of the gel. Load the entire contents (6 μl) of each ligation tube to wells 3 through 10 as follows:

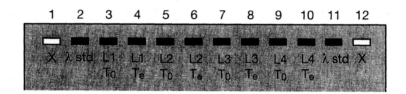

8. Electrophorese at 100–120 V until the bromphenol blue (the leading, darker dye) has migrated one-half to three-quarters of the way to the end of the gel. Good separation is not necessary as we only want to verify that the ligation reactions worked.

9. When the electrophoresis is complete, turn the voltage down to the lowest level and switch the power source off. Remove the lid to the gel box so that the gel can be removed.

B. Staining and Photodocumentation of Agarose Gel

1. Label a staining tray with your group number. Remove the tray supporting the gel from the electrophoresis box and slide the gel into the staining tray.

 CAUTION: Ethidium bromide is a mutagen and suspected carcinogen. Wear gloves, eye protection, and a lab coat, and work over plastic-backed absorbent paper at all times when handling it. Carefully read precautions for handling and disposing of ethidium bromide (Appendix XII) before using ethidium bromide stain.

2. Stain the gel with ethidium bromide (1 μg/ml) for 5 to 10 minutes and rinse several times with tap or distilled water.

3. Briefly observe the gel on the transilluminator and photograph.

 CAUTION: UV light is hazardous and can damage your eyes. *Never* look at an unshielded UV light. Always view through a UV blocking shield or wear UV blocking glasses. Some transilluminators have a switch on the lid so that the UV light switches off when the lid is opened, but many transilluminators do not have this safety feature.

4. Tape your gel photograph in your notebook below:

Results and Discussion

1. Examine the photograph of your gel. Did all of your ligations work? How does the agarose gel verify that the ligation was successful or unsuccessful?

2. How many different ligation products were produced and what is their size range? Are they what you expected?

3. Using two different colored pencils (for vector and insert), diagram the possible ligation products in your reactions. Which do you think are most prevalent? Why?

4. Will all the ligation products formed be replicated if transformed into *E. coli*? Why or why not?

5. Based on your understanding of ligation reactions, would you prefer to use vector and insert cut with restriction enzymes that yield blunt or sticky ends? Why?

6. Under which of the following conditions would it be most advantageous to treat your cut vector with alkaline phosphatase to remove terminal 5'-Ps?

   a. The vector was cut with one restriction enzyme that yields sticky ends.

   b. The vector was cut with one restriction enzyme that yields blunt ends.

   c. The vector was cut with two different restriction enzymes that yield one blunt and one sticky end.

   d. The vector was cut with two different restriction enzymes that each yield sticky ends.

Literature Cited

Dugaiczyk, A., H. W. Boyer, and H. M. Goodman. 1975. Ligation of *Eco*R I endonuclease-generated DNA fragments into linear and circular structures. *J. Mol. Biol.* 96:171.

Sambrook, J., E. F. Fritsch, and T. Maniatis. 1989. *Molecular cloning: A laboratory manual.* 2d ed. New York: Cold Spring Harbor Laboratory Press.

EXERCISE 11
Preparation of Competent
Escherichia coli DH5α

Once a recombinant DNA molecule is formed, it is necessary to transfer the molecule into a living organism in order for it to be replicated. The production of many identical copies of a recombinant DNA molecule by such a host is referred to as "cloning" DNA. Because most cloning vectors are either bacterial plasmids or bacteriophages (viruses that infect bacteria), a means to transfer DNA into a bacterial host is required. Bacteriophage vectors can readily be transferred to a host by simply infecting a bacterium with intact phage particles that contain recombinant phage DNA.

In the early 1970s, an obvious mechanism to transfer recombinant plasmids into a bacterial host was not available. Some bacteria were known to take up free DNA from solution by a natural process known as transformation. However, this is a rather rare event, limited to only a few genera of bacteria, and occurs only when the bacteria are in a state referred to as "competent." *Escherichia coli,* which has long been the genetic workhorse and the host of choice for gene cloning, does not undergo natural transformation. Fortunately, however, in 1970 Mandel and Higa (1970) demonstrated that "nontransformable" bacteria (such as *E. coli*) could take up free DNA after being treated with a solution of ice-cold calcium chloride and then subjected to a brief heat shock. Thus, it is possible to artificially induce competence in nontransformable bacteria, allowing them to successfully take up recombinant DNA molecules. The exact mechanism of transformation in artificially induced competent cells is unknown, but it seems to involve damage to the cell envelope, which allows DNA to pass into the cell.

The original procedure of Mandel and Higa is currently the basis for most transformation procedures used today, although other cations may be used that give equivalent or better transformation efficiencies. In addition, other methods, such as electroporation, are now available for transferring recombinant DNA into bacteria. Electroporation has proven very useful with many organisms that are difficult to transform with the calcium procedure.

When transforming recombinant DNA, it is important that the bacterial host has certain genetic attributes. First, it is essential that the host lack restriction endonucleases so that it does not degrade the recombinant DNA molecules. Most bacterial hosts are also *rec*A mutants. The RecA protein is responsible for homologous recombination in a cell. Thus, if a host had a functional RecA protein, foreign DNA inserted into plasmids could undergo homologous recombination and be incorporated into the host chromosome. This would prevent the cloned DNA from being recovered. Bacterial hosts used in recombinant DNA research also possess mutations resulting in modified cell envelopes that decrease their survival outside of the lab.

Hosts for vectors derived from pUC-type plasmids (such as pGEM™) contain multiple cloning sites in the α-fragment of the β-galactosidase gene (*lacZ*α; see Exercise 8) and must also harbor a *lacZ* deletion (*lacZ*ΔM15) that results in the synthesis of a nonfunctional β-galactosidase. The mutant host protein, however, can become functional in the presence of the α-fragment of β-galactosidase encoded by the cloning vector (this is referred to as α-complementation). Thus, an *E. coli* (*lacZ*ΔM15) strain containing a pGEM™ vector will produce functional β-galactosidase when the *lac* operon is induced. Because β-galactosidase can cleave a variety of chromogenic substrates to form colored products, such a substrate can be added to media which will result in the formation of colored colonies if β-galactosidase is made. For example, X-gal (5-bromo-4-chloro-3-indoyl-β-D-galactoside) is a nontoxic substrate that is hydrolyzed by β-galactosidase to form a blue dye and is commonly added to bacteriological media to visualize colonies with β-galactosidase activity.

α-complementation is a simple, rapid, and elegant system to screen for transformants containing inserts cloned into pUC-like plasmids. Because the transformed cells are plated on media containing an antibiotic, only cells that have taken up vector DNA, which confers antibiotic resistance, will grow. If a host with the *lacZ*ΔM15 mutation takes up a vector **without** an insert, α-complementation will occur and blue colonies will result. However, if the host takes up a plasmid that **has an insert** ligated into the multiple-cloning site of pGEM™, the α-fragment will not be synthesized. These cells will lack β-galactosidase activity and thus form white colonies. This eliminates the tedious and time-consuming process of replica plating colonies to determine which ones contain inserts, as was required with earlier cloning vectors such as pBR322.

In this exercise, you will use the calcium chloride procedure to artificially induce competency in *E. coli* strain DH5α. DH5α is a suitable host for pGEM™ vectors as it lacks restriction endonuclease activity (r_k^-) and contains the *lacZ*ΔM15 deletion. We will test the transformation efficiency of the cells, and they will then be transformed in Exercise 12 with the *V. fischeri* DNA that you ligated into pGEM™-3Zf(+). Remaining competent cells may be frozen for use at a later date.

PART I. PREPARATION AND TESTING COMPETENT CELLS

Cultures

☐ *E. coli* DH5αF⁻, ø80d*lacZ*ΔM15, *end*A1, *rec*A1, *hsd*R17 (r$_k$⁻, m$_k$⁻), *sup*E44, *thi*-1, *gyr*A96, *rel*A1, Δ(*lacZYA-arg*F), U169, λ⁻ (streaked on an LB plate)

Media

☐ 25 × 150-mm culture tube with 5 ml LB broth (1/group)
☐ 15-ml tube with 4 ml LB broth

☐ 250-ml shake flask with 50 ml LB broth (1/group)
☐ LB plates + 50 µg/ml ampicillin (LB/Ap⁵⁰; 4/group)

Reagents

☐ sterile 100 mM CaCl$_2$ (on ice)
☐ pGEM™-3Zf(+), 0.1 ng/µl (on ice)

☐ sterile TE buffer (on ice)

Supplies and Equipment

☐ ice chests
☐ sterile 50-ml polycarbonate or polyallomer Oak Ridge tube
☐ trip balance with 100-ml beaker on each pan
☐ 0.5–10 µl micropipetter
☐ 10–100 µl micropipetter
☐ 100–1000 µl micropipetter
☐ sterile 10-µl micropipet tips
☐ sterile 100-µl micropipet tips
☐ sterile 1000-µl micropipet tips
☐ micropipet tip discard beaker
☐ sterile 5- and 10-ml pipets
☐ discard beaker containing disinfectant solution

☐ spectrophotometer
☐ disposable cuvettes
☐ cuvette rack
☐ refrigerated centrifuge with rotor for 50-ml tubes (prechilled to 4°C)
☐ 37°C shaking incubator
☐ 37°C shaking water bath
☐ 42°C water bath
☐ sterile 15-ml disposable culture tube (2/group)
☐ alcohol jars for flaming
☐ spreading triangles
☐ turntables

➡ Procedure

A. Preparation of Competent Cells

Note: It is important that careful aseptic technique be used throughout this entire exercise.

The day before lab, each group should inoculate a 25 × 150-mm culture tube containing 5 ml of LB broth with a single colony of *E. coli* DH5α streaked on an LB plate. Label the tube with your group number and incubate the tube in a 37°C shaking water bath overnight.

1. Inoculate the 250-ml shake flask containing 50 ml of LB broth with 1 ml (2% inoculum) of the overnight culture of *E. coli* DH5α. Place in the 37°C shaking incubator at 300 rpm.

2. After 2 to 2.5 hours, aseptically remove 2 ml of the culture and check the absorbance at 600 nm on the spectrophotometer. Use the tube of sterile LB broth to zero the spectrophotometer. For efficient transformation, the cell density should be no more than 10^8 cells/ml. This corresponds to an A_{600} of about 0.4 to 0.6. Continue to check the absorbance until the culture is in this range.

3. When the A_{600} is between 0.4 and 0.6, remove your flask from the shaker and place it in your ice chest. Swirl periodically for 5 to 10 minutes to thoroughly chill the cells. Place the sterile 50-ml Oak Ridge centrifuge tube into the ice chest.

   It is crucial that the cells be kept at 0°–4°C for the remainder of this procedure, or their competency will be reduced significantly.

4. Aseptically pipet 38 ml of the chilled culture into the prechilled Oak Ridge centrifuge tube. Label the tube by placing a small piece of tape with your group number on the cap. **DO NOT** write on Oak Ridge tubes with a permanent marker! Balance your tube with another group's on the trip balance. **Aseptically** add or remove culture to balance tubes to ±0.5 g.

5. Centrifuge the cells at 3000 × *g* (5000 rpm in an SS34 rotor) for 10 minutes at 4°C.

6. Pour the supernatant into the discard beaker containing disinfectant solution. Use a sterile pipet to remove traces of medium.

7. Add 10 ml of sterile, ice-cold 100 mM $CaCl_2$ (one-fourth the original culture volume) and **gently** swirl to resuspend the pellet. After the pellet is resuspended, add an additional 10 ml of ice-cold 100 mM $CaCl_2$ and mix gently. It is important from this point on to handle the cells very gently as they become fragile in the $CaCl_2$. Place the tube in your ice chest and incubate on ice for 20 minutes.

8. Dry off the outside of the Oak Ridge tube and centrifuge the $CaCl_2$ suspension at 3000 × *g* (5000 rpm in an SS34 rotor) for 10 minutes at 4°C.

9. Carefully pour the supernatant into the discard beaker and resuspend the pellet in 4.0 ml (one-tenth of the original culture volume) of ice-cold $CaCl_2$. **Do not vortex!** Cells are very fragile after the $CaCl_2$ treatment and should be resuspended by **gently** swirling the tube.

10. Wrap a piece of tape **completely** around the upper portion of the tube and label with your group number. Store in an ice slurry. Competent cells prepared with the $CaCl_2$ procedure may be used immediately, but the transformation efficiency will increase up to sixfold after 24 hours on ice. By 48 hours, the efficiency will have decreased to the original level.

B. Test Transformation of Competent Cells

To test the competency of the cells, we will do a test transformation (see Exercise 12) to determine how efficiently your cells transform DNA.

1. Wrap tape completely around two sterile 15-ml disposable culture tubes and label each with a permanent marker as follows:

   A. TE (negative control)
   B. pGEM™ (positive control)

   Place tubes on ice.

2. Add 200 µl of the competent cells to each tube. Use the 100-1000 µl micropipetter and a **sterile** tip. Swirl the cells gently before removing the aliquot. Keep the competent cells on ice at all times.

3. Using the 0.5–10 µl pipetter with a sterile tip, transfer 10 µl of sterile TE to tube A as a negative control. Be sure you add the TE directly to the competent cell solution.

4. Add 10 µl of the pGEM™-3Zf(+) stock (0.1 ng/µl) to tube B as a positive control. Be sure that you add the DNA **directly** into the competent cell solution. Mix each tube gently and incubate tubes on ice for 20 minutes.

5. Heat-shock cells by placing them in the 42°C water bath for **exactly** 90 seconds. Return the tubes to the rack on your bench.

6. Add 1 ml of sterile LB broth (room temperature) to each tube and mix gently. Incubate the tubes at 37°C, (preferably with gentle shaking) for 45 to 60 minutes (be sure your group number is on your tubes). Why is this incubation necessary?

7. While your tubes are incubating, collect four LB plus ampicillin (LB/Ap^{50}) plates and label them appropriately. You will spread the transformed cells on these plates as indicated in the following table:

Tube	Addition	# of plates	Volume plated
A	TE buffer	1	100 µl
B	pGEM™-3Zf(+)	3	10 µl, 50 µl, 200 µl

8. After the 45- to 60-minute incubation with the LB broth is complete, spread the transformed cell suspensions on the LB/Ap^{50} plates. Plate the volumes indicated in the above table (use the spread plate technique described in Exercise 1; be sure to flame-sterilize your spreading triangle). For volumes less than 100 µl, first add 100 µl of sterile LB broth to the plates and add the transformed cells to the drop of LB. The additional LB broth ensures that the small volume of transformation mix will be evenly spread on the plate.

9. Leave the plates upright for 5 to 10 minutes after spreading to let the fluid soak into the plates. When dry, invert and incubate for 18 to 24 hours at 37°C.

10. Place your tube of competent cells in a rack in an ice chest with an ice slurry. They will be incubated on ice overnight.

PART II. DETERMINATION OF EFFICIENCY & STORAGE OF COMPETENT CELLS

Reagents

☐ sterile 50% (v/v) glycerol
☐ dry ice

☐ ethanol

Supplies and Equipment

☐ sterile 2.0-ml cryovials
☐ foam float for 2.0-ml cryovials
☐ 1-liter beaker

☐ 100–1000 µl micropipetter
☐ sterile 1000-µl micropipet tips
☐ ultracold (−70 to −80°C) freezer

▶ Procedure

A. Determination of Transformation Efficiency of Competent Cells

1. Examine your plates and count the number of colonies on each. Record your results in the following table:

Plate	Volume plated	# Colonies/plate
TE control	100 µl	
pGEM™-3Zf(+)	10 µl	
pGEM™-3Zf(+)	50 µl	
pGEM™-3Zf(+)	200 µl	

2. Based on the number of colonies on a countable plate (generally between 30 and 300 colonies), calculate the transformation efficiency. Transformation efficiencies are traditionally expressed as the number of transformants per µg of supercoiled plasmid DNA (such as pGEM™ or pUC plasmids). Because you know the amount of DNA added to the competent cells and the total volume of LB culture that the volumes plated were removed from, you can calculate the number of transformants per µg of pGEM™ DNA.

Transformation efficiency = _____ transformants/µg pGEM™ DNA

B. Freezing Competent Cells for Long-Term Storage

1. If competent cells yield a high transformation efficiency (>2 × 10⁵/µg pGEM™) based on the test transformation, excess cells may be frozen in an ultracold freezer and used at a later date. Freeze cells as follows:

   CAUTION: Dry ice-ethanol baths are extremely cold (−70°C) and can freeze skin. Wear autoclave gloves when handling dry ice and the dry ice-ethanol bath.

   a. Prepare a dry ice-ethanol bath by placing 100-200 g of dry ice pellets in the bottom of a 1-liter beaker. Add ethanol until all the dry ice is covered. Be careful—the beaker will be extremely cold.

   b. Add 400 µl of competent cells and 100 µl of 50% (v/v) sterile glycerol to a sterile cryovial or microcentrifuge tube (this yields a final concentration of 10% glycerol). Cap and mix **gently** by inversion several times. Label the cryovial or tube with **pencil** (markers will come off in the ethanol bath).

   c. Quick-freeze vials by placing in a foam float and submerging in the dry ice-ethanol bath. Cells will freeze in about one minute and may be stored in a −70°C freezer indefinitely. Transfer them immediately to a −70°C freezer or place them on dry ice to prevent the frozen vials from thawing.

Results and Discussion

1. Think about the size of a recombinant DNA plasmid and the nature of the bacterial cell envelope. Do you think it would be difficult for a molecule of DNA to cross the envelope? What might the calcium be doing to allow this to occur?

2. Which of the following do you think would transform with the greatest efficiency: (i) a 5 kb linear plasmid (yes, they do exist!); (ii) a 5 kb supercoiled plasmid; or (iii) a 5 kb nicked (relaxed) plasmid? Justify your answer.

3. Freeze-thaw cycles and organic solvents have also been shown to induce competency in some species of bacteria. Speculate on how these treatments may work.

Literature Cited

Mandel, M., and A. Higa. 1970. Calcium-dependent bacteriophage DNA infection. *J. Mol. Biol.* 53:159.

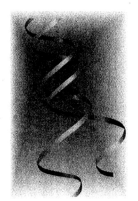

EXERCISE 12
Transformation of Competent *Escherichia coli* DH5α with Recombinant Plasmids

Perhaps the most widely used method to introduce recombinant plasmids into a host is via transformation of a host organism that has been treated with calcium chloride or other cations to induce competency. The transformed cells are plated on a medium containing an antibiotic that only permits growth of cells that have taken up the plasmid (recall that cloning vectors have a gene coding for antibiotic resistance). The resulting transformants will contain one of two types of plasmids: (i) plasmid vectors containing insert DNA from *Vibrio fischeri;* or (ii) intact vector that was either not cut during the restriction endonuclease digestion or that has ligated to itself. Prior to screening the transformants for clones containing the gene(s) of interest, one must first distinguish between these two possibilities. With early plasmid vectors such as pBR322, inserts were ligated into one of two different antibiotic resistance genes and transformants plated on media containing the other antibiotic. Clones with inserts were then identified by loss of antibiotic resistance due to insertional inactivation. This involved replica plating each bacterial colony onto a medium containing both antibiotics. Clones that would not grow on the replica plate could then be identified as having inserts.

Screening by insertional inactivation is a time-consuming process. Modern plasmid vectors, such as pGEM™-3Zf(+), allow for immediate screening of transformants by α-complementation (see Exercise 11). To distinguish between transformants with and without inserts, the plating medium contains the selective antibiotic and also isopropyl-β-D-thiogalactoside (IPTG) and 5-bromo-4-chloro-3-indoyl-β-D-galactopyranoside (X-gal). IPTG is a "gratuitous" inducer of the lactose operon (it induces transcription of the *lac* operon but is not metabolized by the cell). Once the operon is induced, the *lac*Z gene product (β-galactosidase) is made. If a transformed cell contains vector without an insert, the host is able to synthesize the mutated β-galactosidase protein (from the *lac*ZΔM15 gene) and the *lac* promoter in the vector synthesizes the α-fragment of β-galactosidase. These two nonfunctional peptide fragments then associate to form a functional protein that cleaves X-gal, resulting in blue colonies. If the vector contains an insert, which is ligated just downstream of the *lac* promoter, transcription of the **insert DNA** will occur and the α-fragment will not be synthesized. Thus, no functional β-galactosidase is formed and colonies with inserts will appear white on an X-gal/IPTG plate.

In this exercise, you will use the recombinant plasmids constructed in Exercise 10 to transform the competent *E. coli* DH5α produced in Exercise 11. The transformed cells will be plated on a medium containing ampicillin and X-gal (IPTG can be omitted when using pUC-type vectors; Sunday 1987). After determining the plating volume required to yield a reasonable number of transformants per plate, each group will plate the transformation mixture on several dozen plates to generate a *V. fischeri* library. The library will then be screened for clones containing the *lux* operon in Exercise 13.

PART I. INITIAL TRANSFORMATION

Cultures

❑ competent *E. coli* DH5α cells (on ice; prepared in Exercise 11)

Media

❑ sterile LB broth (25 ml/group)
❑ LB plates + 50 µg/ml ampicillin + X-gal (LB/Ap^{50}/X-gal; 15/group)

Reagents

❑ ligation vector/insert mixtures (on ice; from Exercise 10)
❑ *Sal*™ I digested pGEM™-3Zf(+) (on ice; from Exercise 8)
❑ pGEM™-3Zf(+), 0.1 ng/µl (on ice)
❑ sterile TE buffer (on ice)

Supplies and Equipment

❑ ice chests
❑ sterile 15-ml disposable culture tubes (7/group)
❑ 0.5–10 µl micropipetter
❑ 10–100 µl micropipetter
❑ 100–1000 µl micropipetter
❑ sterile 10-µl micropipet tips
❑ sterile 100-µl micropipet tips
❑ sterile 1000-µl micropipet tips
❑ micropipet tip discard beaker
❑ 42°C water bath
❑ 37°C shaking water bath
❑ alcohol jars for flaming
❑ spreading triangles
❑ turntables

Procedure

Note: Use careful aseptic techniques throughout this procedure.

1. Collect your frozen ligation mixes from Exercise 10 (tubes L1 through L4). Spin in a microcentrifuge for 2 to 3 seconds to pool the contents in the bottom of the tubes.

2. Collect the *Sal* I digested vector (tube D from Exercise 8). Label a sterile microcentrifuge tube "8-D" and add 1 µl of the digested vector and 9 µl of sterile TE to it. Mix and spin in a microcentrifuge for 2 to 3 seconds to pool the contents. Place all tubes in your ice chest.

3. Wrap tape completely around each of seven sterile 15-ml disposable culture tubes and label each with a permanent marker as follows:

   A. TE (negative control)
   B. pGEM™- (positive control)

        C. digested pGEM™

        D. L1 (1:1)

        E. L2 (2:1)

        F. L3 (3:1)

        G. L4 (4:1)

Place tubes on ice.

4. To each tube, add 200 μl of the competent cells you prepared in Exercise 11. These have set on ice overnight and should have increased in competency since they were prepared. Use the 100–1000 μl micropipetter and a **sterile** tip. Leave tubes on ice.

5. Using a 0.5–10 μl micropipetter with a sterile tip, transfer 10 μl of sterile TE to tube A as a negative control. Add 10 μl of the uncut pGEM™-3Zf(+) stock (0.1 ng/μl) to tube B as a positive control. Be sure that you add the DNA **directly** into the competent cell solution.

6. To tube C, add the 10 μl of the digested vector dilution you prepared in step 2 (tube 8-D). This is a control to examine if the vector digestion was complete.

7. Add 10 μl of each of the ligation mixes to tubes D through G. Mix gently and incubate tubes on ice for 20 minutes. Be sure that you add the DNA **directly** into the competent cell solution.

8. Heat-shock cells by placing them in the 42°C water bath for **exactly** 90 seconds. Return them to the rack on your bench.

9. Add 2 ml of LB broth (room temperature) to each tube and mix gently. Incubate the tubes at 37°C (preferably with gentle shaking) for 45 to 60 minutes (be sure your group number is on your tubes). Think about what is happening to the cells during this incubation and why it is required.

10. While your tubes are incubating, collect 15 LB/Ap[50]/X-gal plates and label them appropriately. You will spread the transformed cells on these plates as indicated in the following table:

Tube	Addition	# of plates	Volume plated
A	TE buffer	1	100 μl
B	pGEM™-3Zf(+)	1	100 μl
C	digested pGEM™-3Zf(+)	1	100 μl
D	L1 (1:1)	3	10 μl, 50 μl, 200 μl
E	L2 (2:1)	3	10 μl, 50 μl, 200 μl
F	L3 (3:1)	3	10 μl, 50 μl, 200 μl
G	L4 (4:1)	3	10 μl, 50 μl, 200 μl

11. After the 45- to 60-minute incubation with the LB broth is complete, spread the transformed cell suspensions on the LB/Ap^{50}/X-gal plates. Plate the volumes indicated in the above table (use the spread plate technique described in Exercise 1; be sure to flame-sterilize your spreading triangle). For volumes less than 100 μl, first add 100 μl of sterile LB broth to the plates and add the transformed cells to this. The additional LB broth ensures that the small volume of transformation mix will be evenly spread on the plate.

12. Leave the plates upright for 5 to 10 minutes after spreading to let the fluid soak into the agar. When dry, invert and incubate for 18 to 20 hours at 37°C. **Note: Do not let the plates incubate for longer than this at 37°C.**

13. Immediately after spreading the plates, place the tubes of transformed cells (transformation mix) on ice. Store them in the refrigerator. We do not want them to continue growing, or we will not be able to determine the appropriate volumes to plate tomorrow. Be sure that each tube is clearly labeled with your group number and the identity of the contents.

You will spread more of the transformed cells on a large number of plates in the next period. However, you must first determine which insert:vector ratios give the highest number of transformants with inserts and which plating volume gives an appropriate number of colonies to screen (approximately 200 colonies per plate).

PART II. CREATION OF GENOMIC LIBRARY

Cultures

☐ refrigerated transformation mixtures (from Exercise 12, Part I)

Media

☐ sterile LB broth (25 ml/group)
☐ LB plates + 50 µg/ml ampicillin + X-gal (LB/Ap^{50}/X-gal; 20/group)

Supplies and Equipment

☐ 100–1000 µl micropipetter
☐ sterile 1000-µl micropipet tips
☐ micropipet tip discard beaker

☐ alcohol jars for flaming
☐ spreading triangles
☐ turntables

➡ Procedure

1. After 18 to 20 hours, remove your transformation plates from the 37°C incubator and incubate for 2 to 3 hours at room temperature (20° to 26°C). This is necessary because bioluminescence does not occur at 37°C, even though this is the optimum temperature for *E. coli*.

2. Organize your plates according to the table from Period 1. First, examine your control plates. You should have no transformants on the negative control plate (A) and numerous colonies on the positive control plate (B). What color are the colonies on the positive control? Can you explain why?

3. Examine plate C. Are there colonies on it? What color are they? Think about what this control tells you and why it is included.

4. Now examine the three plates prepared from each of the transformation mixes. You should have a mixture of white and blue colonies. The objective of this preliminary transformation was to determine which ligation mixture results in the greatest frequency of white colonies (clones with insert DNA) and to determine a volume of the transformation mix that yields a reasonable number of colonies per plate (200). It is important that the density not be too great or it will be difficult to pick clones due to crowding. In addition, luminescent clones must reach a certain size before they will begin to bioluminesce, and the colonies may be too small if the plates are crowded.

5. You may already have a clone that contains the *lux* genes. Because the screening is easy with such a readily observable phenotype, take your plates transformed with the ligation mixtures into a dark room and check to see if you have a bioluminescent clone. If you do, congratulations! You have just cloned the *lux* operon and genetically engineered *E. coli* to glow in the dark! See Exercise 13 for tips on screening and how to pick a bioluminescent clone.

6. You will now create a genomic library with your saved transformation mixes (tubes D through G). Determine which insert:vector ratio gave the highest frequency of white colonies (transformants with inserts) and record your data below:

Tube	I:V	Volume plated	# of blue colonies	# of white colonies	% whites	Volume LB to add
D (L1)						
E (L2)						
F (L3)						
G (L4)						

Calculate the amount of sterile LB to add to the best transformation (you may use more than one) so that plating 100 µl will yield approximately 200 colonies. Record your volumes in the table above and verify your calculations with your instructor before proceeding.

7. Spread each of 20 LB/Ap^{50}/X-gal plates with 100 µl of this transformation mixture and allow the plates to dry. Be sure each plate is labeled with your group number and the transformation mixture used.

8. Incubate all plates for 18 to 20 hours at 37°C. Your instructor will transfer the plates to room temperature 3 to 4 hours before you screen the library (see Exercise 13) to allow the bioluminescence genes to be expressed.

Results and Discussion

1. As in all scientific experiments, the controls set up in this exercise are very important. Think about the rationale for each control and what the results tell you. Record your thoughts and results below:

Control	Rationale	Results
TE buffer		
Uncut pGEM™		
Digested pGEM™		

2. Based on the number of transformants found on your uncut pGEM™ control, calculate the transformation efficiency of your cells (see Exercise 11). Did the efficiency increase? By how much? Propose a possible explanation.

3. Did you observe transformants in the control with the digested pGEM™ vector? If so, what does this tell you?

4. What types of misleading results could you obtain if you omitted the controls? Why were different volumes plated for different controls?

5. Did you obtain different percentages of transformants containing inserts with the different insert:vector ratios? If so, what is a possible explanation? Should other ratios have been tested? If so, suggest which ones.

6. Note that we did not include IPTG in the plates that we used in this exercise. In wild-type *E. coli*, β-galactosidase is not expressed in the absence of an inducer such as IPTG. Why do you think we can get blue colonies of *E. coli* DH5α (which indicates that the mutant β-galactosidase from the genome and the α-fragment were both synthesized) in the absence of IPTG? Hint: Refer to the map of pGEM™ and review the requirements for expression of the *lac* operon.

Literature Cited

Sunday, G. J. 1987. Use of IPTG with M13mp and pUC hosts. *Gibco BRL Focus.* 9:16.

EXERCISE 13
Screening the *Vibrio fischeri* Genomic Library for Light Producing Clones

There are numerous concerns to be addressed when creating a genomic library. One of the most important is, "How many clones need to be screened in order to find the target gene(s)?" This question is based on the premise that the clones represent random pieces of the chromosome such that no segment of DNA was systematically excluded during the preparation of the recombinant plasmids. Assuming that this premise is met, the number of clones needed will depend on several factors: (i) the genome size of the organism you are working with; (ii) the average size of the insert DNA in your clones; and (iii) the frequency of occurrence of the target gene in the genome of the organism.

Assuming that the gene of interest occurs once in the genome of an organism, the probability that the gene is represented in a given library can be estimated statistically from calculations based on the Poisson distribution (Clarke and Carbon 1976). The number of clones N that must be screened to isolate a given sequence with a probability of P is given by the following equation:

$$N = \frac{\ln(1 - P)}{\ln(1 - I/G)}$$

where I is the average number of base pairs per cloned fragment and G is the number of base pairs in the target genome. For example, if a library is being prepared from the *Escherichia coli* genome (approximately 4×10^6 bp) and the average fragment size is 9 kb, 2275 clones would need to be screened to achieve a 0.99 (99%) probability that all genes would be represented. An easy rule of thumb is that in order to achieve a 99% probability of finding a given sequence in a library, the total number of base pairs present in the clones screened must be about five times greater than the total genome size of the target organism. Therefore, the larger the average clone size, the fewer clones you will need to screen. This relationship also shows that the larger the genome of the target organism, the larger the number of clones that must be screened in order to achieve a high probability of finding a given DNA sequence. This becomes a major problem when making genomic libraries from eukaryotic organisms. For example, the size of the human genome is almost 1000 times larger than that of *E. coli*. Thus, more than 2 million clones would have to be screened in a human genomic library with an average fragment size of 9 kb in order to ensure a 99% probability of obtaining a target sequence. For this reason, much of the work on the Human Genome Project uses Bacterial Artificial Chromosomes (BACs) and Yeast Artificial Chromosomes (YACs) that have inserts ranging from 100 to more than 1000 kb.

The steps involved in producing a genomic library are often complex and time-consuming—but the techniques that are used are well-established and many of the pitfalls are known. The most difficult and crucial step is often the screening of the recombinants. A selection or screening method that is insensitive or inappropriate can negate the entire effort required to create the library.

A selection or screening procedure represents an assay for the specific gene or genes of interest. Many strategies can be used for identifying desired clones, but most fall into one of a few categories. A simple, direct selection strategy is the complementation of a genetic defect. For example, if you want to clone the genes for proline synthesis from a species of *Vibrio,* you could isolate or obtain an *E. coli* mutant that is deficient for proline synthesis and use this mutant as the host for the recombinant plasmids. If the host cells are plated on minimal medium after transformation, the only organisms that will grow are those that have acquired the ability to synthesize the essential amino acid proline. If a mutant *E. coli* host acquired the ability to produce proline, it would be safe to conclude that the recombinant plasmid that it carries encodes the *Vibrio* genes for proline synthesis. Such direct methods eliminate the need to screen numerous clones and save tremendous amounts of time.

Another approach used to identify clones containing a desired gene is to screen for the product of that gene with polyclonal or monoclonal antibodies. Obviously, this strategy requires that the cloned gene be expressed in the host organism. This can be accomplished if the host recognizes the promoter of the target organism's gene (which is often the case with diverse bacterial species) or by cloning the genes in an expression vector where a bacterial promoter initiates transcription of the insert DNA. A second requirement for antibody screening is that the cells must be lysed (unless the protein associates with the host's outer membrane) to release the gene product. Cells are lysed with a detergent followed by binding the cell lysates onto a solid support like nitrocellulose or nylon membranes. The membrane is then immersed in a solution that contains antibodies to the target protein. After allowing sufficient time for the antibodies to bind to the target (antigen), the excess antibodies are washed away. The presence of the primary antibody on the membrane is detected by a second antibody that carries a radioactive, fluorescent, or enzymatic label that allows the clone producing the target gene product to be identified.

When the object of the cloning is a gene encoding an enzyme, it is often easy to screen for enzymatic activity. There are a wide range of commercially available compounds that are enzyme substrates or analogs of enzyme substrates. These are often conjugated to molecules that become chromogenic or fluorogenic upon cleavage of the substrate. Methyl umbelliferone and o-nitrophenol compounds are commonly used for such purposes. Some enzymes, such as many oxygenases, produce colored compounds from their native substrates, and these can be screened without the use of synthetic substrates. Application of the substrate or conjugated substrate is readily done by spraying the clones with a fine mist of a solution containing the substrate.

All of the previous methods require the expression of a gene in order to identify the clone. There are many reasons why a given gene might **not** be expressed in a foreign host. If there is no promoter on the insert and the insert is not in the proper orientation or reading frame to be transcribed by a promoter on the vector, a gene product (protein) will not be made even if the entire gene is present. If there is a promoter present, it is possible that the promoter might not be recognized by the host's transcriptional machinery. Failure to detect a gene product could also occur because organisms frequently process their proteins in a unique fashion: an improperly folded peptide might not be detected by an antibody that would detect the native protein. There are also reasons why expression of the desired gene might not be desirable. For example, if the gene product is toxic to the host, all clones with that gene would die if it is expressed. For these reasons, it is often ideal to detect the target DNA sequence directly. This can be done by nucleic acid **hybridization.**

The most commonly used hybridization procedure for screening bacterial colonies is called the colony blot (Grunstein and Hogness 1975). It involves transferring bacterial cells to a solid support (nitrocellulose or nylon) where they are lysed and their DNA is immobilized for hybridization. This procedure will be used in Exercise 21 to screen for subclones containing the *lux*A gene but that do not bioluminesce! Principles of hybridization are given in Appendix VI.

In this exercise, you will screen the library constructed in the preceding exercises. The screening will be done by an inexpensive optical method—you will go into a dark room and eyeball the plates to find clones that bioluminesce! The *lux* operon consists of five genes which include the genes for the α and β subunits of luciferase as well as three genes responsible for the synthesis of the long-chain aldehyde required for light production (see Introduction). Because the luminescent colonies contain the entire five-gene *lux* operon, and *E. coli* recognizes the *V. fischeri lux* promoter, clones containing the operon will produce light. After purifying the clones on LB plates with ampicillin and X-gal, you will begin an extensive characterization of your clones that culminates in the sequencing and analysis of a subcloned portion of the *lux* operon from a luminescent clone.

Cultures

☐ plates from Exercise 12 containing transformants

Media

☐ LB plates + 50 µg/ml ampicillin + X-gal (LB/Ap^{50}/X-gal)

Supplies

☐ sterile toothpicks   ☐ inoculating loop
☐ toothpick discard beakers   ☐ dark room

➡ Procedure

1. Your instructor will have taken your plates out of the 37°C incubator 3 to 4 hours before the laboratory period and allowed them to incubate at room temperature. This will allow any clones containing the *lux* operon to become bioluminescent.

2. Examine your plates. You should have approximately 200 colonies per plate and a fairly high proportion (5–15%) of transformants containing inserts (white colonies).

3. Take all of your plates into a dark room and arrange them in an orderly pattern.

4. Turn off the lights and allow your eyes to dark adapt for 5 to 10 minutes. Carefully scan each plate, looking for a bioluminescent clone. If you find one, congratulations! You have just cloned the *lux* genes into *E. coli!* Keep looking as you may have more than one clone.

5. If you have a bioluminescent clone, you must carefully circle it with a permanent marker. This is rather difficult in the dark. However, if you rapidly flick the room light on and off in a strobe-like manner, you will be able to see the colony when the light is on, but by immediately turning it off, your eyes will still be dark adapted and you should be able to see the bioluminescence. Strobe lights are often used by researchers who clone bioluminescence genes.

6. Once you have circled all bioluminescent clones, return to the laboratory and examine your plate(s). What color is the bioluminescent colony(s)? Is it the color you expected?

7. Because you isolated the clone(s) yourself, you have the privilege of naming the strain(s). Strains are often named with two letters indicating the initials of the person who isolated or constructed the strain, followed by a number indicating the number of strains that person has isolated. For example, *E. coli* JM109 was constructed by Joachim Messing. However, you may use any designation you wish in naming your strain. Feel free to be creative. Label an LB/Ap^{50}/X-gal plate with the name of your strain(s) and indicate the phenotype (i.e., Lux$^+$).

8. With a sterile toothpick, carefully pick the bioluminescent colony and patch it in several short lines near the edge of the plate. Dispose of the toothpick in the toothpick discard beaker containing disinfectant. Streak the plate for isolation (see Exercise 1) and incubate at 37°C.

9. You will also pick several other clones containing inserts to determine sizes of the other cloned inserts. To save plates, we will streak several clones on the same plate. Divide an LB/Ap^{50}/X-gal plate into quarters by drawing crossed lines on the bottom of the plate with a permanent marker.

10. Pick a nonbioluminescent white colony from one of your plates and patch it near the edge of one quadrant of the plate as indicated in Figure 13.1. With a flame-sterilized loop (cooled by touching the sterile agar), make a short pass through the first patch and about 1 cm toward the center of the plate. Reflame the loop and streak back and forth in a tight pattern to the end of the quadrant. Repeat with two other white colonies and one blue colony. Assign each clone a strain designation and note the phenotype. You may wish to start a strain notebook with entries of each strain. You will obtain additional information on each as you analyze the clones.

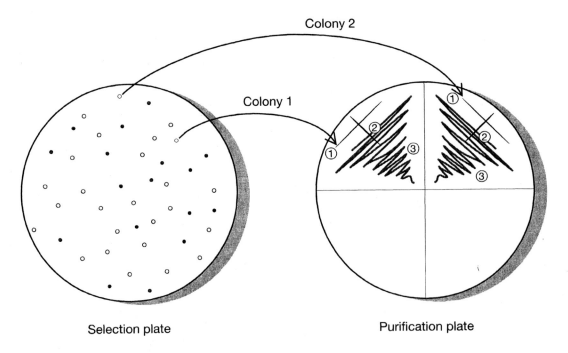

Figure 13.1
Streaking for isolated colonies on quadrant plates.

11. Incubate your plates at 37°C for 18 to 20 hours. To verify the bioluminescent phenotype of your clone(s), remove the plates after this time and continue incubating at room temperature. Luminescent clones of *E. coli* will continue to glow for several days at room temperature. After your plates have grown up, you will isolate the plasmids from several of the clones by a plasmid mini-prep procedure in the next exercise.

Results and Discussion

1. Estimate the average insert size of the DNA fragments in your total genomic digest (from Exercise 8). Assuming that the *V. fischeri* genome is 4×10^6 base pairs, how many clones would you need to screen to ensure that you would obtain a clone containing the *lux* genes? Count the total number of white clones you obtained from your cloning. Were there enough clones to ensure success in obtaining the *lux* clone? If not, did the entire class obtain enough clones?

2. Based on the entire class's data for the total number of white clones and the number of bioluminescent clones isolated, calculate the frequency of cloning the *lux* genes. Does this support your predictions? If different than predicted, can you suggest an explanation?

3. Can you think of any ways that you could decrease the number of plates you would need to screen a library? Think about the advantages of minimizing the number of plates used when cloning genomes from higher organisms that may be 10 to 1000 times larger than the *V. fischeri* genome.

Literature Cited

Clarke, L., and J. Carbon. 1976. A colony bank containing synthetic ColE1 hybrid plasmids representative of the entire *E. coli* genome. *Cell* 9:91–99.

Grunstein, M., and D. S. Hogness. 1975. Colony hybridization: A method for the isolation of cloned DNAs that contain a specific gene. *Proc. Natl. Acad. Sci.* 72:3961–3965.

IV
Restriction Mapping and Southern Blotting

EXERCISE 14
Small-Scale Plasmid Isolations (Mini-Preps) from Bioluminescent Clones

The ability to isolate plasmid DNA is crucial to recombinant DNA research and many other applications in molecular biology. Large-scale plasmid purifications (see, for example, Exercise 6) often require time-consuming procedures such as cesium chloride density gradient centrifugation, gel filtration columns, or affinity columns. These techniques can produce highly purified plasmid DNA, but consume considerable amounts of time and materials. However, not all manipulations involving plasmids require highly purified material. For example, when screening a large number of clones, it is more important to be able to rapidly isolate small amounts of moderately pure plasmid than to spend the time to prepare highly purified DNA. Several rapid small-scale plasmid isolation procedures have evolved and are widely used procedures in molecular biology. Because small amounts of bacterial cells are used and small amounts of plasmid are isolated, the procedures are referred to as mini-preps. Despite the small quantities of cells used, the yield of plasmid is quite impressive—usually in the range of 1 to 5 µg.

Plasmid purification procedures selectively enrich plasmid DNA over chromosomal DNA, which is present in the cell in much greater quantities. The two most widely used mini-preps are the alkaline lysis (Birnboim and Doly 1979) and rapid boiling (Holmes and Quigley 1981) procedures. Both protocols involve a precipitation of cell debris and protein and take advantage of physical properties of plasmids to separate them from the chromosomal DNA. Either of these procedures can be done in 1 to 2 hours, allowing rapid screening of many different plasmids. The plasmid DNA produced in these protocols is relatively clean and can be used in restriction digestions, transformations, and even DNA sequencing.

In both procedures, cells are pelleted rapidly in a microcentrifuge and the pellet is resuspended in a buffered medium. In the alkaline lysis procedure (similar to the large-scale procedure described in Exercise 6), cells are lysed with a solution of SDS and NaOH. In addition to lysing the cells, these components serve several other valuable functions. The SDS and the NaOH solubilize and denature cellular constituents, and the elevated pH will begin to degrade RNA. The alkaline conditions also result in DNA denaturation (strand separation). Because the chromosomal DNA is very large, it is broken by shear forces when the cells lyse, while the small plasmids remain intact. Thus, when denatured, the plasmids remain linked to their complementary strand—much as two links of a chain are linked together. When the lysate is neutralized with potassium acetate, the hydrogen bonds reform. Because the complementary plasmid DNA strands are held in close proximity to each other by the linked strands, they will immediately find their complementary strand and base pair properly to form an intact double-stranded plasmid. However, the chromosomal DNA will not be near its complementary strand. When the solution is neutralized, the DNA will base pair, but not to its complementary strand. The result will be a rather large, interlinked mass of DNA strands. The elevated salt concentration resulting from the potassium acetate addition will cause the SDS and protein to form a flocculent precipitate. The precipitated SDS and protein trap the chromosomal DNA fragments and are removed by centrifugation. The plasmids (and RNA) in the supernatant are then precipitated with alcohol and dissolved in a Tris/EDTA (TE) buffer.

The alkaline lysis procedure works well not only with *Escherichia coli* but also with a wide variety of other bacterial genera and a wide range of plasmid sizes. The rapid boiling procedure is even faster, but it seems to work best with *E. coli*. Many other organisms are not lysed as efficiently by boiling as they are with the alkaline lysis procedure. In addition, the rapid boiling technique does not recover large plasmids as well as does the alkaline procedure. However, because most recombinant work is done in *E. coli*, it is widely used for analyzing recombinant plasmids.

The boiling procedure is similar to the alkaline lysis procedure in that the plasmid DNA is enriched by the selective removal of protein and chromosome. However, in the boiling protocol, cells are lysed by lysozyme, heat, and a non-ionic detergent. The detergent, Triton X-100, is not as disruptive to membranes and proteins and results in less inactivation of nucleases than lysis with SDS. However, nucleases are generally inactivated by the heat treatment. The cells are lysed by placing them in a boiling water bath for a short period of time. This results in almost simultaneous lysis and protein coagulation. The heat denatures the DNA, and, because the plasmids are not sheared, they remain interlocked. When the solution is cooled, the interlocked plasmid strands reanneal with their complementary strand, forming an intact plasmid. The chromosomal DNA, which is sheared upon lysis, is trapped with the coagulated proteins and may be removed by precipitation. The plasmids remain in solution and may be ethanol precipitated and resuspended in TE buffer.

Plasmids purified by mini-prep procedures contain significant amounts of RNA. The alkaline lysis procedure results in a plasmid solution that contains less RNA than in the boiling procedure because RNA is partially degraded under alkaline conditions. This RNA will not affect transformation of the plasmids, but it will mask small restriction fragments on agarose gels. If fragments smaller than 2 kb are to be visualized by agarose gel electrophoresis, the RNA may easily be removed by RNase treatment.

In this exercise, we will use a modification of the rapid boiling method to prepare plasmid mini-preps from clones containing the *lux* operon and other portions of the *Vibrio fischeri* chromosome. (A good modification of the alkaline lysis procedure is provided in Exercise 22 and may also be substituted.) The plasmids will then be digested with *Sal* I and the digests analyzed by agarose gel electrophoresis to determine the insert size.

Cultures

☐ overnight cultures of *E. coli* DH5α (containing recombinant plasmids)

Reagents

☐ STET buffer (8% sucrose, 0.5% Triton X-100, 50 mM EDTA, 50 mM Tris [pH 8.0])
☐ lysozyme solution (20 mg/ml in 10 mM Tris [pH 8.0]; on ice)
☐ RNase A (1 mg/ml; boiled to inactivate DNase; on ice)
☐ isopropanol (room temperature)
☐ 70% and 95% ethanol (optional; on ice)
☐ TE buffer (10 mM Tris [pH 8.0], 1 mM EDTA; on ice)

☐ 3 M sodium acetate (pH 5.2) (optional; on ice)
☐ *Sal* I (on ice; 5 μl/group)
☐ *Xba* I (on ice; 2 μl/group)
☐ 10X restriction buffer (on ice)
☐ sterile deionized water (on ice)
☐ loading dye (room temperature)
☐ 1X TAE buffer
☐ LE agarose (low EEO)
☐ ethidium bromide stain (1.0 μg/ml)

Supplies and Equipment

☐ mini ice buckets (1/group)
☐ sterile 1.5-ml microcentrifuge tubes
☐ microcentrifuge tube opener (1/group)
☐ microcentrifuge tube rack (1/group)
☐ boilable microcentrifuge tube float (1/group)
☐ 0.5–10 μl micropipetter
☐ 10–100 μl micropipetter
☐ 100–1000 μl micropipetter
☐ sterile 10-μl micropipet tips
☐ sterile 100-μl micropipet tips
☐ sterile 1000-μl micropipet tips
☐ pipet tip discard beakers
☐ boiling water bath
☐ 37°C water bath

☐ 65°C water bath
☐ vortex mixer
☐ microcentrifuge
☐ 250-ml Erlenmeyer flask (1 per 2 groups)
☐ 25-ml Erlenmeyer flask (1 per 2 groups)
☐ microwave oven
☐ top loading balance
☐ horizontal mini gel electrophoresis chamber
☐ power source
☐ gel casting tray with 12-tooth comb
☐ gel staining tray
☐ spatula
☐ transilluminator
☐ Polaroid camera and type 667 film
☐ 3-cycle semi-log paper

Procedure

A. Rapid Boiling Plasmid Mini-Prep

The evening before the lab you should inoculate the clones you wish to do mini-preps on (see table in step 1 below) into 50-ml sterile tubes with 5 ml of LB broth (+ 50 µg/ml ampicillin). Incubate in a shaking water bath overnight at 37°C. (If you wish to observe bioluminescence in the broths containing *lux* clones, you should incubate at 28°C or less.) Alternately, you may do mini-preps by scraping a small amount of bacterial growth off a Petri plate and resuspending it in saline or LB broth. Treat this suspension as you would the culture in the following procedure. This method generally results in a lower yield of plasmid than broth cultures, but allows rapid plasmid preps directly from plates.

Note: Start the boiling water bath before beginning.

1. Label six **clear** microcentrifuge tubes A through F with a permanent marker. Transfer 1.5 ml of your overnight cultures to the tubes as follows:

Tube	Clone
A	pGEM (blue)
B	Lux⁻ white #1
C	Lux⁻ white #2
D	Lux⁺ (white)
E	Lux⁺ (white)
F	Lux⁺ (white)

Tube A contains culture from a blue clone that should contain pGEM™ that has ligated to itself. Tubes B and C contain Lux⁻ clones, which will give you an indication of the sizes of inserts ligated into the recombinant plasmids. Tubes D through F are replicate mini-preps from the **same** bioluminescent clone (tube). This will provide enough DNA for all the restriction digestions in the restriction mapping exercise (see Exercise 15).

2. Centrifuge at 8–10,000 × *g* (10–11,000 rpm) in a microcentrifuge at room temperature for 60 seconds. It is best not to centrifuge too long or too hard or you will have difficulty resuspending the pellet. Discard the supernatant by pouring it into the discard beaker. Remove traces of supernatant with a micropipetter.

3. Vortex the tube vigorously to loosen the pellet. Resuspend the pellet in 350 µl of STET buffer by pipetting repeatedly in and out.

4. Add 25 μl of 20 mg/ml lysozyme and vortex to mix. Incubate at room temperature for 3 to 5 minutes. **Note:** the lysozyme step may be skipped with most host strains of *E. coli*, as they lyse easily.

 CAUTION: In the next step, you will be using a boiling water bath. Boiling water is hazardous in the lab, especially when elevated. Use caution when adding and removing tubes from the water bath and when working in the vicinity of boiling water.

5. Put the tubes into a plastic microcentrifuge tube float and place them in a boiling water bath for 60 to 70 seconds.

6. Remove the tubes from the water bath and immediately centrifuge at 12–14,000 × *g* (full speed in a microcentrifuge) for 10 minutes at room temperature. Wait for the other groups sharing your centrifuge before beginning.

7. Remove gelatinous pellets in each tube with a sterile toothpick. To do this, stab the pellet with the toothpick and carefully withdraw, trying not to smear any of the pellet on the tube wall. Place the toothpick in the toothpick discard beaker.

Note: If you encounter nuclease problems in this mini-prep procedure, you may extract the aqueous solution with chloroform at this point. However, this is usually not necessary.

8. Add 400 μl of isopropanol to each tube and mix by inversion. **Immediately** go to the next step. The tubes should be at room temperature for a **maximum** of 5 minutes before centrifugation. Longer incubations will result in significant protein precipitation.

9. Centrifuge at 12–14,000 × *g* in a microcentrifuge for 10 minutes at room temperature. Be sure that the hinges of the microcentrifuge tubes are pointing up in the microcentrifuge.

10. Pour out the supernatants and remove any last traces with a micropipetter. You may not see a pellet—be careful not to touch the tip to the side where the pellet should be.

Note: Steps 11–13 are optional but will yield cleaner mini-prep DNA and minimize chances of nuclease contamination. Protein co-precipitates more with isopropanol than ethanol, so the ethanol-precipitated plasmid will be less likely to contain nucleases. If you do not do these, go to step 14.

11. Dissolve the precipitate in 100 μl of TE, add 10 μl 3 M sodium acetate (pH 5.2), and mix. Add 2 volumes of ice-cold 95% ethanol, mix, and incubate on ice for 10 minutes to precipitate the nucleic acids.

12. Centrifuge at 12–14,000 × *g* for 10 minutes at room temperature. Pour out the supernatants and remove any last traces with a micropipetter.

13. Wash the precipitate with 1 ml of 70% ethanol. This washes out the salt and iso-propanol, which will decrease the solubility of DNA. Centrifuge at 12–14,000 × *g* for 5 minutes at room temperature (this centrifugation can be skipped if you have a visible pellet that stays on the wall of the tube). Pour the supernatant into the discard beaker and remove the last traces with a micropipetter.

14. Set the tubes on your bench for 5 to 10 minutes to dry (or dry pellets by care-fully blowing warm air from a hair dryer over the tubes). Verify that all ethanol has been removed by sniffing the tubes for traces of alcohol odor.

15. Resuspend each pellet in 50 µl of TE buffer (for lower copy number plasmids, such as pBR322, resuspend in 20 µl). Combine tubes D, E, and F (containing the *lux* clones) into tube D. Perform restriction digestions on 2 to 4 µl of these sus-pensions.

16. The mini-prep DNA will also contain large amounts of RNA at this point. Add 2 µl of a 1 mg/ml RNase A stock to tubes B, C, and D and incubate for 10 to 20 minutes at 37°C. Removing RNA will allow you to visualize any small restriction fragments on an agarose gel that would otherwise be masked by the smear of RNA.

   Do not add RNase to tube A, which is a mini-prep from a blue clone that should contain pGEM™-3Zf(+). This will allow you to see what the RNA looks like when you run the gel. Because pGEM™-3Zf(+) is about 3 kb, the RNA will not interfere with visualization of the plasmid.

B. Restriction Endonuclease Digestion

1. Fill your mini ice bucket with crushed ice and place each of the required enzymes, buffers, and so on, in it. Each group will be provided with its own set of enzymes and reagents containing slightly more than the required volume.

2. Using a permanent marker, label four sterile 1.5-ml microcentrifuge tubes "A/Sal," "B/Sal," "C/Sal," and "D/Sal." Label a fifth tube "D/Xba." Place them in your micro-centrifuge tube rack.

3. Calculate the amount of water to add to each tube to give a final volume of 20 µl. Using micropipetters and sterile tips, add the following reagents to each of the tubes. Be sure to read the Reminders following the table before proceeding.

Tube	Clone	10X buffer	Mini-prep DNA	H₂O	*Sal* I	*Xba* I
A/*Sal*	blue	2 µl	4 µl	___µl	1 µl	—
B/*Sal*	white	2 µl	4 µl	___µl	1 µl	—
C/*Sal*	white	2 µl	4 µl	___µl	1 µl	—
D/*Sal*	Lux⁺	2 µl	4 µl	___µl	1 µl	—
D/*Xba*	Lux⁺	2 µl	4 µl	___µl	—	1 µl

Reminders:

- Always add buffer, water, and DNA to tubes prior to adding the enzymes.

- Touch the pipet tip to the side of the tube when adding each reagent. Be sure you **see** the liquid transferred to the side of the tube, especially with volumes <2 µl.

- You may use the same tip to add the same reagent to different tubes, but be sure to use a new tip for each reagent.

- Check off each addition in the table as you add it. Omissions or incorrect volumes will usually prevent the restriction digestion from working.

4. After adding all ingredients, cap the tubes and vortex to mix. Pool the reagents in the bottom of the microcentrifuge tube by spinning the tubes in the microcentrifuge for 2 to 3 seconds. Be sure that the tubes are in a balanced configuration. Vortex or tap each tube to mix and recentrifuge to pool the digestion mix in the bottom of the tube.

5. Push your tubes into the foam microcentrifuge tube holder and float them in a 37°C water bath. Incubate for 30 to 45 minutes. If you did not add RNase to the mini-prep DNA in Part A, add 1 µl of 1 mg/ml RNase to restriction digestions 10 to 15 minutes before the digestion is done. **Do not add RNase to tube A** (with the pGEM™-3Zf[+] DNA). While the restriction digestions are proceeding, prepare and cast an agarose gel.

Note:
Digests can be frozen at this point and analyzed at a later date.

C. Agarose Gel Electrophoresis

Each group will run an agarose gel on the digestions of its mini-preps to characterize each plasmid. See Exercise 4 for details on preparing, running, staining, and visualizing agarose gels.

1. Prepare one batch of **0.8% agarose** in 1X TAE buffer per two groups.

CAUTION: In the next step, you will be heating agarose to boiling. Agarose can become superheated and boil over in the microwave or when swirled. Wear gloves and eye protection when removing the agarose and swirling heated flasks.

2. Melt the agarose in a microwave oven and cool to about 60°C or place in the 60°C water bath until ready to use. Cast a gel with a **12-tooth** comb. When solidified, place the gel in the electrophoresis chamber.

3. While the digestion is proceeding, label four tubes A/UC, B/UC, C/UC, and D/UC. Add 2 µl of the uncut plasmid (from the mini-preps) to each of these tubes. Add

6 µl of TE buffer and 2 µl of loading dye to each tube and pulse for 2 to 3 seconds in a microcentrifuge to pool the ingredients. These will serve as uncut controls for each restriction digestion.

4. After the restriction digestion has incubated for 30 to 45 minutes, remove the tubes from the water bath and place them in the microcentrifuge rack.

5. Add 2 µl of loading dye to each tube. Pulse each tube for 2 to 3 seconds in the microcentrifuge to mix the dye with the restriction digestion.

6. Remove the *Hin*d III digest of lambda DNA from your freezer box and heat the standards in the 65°C water bath for 2 to 3 minutes to melt the *cos* sites in lambda.

7. Load 10 µl of the *Hin*d III digest of lambda as a size standard into wells 2 and 12. Load 10 µl of each digestion and the undigested plasmids into the wells in the gel as follows:

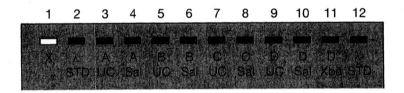

8. Electrophorese at 100 V until the bromphenol blue (the leading, darker dye) has migrated about three-quarters of the way to the end of the gel.

9. When the electrophoresis is complete, turn the voltage down to the lowest level and switch the power source off. Remove the lid to the gel box so that the gel can be removed.

D. Staining and Photodocumentation of Agarose Gel

1. Label a staining tray with your group number. Remove the tray supporting the gel from the electrophoresis box and slide the gel into the staining tray.

CAUTION: Ethidium bromide is a mutagen and suspected carcinogen. Wear gloves, eye protection, and a lab coat, and work over plastic-backed absorbent paper at all times when handling it. Carefully read precautions for handling and disposing of ethidium bromide (Appendix XII) before using ethidium bromide stain.

2. Stain the gel with ethidium bromide (1 µg/ml) for 5 to 10 minutes and rinse several times with tap or distilled water.

3. Briefly observe the gel on the transilluminator and photograph.

> **CAUTION:** UV light is hazardous and can damage your eyes. *Never* look at an unshielded UV light. Always view through a UV blocking shield or wear UV blocking glasses. Some transilluminators have a switch on the lid so that the UV light switches off when the lid is opened, but many transilluminators do not have this safety feature.

4. Tape your gel photograph in your notebook below:

Results and Discussion

1. Carefully examine the photograph of your agarose gel. Prepare a standard curve from the *Hin*d III restriction fragments of lambda DNA on the 3-cycle semi-log paper provided (see Exercise 4). Measure the distance migrated by each uncut plasmid and each restriction fragment and calculate the size (in kb).

2. How does the size of the uncut plasmid from the blue colony compare with the size of the uncut plasmids from white colonies? Is this what you would expect? What is the apparent size of the uncut pGEM™-3Zf(+) (from mini-prep A)? Compare with the known size of pGEM™-3Zf(+) (Appendix XI). Can you explain any differences?

3. Based on your response to questions 1 and 2, what do you think the purpose of running the uncut plasmids on a gel is?

4. Because the DNA cloned and vector were digested with *Sal* I, cutting the plasmids with *Sal* I should remove the insert from the vector. Examine the sizes of inserts from the nonbioluminescent clones. Compare the insert sizes you obtained with those from the rest of the class. What insert sizes are most commonly found in the clones from the total genomic digest? Is there ever more than one insert? If not, under what conditions (if any) would you expect this to be possible?

5. What is the size of the insert from the bioluminescent clone? Do all of the bioluminescent clones have the same size inserts? Compare photographs of other gels in the class.

6. Do you think that all the recombinant plasmids in the bioluminescent clones are identical? What does the *Xba* I digest of the *lux* plasmids indicate? Compare the results of your *Xba* I digest with others from the rest of the class. What does the *Xba* I digest allow you to determine about the luminescent clone?

Literature Cited

Birnboim, H. C., and J. Doly. 1979. A rapid alkaline extraction procedure for screening recombinant plasmid DNA. *Nucleic Acids Res.* 7:1513.

Holmes, D. S., and M. Quigley. 1981. A rapid boiling method for the preparation of bacterial plasmids. *Anal. Biochem.* 114:193.

Sambrook, J., E. F. Fritsch, and T. Maniatis. 1989. Molecular cloning: A laboratory manual. 2d ed. New York: Cold Spring Harbor Laboratory Press.

10
9
8
7
6
5
4
3
2
1
9
8
7
6
5
4
3
2
1
9
8
7
6
5
4
3
2
1

Semi Log 3 x 10

EXERCISE 15
Restriction Mapping of Plasmids from Bioluminescent Clones

The construction of an accurate map indicating the sites where restriction endonucleases cut a cloned DNA molecule is essential for most subsequent manipulations of that DNA. This procedure, called **restriction mapping**, involves the digestion of a given piece of DNA with a series of restriction enzymes and the subsequent determination of the size and the order of the resultant fragments by agarose gel electrophoresis.

The information gained by placing the recognition sites of various restriction enzymes on a restriction map has many applications in modern molecular biology. Subcloning, the process of cloning a fragment of DNA from a piece of DNA already cloned (see Exercises 17–22), generally requires a detailed restriction map of the larger cloned DNA. Subcloned DNA is often used as a template for DNA sequencing or for the production of nucleic acid "probes." Such probes can be used to identify specific genes in genomic libraries, or in restriction digests separated by agarose gel electrophoresis. Restriction maps of cloned DNA are also used to compare genes of different organisms or individuals. This provides evidence of the evolutionary relationships between species at the molecular level without having to determine the exact nucleotide sequence.

There are many different strategies that can be used to construct restriction maps. The simplest procedure involves use of a series of enzymes that are used to digest a given DNA singly and then in combination with each other. After you determine the sizes of the resulting restriction fragments, you can often deduce the order of the fragments in the original plasmid. Additional strategies that may be used include use of partial digests with a given enzyme, radiolabeling, or purification of a restriction fragment from a gel followed by redigestion with a second enzyme.

The most useful restriction endonucleases for generating restriction maps are those that cut DNA infrequently, as the production of too many fragments makes the interpretation difficult. Therefore, the first steps in a mapping project are generally carried out with restriction enzymes that have six base pair recognition sequences (see Exercise 4). The enzymes that are most useful initially cut the DNA only once. Using these single-cut enzymes, digests are performed individually and in combination with other enzymes. This usually allows the unambiguous placement of a few restriction sites. Map construction is generally done in a step-wise, cumulative fashion. After the sites for enzymes that cut infrequently are determined, the sites for enzymes that cut more frequently can be accurately located.

To determine the size of restriction fragments, the fragments in a digest are separated by electrophoresis in an agarose gel that also includes DNA molecules of known size as standards. Because the migration of DNA fragments is inversely proportional to the log of their size (see Exercise 4), the DNA standards allow you to construct a standard curve on semi-log paper relating size to mobility.

When digesting plasmid DNA, the number of fragments produced is a function of the number of times the circular DNA molecule is cut. A single cut will produce one linear DNA fragment, two cuts will produce two fragments, and so on. When two enzymes are used simultaneously to digest a given DNA molecule, the resultant number of fragments should be equal to the sum of the fragments present in the single-enzyme digests. When setting up digests for mapping, it is important to include an uncut plasmid control. Due to the different mobilities of the circular and linear forms of the plasmid (see Exercise 4), this control will allow you to determine whether an enzyme makes a single cut or does not cut the plasmid at all.

For the type of map construction that you will be doing, it is important that the digestion of the DNA be complete. When a digest is complete, the resulting fragments should be present in **equimolar** quantities, but the amount of mass of each fragment will vary, depending on its size. The ethidium bromide fluorescence is proportional to the size of the fragment. Therefore, when looking at a digest, the intensity of each band should decrease as you proceed from the largest fragments to the smallest. If a band appears lighter than its size would suggest, it could be the product of a **partial digest**. If a band appears brighter than expected, it could be the result of two separate fragments from different regions of the plasmid that are essentially the same size (a **doublet**). It is important to be able to recognize these patterns in gels when restriction mapping. Further guidelines on restriction mapping, examples, and restriction mapping problems are provided in Appendix XIX.

In this exercise, you will map the location of several restriction sites in the *Sal* I fragment of a bioluminescent clone. Your task in mapping is simplified because you have a map of pGEM™-3Zf(+) (Appendix XI) that shows if and where the enzymes you will use cut the vector. You will only have to map the sites that are located in the insert DNA. You will also locate the restriction fragment that contains the gene for the α-subunit of luciferase (*lux*A) by DNA hybridization with an oligonucleotide probe (see Exercise 16).

Reagents

- ☐ pUWL500 or pUWL501 (0.1 µg/µl) (or mini-prep DNA of bioluminescent clone from Exercise 14)
- ☐ BRL 1 kb ladder (0.05 µg/µl)
- ☐ various restriction enzymes (on ice)
- ☐ 10X restriction buffer (compatible with the assigned enzymes; on ice)

- ☐ sterile deionized water (on ice)
- ☐ loading dye (room temperature)
- ☐ 1X TAE buffer
- ☐ LE agarose (low EEO)
- ☐ ethidium bromide stain (1.0 µg/ml)

Supplies and Equipment

- ☐ mini ice buckets (1/group)
- ☐ sterile 1.5-ml microcentrifuge tubes
- ☐ microcentrifuge tube opener (1/group)
- ☐ microcentrifuge tube rack (1/group)
- ☐ microcentrifuge tube float (1/group)
- ☐ 0.5–10 µl micropipetter
- ☐ 10–100 µl micropipetter
- ☐ sterile 10-µl micropipet tips
- ☐ sterile 100-µl micropipet tips
- ☐ micropipet tip discard beaker
- ☐ 250-ml Erlenmeyer flask (1 per 2 groups)
- ☐ 25-ml Erlenmeyer flask (1 per 2 groups)
- ☐ microwave oven

- ☐ top loading balance
- ☐ spatula
- ☐ 37°C water bath
- ☐ 65°C water bath
- ☐ microcentrifuge
- ☐ horizontal mini gel electrophoresis chamber
- ☐ gel casting tray
- ☐ 12-tooth comb
- ☐ electrophoresis power source
- ☐ transilluminator
- ☐ Polaroid camera with type 667 film
- ☐ 3-cycle semi-log paper

▶Procedure

A. Restriction Endonuclease Digestion

1. Each group will be assigned either pUWL500 or pUWL501 and three different enzymes. Write your assigned enzymes and DNA in the following table. Fill your mini ice bucket with ice and place each of the required cold enzymes, buffers, and plasmid DNA in it.

2. Label eight sterile 1.5-ml microcentrifuge tubes A through H with a permanent marker and place them in your microcentrifuge rack.

3. Calculate the amount of water to add to each digestion to give final volumes of 20 µl. Using micropipetters and sterile pipet tips, add the following reagents to each of the tubes. Be sure to read the Reminders below the table before proceeding.

Tube	10X buffer	Plasmid		H₂O	Assigned enzymes		
A	2 µl	5 µl		____ µl	1 µl	—	—
B	2 µl	5 µl		____ µl	—	1 µl	—
C	2 µl	5 µl		____ µl	—	—	1 µl
D	2 µl	5 µl		____ µl	1 µl	1 µl	—
E	2 µl	5 µl		____ µl	1 µl	—	1 µl
F	2 µl	5 µl		____ µl	—	1 µl	1 µl
G	2 µl	5 µl		____ µl	0.7 µl	0.7 µl	0.7 µl
H	2 µl	5 µl		____ µl	—	—	—

Reminders:

- Always add buffer, water, and DNA to tubes prior to adding the enzymes.

- Touch the pipet tip to the side of the tube when adding each reagent. Be sure you **see** the liquid transferred to the side of the tube, especially with volumes <2 µl.

- You may use the same tip to add the same reagent to different tubes, but be sure to use a new tip for each reagent.

- Check off each addition in the table as you add it. Omissions or incorrect volumes will usually prevent the restriction digestion from working.

4. After adding all ingredients, cap the tubes and vortex to mix. Pool the reagents in the bottom of the microcentrifuge tube by spinning the tubes in the microcentrifuge for 2 to 3 seconds. Be sure that the tubes are in a balanced configuration. Vortex or tap each tube to mix and then recentrifuge to pool the digestion mix in the bottom of the tube.

5. Push the tubes into the foam microcentrifuge tube holder and incubate in a 37°C water bath for 1 hour. While the restriction digestions are proceeding, cast an agarose gel (see part B).

Note:

Digests can be frozen at −20°C after the incubations are completed and analyzed at a later date.

B. Agarose Gel Electrophoresis

See Exercise 4 for details on preparing, running, staining, and visualizing agarose gels.

1. Prepare one batch of **0.8% agarose** in 1X TAE buffer per two groups.

 CAUTION: In the next step, you will be heating agarose to boiling. Agarose can become superheated and boil over in the microwave or when swirled. Wear gloves and eye protection when removing the agarose and swirling heated flasks.

2. Melt the agarose in a microwave oven and cool to about 60°C or place in the 60°C water bath until ready to use. Cast a gel with a **12-tooth** comb. When solidified, place the gel in the electrophoresis chamber.

3. After the restriction digestions have incubated for 1 hour, remove the tubes from the water bath and place them in the microcentrifuge rack.

4. Add 2 µl of loading dye to each tube. Pulse each tube for 2 to 3 seconds in the microcentrifuge to pool the contents of the tube.

5. Add 5 µl of the BRL DNA ladder into lanes 2 and 11. Avoid loading the first and last wells.

Note: If you wish, you can load your undigested chromosomal DNA (1 µg) into lane 1 and the *Sal* I digested DNA (2 µg) into lane 12. This will allow you to see the position of *lux*A in the digest when you do the Southern Blotting lab (Exercise 16).

Load 10 µl of each of tubes A through H into the remaining wells in the gel as follows:

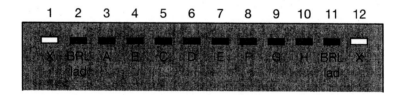

6. Electrophorese at 100 V until the bromphenol blue (the leading, darker dye) has migrated about three-quarters of the way to the end of the gel.

7. When the electrophoresis is complete, turn the voltage down to the lowest level and switch the power source off. Remove the lid to the gel box so that the gel can be removed.

C. Staining and Photodocumentation of Agarose Gel

1. Label a staining tray with your group number. Remove the tray supporting the gel from the electrophoresis box and slide the gel into the staining tray.

CAUTION: Ethidium bromide is a mutagen and suspected carcinogen. Wear gloves, eye protection, and a lab coat, and work over plastic-backed absorbent paper at all times when handling it. Carefully read precautions for handling and disposing of ethidium bromide (Appendix XII) before using ethidium bromide stain.

2. Stain the gel with ethidium bromide (1 µg/ml) for 5 to 10 minutes and rinse several times with tap or distilled water.

3. Briefly observe the gel on the transilluminator and photograph. **Do not discard your gel!** Carefully remove it from the transilluminator and return it to the gel staining tray. You will use this gel to perform a Southern blot in the next exercise.

CAUTION: UV light is hazardous and can damage your eyes. *Never* look at an unshielded UV light. Always view through a UV blocking shield or wear UV blocking glasses. Some transilluminators have a switch on the lid so that the UV light switches off when the lid is opened, but many transilluminators do not have this safety feature.

4. Tape your gel photograph in your notebook below:

Results and Discussion

1. The measurements needed to construct the map are made from the photograph of the agarose gel. The migration distances are measured from the leading edge of the well to the leading edge of the fragment. The sizes of the DNA fragments in the BRL 1 kb ladder (in base pairs) are given in the following table. Carefully measure the distance that the DNA size standards have migrated and record the distance in this table:

Fragment #	Size (kb)	Distance (mm)	Fragment #	Size (kb)	Distance (mm)
1	12.2		8	5.09	
2	11.2		9	4.07	
3	10.2		10	3.05	
4	9.16		11	2.04	
5	8.14		12	1.64	
6	7.13		13	1.02	
7	6.11		14	0.51	

There are additional small fragments in the BRL 1 kb ladder (<0.4 kb) that are not discernable as sharp bands on an 0.8% gel.

2. The accuracy of your map will depend on how evenly the fragments in the gel ran and how accurately you measure the migration distances of your standards and restriction fragments. How can you determine if the fragments in the gel ran evenly (i.e., the same size fragments migrated the same distance no matter which lane they were loaded into)? What can you do if you have evidence that your gel did not run evenly?

3. The rate of migration of linear DNA in an agarose gel is inversely proportional to the $\log_{10}$ of the fragment size (see Exercise 4). Plot the mobility (in mm on the x-axis) versus fragment size (in kb on the log scale) using the provided 3-cycle semi-log paper. The relationship between the mobility and log of the fragment size should be linear over most of the range of the fragment sizes. The linearity breaks down, however, for large fragments. Prepare a standard curve by drawing a straight line through the linear data points and interpolate through additional points.

4. Design a data table and record the mobility (in mm) of the DNA fragments that result from each digest. Be sure you include which plasmid and which enzymes you used for each lane. Use the standard curve to determine the size of each restriction fragment. Are your measurements more accurate in sizing large or small fragments? Why?

TABLE 15.1 Restriction fragment sizes resulting from the digestion of pUWL _____ with the enzymes _____, _____, and _____.

5. Add the sizes of all restriction fragments for each individual digest. The sum of the fragments should be equal to the size of the fragment that results from a restriction digest in which the plasmid is cut once. What do you think is an acceptable margin of error? If your error is greater than what you feel is acceptable, first recheck your standard curve to be sure it was prepared properly.

6. If the sums are low and the error is too great to be attributed to simple measurement inaccuracies, recheck the photograph to determine if you have missed a fragment. Two fragments might be the same size (or nearly the same size) and run as a single band, creating the appearance that there is only one fragment present. These doublets can generally be recognized as their brightness is greater than expected based on the size of the fragment.

7. If the sums are too high, it is likely that you did not have complete digestion of the DNA, and, as a result, products of incomplete digestion were measured. The fragments resulting from partial digestions usually can be identified by an intensity less than expected based on their size.

8. If the data are satisfactory, use the guidelines stated in the introduction of this exercise and the examples in Appendix XIX to construct the restriction map for your assigned enzymes. Remember that you have a very accurate map of the plas-

mid vector (Appendix XI) that will be useful in constructing your map. It is helpful to use the provided linear graph paper to make your map, with each small square representing 0.1 kb. After completing your map, compare your results with other groups that used the same enzymes on a plasmid with the insert in the opposite orientation. Did you obtain similar results?

9. Again, estimate the accuracy of your measurements. Now that you know the approximate sizes of fragments generated from the enzymes that you used, can you suggest additional procedures to generate a more accurate map?

10. You should have determined the location of three enzymes on your restriction map. Now that you know where these enzymes cleave, will it be easier to map additional enzymes? Can you use the data of other groups to improve the accuracy of your map?

Collectively, you will use the class data to generate a restriction map showing the cleavage site of each enzyme. Record the complete map in your notebook.

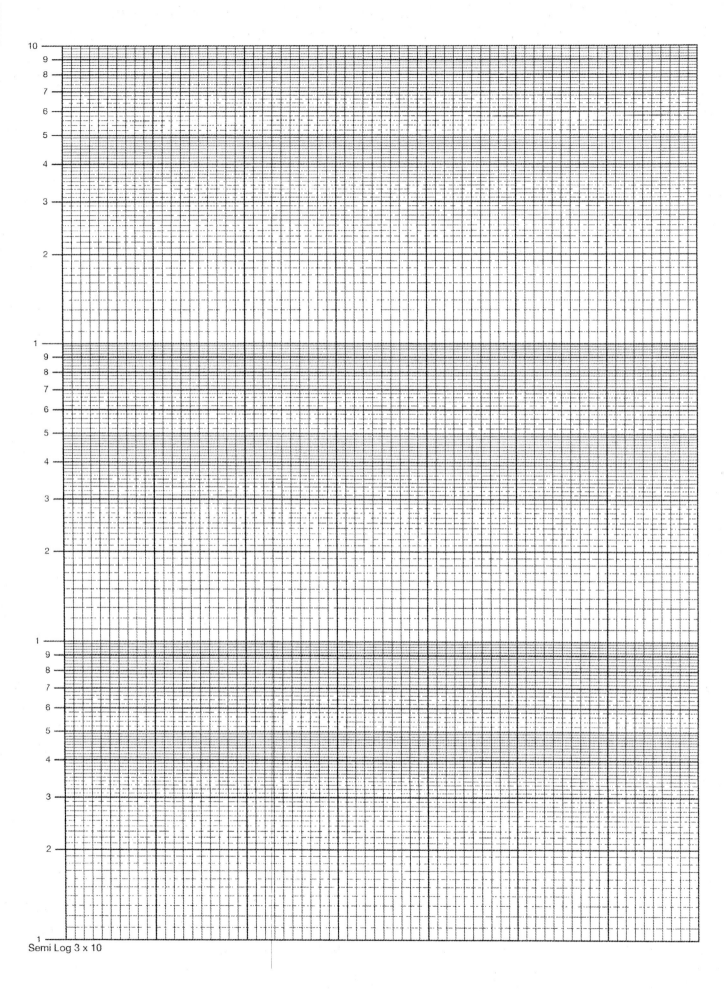

Semi Log 3 x 10

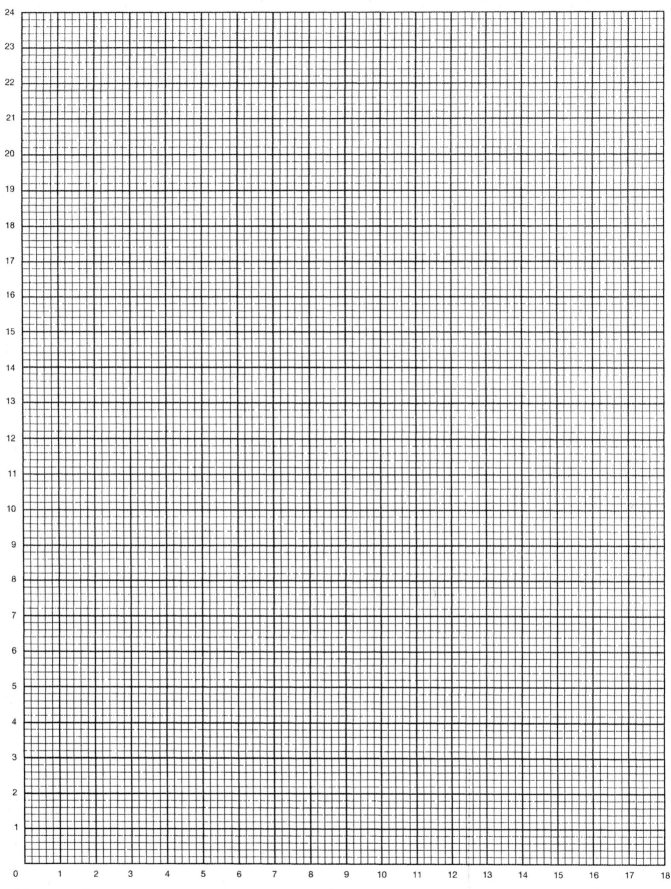

5 Squares to a Centimeter

EXERCISE 16
Southern Blotting and Hybridization to Detect the *lux*A Gene

Nucleic acid hybridization has long been used to detect gross similarities in the nucleotide sequences of different organisms. With the improvements in the isolation and purification of type II restriction enzymes, which recognize and cut specific DNA sequences, genomes could be cut into unique fragments. These restriction fragments, when separated electrophoretically in an agarose gel, produce a distinctive pattern for each organism. It was then possible to produce and isolate specific fragments of the chromosome that could be used in DNA-DNA or DNA-RNA hybridizations. However, DNA embedded in a gel is not readily available for hybridization, and it must first be removed. In the earliest experiments, DNA fragments were painstakingly extracted from the agarose and hybridization was carried out with each purified fragment. Clearly, it would be useful to have a rapid method of detecting fragments in an agarose gel that are complementary to a given nucleic acid probe. In 1975 Edwin Southern published a method (Southern 1975) that permitted the simultaneous hybridization of all DNA fragments in a restriction pattern without prior purification.

In the **Southern blotting** procedure, DNA is cut with one or more restriction enzymes, and the resultant fragments are separated by agarose gel electrophoresis. When adequate fragment separation is achieved, the DNA in the gel is denatured and a nitrocellulose membrane is placed on top of the gel. Buffer is then wicked perpendicularly through the gel, carrying the DNA fragments and transferring them to the membrane where they become chemically bound. DNA fragments transferred to the nitrocellulose are localized in a pattern identical to that in the gel. Thus, a replica of the restriction pattern is created on the membrane. After baking the nitrocellulose, the single-stranded nucleic acids are irreversibly bound to the membrane, and are available to hybridize with a complementary piece of DNA or RNA. Southern blotting, coupled with hybridization, has proved to be an extremely valuable tool with many applications in molecular biology. A variation of Southern blotting is used to transfer RNA molecules separated by electrophoresis to a membrane and is humorously called

northern blotting. Northern blots can also be hybridized with DNA or RNA probes. A second variation on the Southern procedure is used to blot proteins onto membranes and, in keeping with the geographic tradition, is called **western blotting**. Western blots are useful to identify specific proteins by screening with labeled antibodies in a manner similar to screening Southern or northern blots with nucleic acid probes.

A modification of the Southern procedure involves the transfer of DNA to a nylon membrane under alkaline conditions (Reed and Mann 1985). Nylon membranes are more durable, easier to handle, more effective at binding DNA, and able to bind smaller oligonucleotides than nitrocellulose. The alkaline conditions denature the DNA during transfer, eliminating the need for a separate denaturation step. In addition, the alkaline conditions induce covalent binding of DNA to the membrane, eliminating the need to bake the membrane.

Specific genes in the restriction pattern on the membrane can then be identified by hybridization with an RNA or DNA **probe**. A nucleic acid probe is a piece of DNA or RNA complementary to a nucleic acid sequence of interest. The nucleic acid probe may be complementary to a small portion of a specific gene (as small as 15 bases) or to one or more entire genes (up to several kb). In order to locate a specific DNA sequence by hybridization, the DNA or RNA probe is labeled with a reporter group. This reporter is frequently a radioactive compound, but it can also be either an enzyme that produces a colored product or a chemical group that fluoresces or produces chemiluminescence.

The advent of cloning and nucleic acid sequencing techniques, coupled with the introduction of DNA synthesizers, has resulted in tremendous improvements in probe technology. Once the sequence of a given piece of DNA is determined, DNA synthesizers can rapidly produce large amounts of pure oligonucleotides of a specific sequence. These oligonucleotide probes (usually 15 to 50 bases long) allow extremely specific and rapid detection of complementary sequences.

The utility of nucleic acid hybridization is based on the original discovery by Watson and Crick that DNA is a double-stranded molecule held together by hydrogen bonds between complementary bases. Because of the complementarity of the two strands, denatured DNA derived from the same parent molecule can **reanneal** (or hybridize) under appropriate pH, temperature, and ionic strength. Nucleic acid hybridization is usually performed with target DNA that has been immobilized onto membranes (from Southern blots or by spot loading DNA onto membranes) and a probe in a liquid phase. However, hybridization may also be done with both the target DNA and probe in solution (solution hybridization) or even directly in cells (*in* *situ* hybridization). The latter technique allows the identification of the cellular location of specific DNA or RNA sequences.

The specificity of a probe for its target sequence during a hybridization procedure is controlled by adjusting the **stringency**. If the hybridization is performed under conditions of **low stringency**, the probe may bind to DNA sequences on the blot that are not completely complementary to the probe sequence, and under **high stringency** the probe should only bind to exactly complementary sequences. Typically, probes are hybridized to blots under conditions of relatively low stringency and then the stringency of the hybridization is increased by successive washes of the blot in solutions that reduce the tendency of the probe to remain bound to anything but the target DNA sequence (which is exactly complementary to the probe). The stringency may be increased during a washing step by raising the temperature, lowering the ionic strength of the buffer, or adding denaturants to the buffer. Further information on the principles of nucleic acid hybridization and use of probes in molecular biology is given in Appendix VI.

Nucleic acid hybridization is one of the most powerful tools used in molecular biology. Hybridization may be used to quantify nucleic acids, to examine the relatedness of different nucleic acids, or to locate specific DNA sequences in the complex library of genes that makes up an organism's genome. Hybridization has been particularly valuable for recombinant DNA studies, enabling researchers to find specific genes in genomic or cDNA libraries and in restriction fragments. Additional applications of hybridization range from basic research in molecular biology to practical applications such as the clinical diagnosis of bacterial, viral, or genetic diseases, as well as the detection of nonculturable or genetically engineered organisms in the environment. **DNA fingerprinting** is an application of hybridization that is increasingly being used in criminology and forensic medicine. In this procedure, a sample of DNA from an individual or extracted from evidence left at a crime scene is digested with restriction endonucleases and separated on an agarose gel. Following a Southern blot, the membrane is hybridized with specific probes that produce a complex pattern of signals similar to the UPC symbol on a cereal box. This "fingerprint" is different for each individual and is a powerful tool in identifying individuals at the molecular level.

In this exercise, you will transfer the restriction fragments of pUWL500 or pUWL501, which were produced in Exercise 15, to a nylon membrane by Southern blotting. The nucleic acids on this membrane will then be hybridized with a nonradioactively labeled oligonucleotide probe to identify the DNA fragment that contains the *lux*A gene of *Vibrio fischeri*. The probe that will be used in this exercise is labeled with alkaline phosphatase (an enzyme that cleaves phosphate groups off nucleic acids and other molecules) by covalently linking the enzyme via a linker molecule to a base in the probe. This enables the enzyme-linked oligonucleotide probe to be detected colorimetrically after hybridization. Alkaline phosphatase catalyses the removal of the phosphate group from 5-bromo-4-chloro-3-indoylphosphate (BCIP). The resulting product reduces nitro blue tetrazolium (NBT) to form a blue-colored, insoluble formazan dye. The amount of color formed is proportional to the amount of probe fixed. The absence of any color indicates that the target sequence was not present, or present in an insufficient quantity to be detected.

PART I. SOUTHERN BLOTTING

Reagents

- ☐ gel from Exercise 15
- ☐ 0.4 N NaOH

- ☐ 5X SSC
- ☐ deionized water

Supplies and Equipment

- ☐ disposable gloves
- ☐ Rubbermaid™ washing trays (2/group)
- ☐ glass baking dish
- ☐ gel support platform
- ☐ 5-ml glass pipet or glass rod
- ☐ Whatman 3MM paper wicks (cut to exact width of gel and 25 cm long; 2/group)
- ☐ nylon membrane (cut to exact size of gel; 1/group)

- ☐ Whatman 3MM paper blotter (cut to exact size of gel; 4/group)
- ☐ paper towel blotters (cut to exact size of gel; 5 cm high stack/group)
- ☐ masonite board (cut to exact size of gel; 1/group)
- ☐ platform shaker
- ☐ 10 × 12-cm 3MM paper for drying nylon after blotting (4/group)

Procedure

Note: Always wear gloves when handling nylon membranes. Oils from your skin will prevent proper wetting of the membrane and subsequent transfer of the DNA.

1. Pour a small amount of deionized water into the wash tray and wet the membrane (cut to the exact size of the gel) by immersing it in the water. Do not allow the membrane to dry at any time prior to the completion of the transfer procedure.

2. Place the support platform in the glass baking dish [see Figure 16.1(a)]. Add 0.4 N NaOH to a separate wash tray until it is about 1 cm deep. Wet one of the Whatman 3MM wicks with the 0.4 N NaOH and lay it over the support. Repeat with the second wick, being sure that each wick lays flat on the platform. It helps to flood the surface of the first wick in position before adding the next. Roll a pipette or glass rod over the upper surface of the wicks to ensure that any air bubbles are removed from between the wicks.

3. Flood the surface of the wicks with 0.4 N NaOH. Invert the gel and place it on the wicks so that the open wells are facing down toward the wicks. Make sure that there are no air bubbles between the 3MM wicks and the gel as air bubbles prevent the transfer of DNA. Flood the upper surface of the gel with 0.4 N NaOH.

Note: Large fragments of DNA transfer inefficiently in a Southern blot. If complete transfer of fragments greater than 15 kb is required, they may be fragmented by soaking the gel in weak acid followed by incubation in base (Wahl, et al. 1979). This results in acid depurination followed by hydrolysis of the DNA at the depurination sites in alkali. With the plasmids and DNA fragments used in this exercise, however, this procedure is not necessary.

4. Rinse the wetted membrane with 0.4 N NaOH and lay it carefully on the gel [see Figure 16.1(b)]. **Remember to wear gloves!** Align the membrane with two diagonal corners, then gently roll the membrane down onto the gel. Remove any trapped bubbles by gently pushing them to the side or rolling them out with a pipette. Flood the surface of the membrane with additional NaOH.

Note: Do not move the membrane after it has come into contact with the gel! The binding of the DNA to the membrane is rapid and, in the case of nylon, irreversible. Movement of the membrane after placement will result in sets of the restriction pattern being superimposed on the nylon.

5. Wet two pieces of 3MM blotter in the 0.4 N NaOH and place them, one at a time, precisely on top of the membrane. Carefully remove all bubbles. Stack the remaining 3MM blotters (dry) and then the paper towels on top of the blotters. Be sure that the blotters do not contact the wicks—this will cause the buffer to bypass the gel, resulting in inefficient and uneven transfer of the DNA in the gel. Place the masonite board on top of the paper towels and pour the remaining NaOH from the wash dish into the glass baking dish. The assembled apparatus should look like Figure 16.1(c) when complete.

6. The transfer will be complete in 1 to 4 hours. Alternately, the apparatus can be left overnight, but be sure that there is adequate NaOH (2–3 cm deep) in the baking dish to maintain capillary flow if you do an overnight transfer.

7. After the transfer is complete, remove and discard the paper towels and 3MM blotters. Be sure you are wearing gloves and be careful not to disturb the nylon. Remove the nylon and gel as a single unit and lay them (gel side up) on a clean, dry piece of 3MM paper. Mark the position of the wells with a black fine-point permanent marker or soft pencil. Peel off and discard the gel. The transfer of the tracking dyes to the nylon is an indication that the DNA transfer has occurred.

8. Float the nylon, DNA side down, on 5X SSC in a washing tray. Gently agitate at room temperature for 5 minutes.

9. Remove the nylon from the SSC, then place the membrane on 3MM paper to blot dry with the DNA side up. When visible fluid has been absorbed, place the membrane between two pieces of 3MM paper and tape them together. Label with your group number and store it in your drawer or in a desiccator. Dry membranes are stable and may be hybridized at any time.

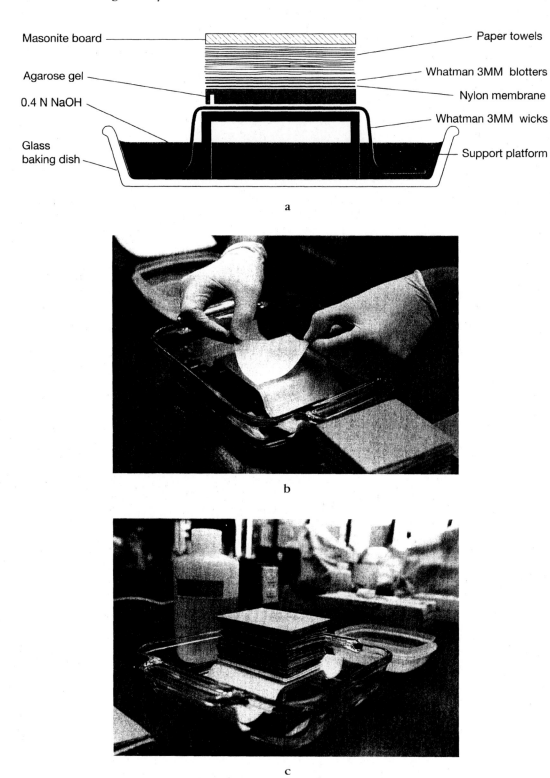

Figure 16.1
Assembling a Southern blot. (a) Schematic of an assembled Southern blot. (b) Laying the membrane onto the gel. (c) A completed Southern blot apparatus.

PART II. HYBRIDIZATION OF TRANSFER MEMBRANE WITH *lux*A PROBE

Reagents

- ❑ prehybridization buffer (5X SSC, 0.5% BSA [fraction V], 1% SDS)
- ❑ hybridization buffer (prehybridization buffer + 0.2 nM probe to *lux*A)
  Probe sequence: 5' GATGGAACTCTATAATGAAATTGC 3'
- ❑ wash solution I (50°C) (0.5% SDS, 0.5% N-lauroylsarcosine, 2X SSC)
- ❑ wash solution II (room temperature) (1X SSC, 0.5% Triton X-100)
- ❑ alkaline phosphatase substrate buffer (containing BCIP and NBT)

Supplies and Equipment

- ❑ dried nylon membrane from Southern blot (part 1)
- ❑ Seal-a-Meal™ heat sealer (or equivalent)
- ❑ heat-sealable polyethylene bags (2 per group)
- ❑ Rubbermaid™ washing tray with lid (1 per group)
- ❑ scissors
- ❑ forceps
- ❑ 5-ml disposable pipets (nonsterile)
- ❑ pipet aids or pipet bulbs
- ❑ aluminum foil
- ❑ 50°C shaking water bath
- ❑ 50°C shaking incubator
- ❑ 44°C water bath
- ❑ platform shaker

➡ Procedure

1. Remove the membrane from between the two sheets of 3MM paper. Remember to wear gloves when handling the membrane.

2. Transfer the membrane to a heat-sealable bag and heat-seal the bag so that the bag is approximately 1 cm larger than the membrane on three sides and 3–4 cm larger on one side (see Figure 16.2). Cut off one corner so there is an opening large enough to insert a pipet (about 1 cm).

3. Add 200 μl of the prehybridization buffer to the bag for every square cm of membrane. Carefully squeeze out all of the air bubbles as demonstrated by your instructor (see Figure 16.3). Heat seal the opening and place the bag in a 50°C shaking water bath to prehybridize for 15 minutes.

4. Remove the bag from the water bath and cut off the corner of the bag opposite the first cut (see Figure 16.2). Squeeze out as much of the prehybridization buffer as possible into a sink. Roll a pipet over the bag on a flat surface to remove the remaining prehybridization buffer.

5. Add 100 μl of the hybridization buffer (this is prehybridization buffer containing the probe) to the bag for every square cm of membrane. Squeeze out all of the bubbles and heat seal the bag (try to minimize loss of the hybridization buffer). Return the bag to the 50°C shaking water bath and hybridize for 30 minutes with gentle shaking.

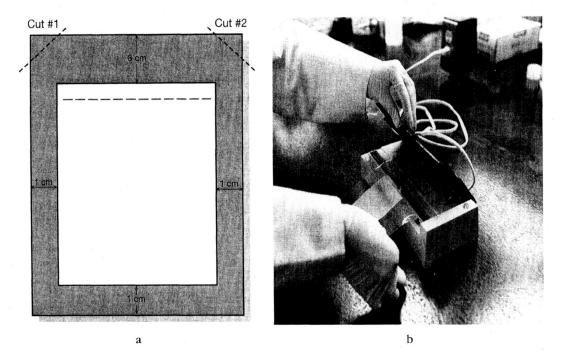

Cut #1   Cut #2

3 cm

1 cm   1 cm

1 cm

a   b

Figure 16.2
Heat sealing a Southern membrane in a heat-sealable bag. (a) Location of membrane in the heat-sealable bag. Note the increased space at the top of the bag to allow room to remove bubbles after adding the prehybridization and hybridization buffer. (b) Use of a Seal-a-Meal™ heat sealer.

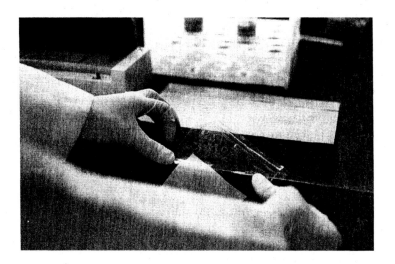

Figure 16.3
Removal of bubbles from a hybridization bag.

6. Cut off a third corner of the heat-sealable bag and discard the hybridization buffer in the sink. Then cut the bag open on three sides, remove the membrane with a pair of forceps, and place it in a plastic washing dish. Be sure that you do not allow the membrane to dry at any stage throughout the washing or color development procedure.

7. Add approximately 200 ml of wash solution I that has been preheated to 50°C to the wash dish. Replace the lid and put the container in a 50°C incubator. Wash the membrane for 20 minutes with gentle agitation.

8. When the first wash is complete, pour out the wash solution and replace it with approximately 200 ml of wash solution II. Wash the membrane at room temperature for 10 minutes with gentle agitation.

9. Transfer the washed, hybridized membrane to a second heat-sealable bag and heat seal so that the bag is approximately 1 cm larger than the membrane on all sides.

10. Add 100 μl of the alkaline phosphatase substrate buffer (containing NBT and BCIP) for every square cm of membrane. Squeeze out air bubbles, heat seal the bag, and wrap the bag in aluminum foil (the color development must be done in the dark).

11. Submerge the foil-wrapped bag in the 44°C water bath and incubate for 15 to 30 minutes (additional sensitivity can often be achieved with a longer incubation).

12. Check your membrane periodically for color development. When you observe the presence of purple-blue bands, stop the reaction by removing the membrane and rinsing it thoroughly in deionized water. Blot the membrane dry and store it between two pieces of Whatman 3MM paper.

Results and Discussion

1. Regions of the membrane that bound DNA fragments complementary to the probe will be indicated by a purple-blue dye. Examine each lane of your membrane for the presence of bands identified by the probe. Compare these bands with those on the photograph of the gel (from Exercise 15). How many bands on the membrane do you observe in each lane? How many do you expect to observe? Explain your rationale.

2. Are all of the bands on the nylon membrane of the same intensity? How might you explain variations in the intensity? What could give rise to this? Recall that the intensity of the ethidium bromide fluorescence was not the same for all of the fragments. Is there any relationship between the intensity of the bands on your photograph and those on the nylon membrane?

3. Are there bands that developed on the hybridized membrane that are not visible on the photograph of the agarose gel? If so, what does this suggest about the relative sensitivity of DNA hybridization compared to ethidium bromide fluorescence?

4. Compare the photograph of your gel with the developed, hybridized nylon membrane. By measuring the distance from the marks on the membrane (corresponding to the wells on the gel) to the blue bands, you should be able to determine which restriction fragments on the photograph contain the target sequence for the *lux*A probe. Use this to identify the location of the *lux*A gene on the restric-

tion map of pUWL500 and pUWL501 prepared in Exercise 15. Does being able to identify the location of a specific DNA fragment help you in your effort to map the plasmid? Refer to problem 5 in Appendix XIX.

Literature Cited

Reed, K. C., and D. A. Mann. 1985. Rapid transfer of DNA from agarose gels to nylon membranes. *Nucleic Acids Res.* 13:7207–7221.

Southern, E. M. 1975. Detection of specific sequences among DNA fragments separated by gel electrophoresis. *J. Mol. Biol.* 98:503–517.

Wahl, G. M., M. Stern, and G. R. Stark. 1979. Efficient transfer of large DNA fragments from agarose gels to diazobenzyloxymethyl-paper and rapid hybridization by using dextran sulfate. *Proc. Natl. Acad. Sci.* 76:3683.

PART FIVE

Subcloning the *lux*A Gene

EXERCISE 17
Restriction Digestion of *lux* Plasmids and Cloning Vector for Subcloning *lux*A

Most genomic libraries consist of clones that contain fairly large inserts in order to reduce the number of clones that must be screened to find the gene(s) of interest. For example, plasmid libraries may contain inserts up to 10 or 15 kb, and lambda and cosmid libraries often contain even larger inserts. Generally, there are significant amounts of DNA in the clone other than the gene or genes of interest. Thus, after one has cloned a desired sequence of DNA from an organism's genome, it is often necessary to **subclone** a smaller portion of the insert DNA into another vector for further analysis. This allows you to eliminate additional DNA not related to the gene of interest and makes further manipulation of the DNA easier. In addition, subcloning is often required if you wish to prepare probes to the cloned gene(s) or sequence the DNA.

Because cloning in plasmids is reliable and allows easy screening of clones containing inserts, plasmid vectors are most commonly used for subcloning. In addition, many modern cloning vectors (like the pGEM™ plasmids) have promoters flanking multiple cloning sites so that RNA probes can be made by transcribing the cloned DNA. Some vectors (such as pGEM™-3Zf[+]) also have an origin of replication for a single-stranded bacteriophage that can be used to generate single-stranded DNA for DNA sequencing. Many of the commonly used vectors are available in pairs (such as pUC18 and pUC19) with the sequence of the multiple cloning site inserted into the *lacZα* gene in opposite orientations in the two plasmids. This is often an important consideration if one wishes to express the cloned insert DNA. By having opposite orientations of the multiple cloning site available, such vectors allow insertion of a gene lacking its own promoter in an orientation with respect to the *lac* promoter that will allow proper transcription of the cloned gene.

Prior to subcloning a portion of an insert DNA, it is first necessary to generate a restriction map of the insert (see Exercise 15) and to identify which restriction fragments contain the sequence of interest (see Exercise 16). Once this is done, it is possible to remove the fragment and ligate it into a vector of choice. In this exercise, you will digest vector and insert DNA so that you can subclone a restriction fragment containing the *lux*A and *lux*B genes into pUC19 (see Exercise 19). We will also excise the restriction fragment of interest from an agarose gel and purify it from the agarose (see Exercise 18). Ligating the purified fragment into a vector should virtually guarantee obtaining the correct clone.

Because this clone will lack the genes involved in the synthesis of the aldehyde required for bioluminescence, it will not produce light. However, as it will contain both genes required to synthesize luciferase (the product of the *lux*A and *lux*B genes), the clones should be able to bioluminesce if *lux*A and *lux*B are expressed and an exogenous aldehyde is supplied. The correct clone can also be identified by colony hybridization or verified by performing plasmid mini-preps on the clones. The insert to be subcloned will be excised from pUWL500 or pUWL501 with *Sst* I (which produces sticky ends) and *Eco*R V (which produces blunt ends). The vector will also be cut with *Sst* I and *Hin*c II. The vector does not contain a site for *Eco*R V in the multiple cloning site, but because *Eco*R V (used to cut the insert DNA) yields blunt ends, any other enzyme that produces blunt ends may be used to cut the vector. Because the ends generated by the two enzymes used to digest the vector are not compatible, the vector cannot religate to itself, which reduces the number of clones that will contain vector. In addition, the insert can only be ligated in one orientation in the vector, which should allow expression of *lux*A and *lux*B from the *lac* promotor in the vector. Because the insert can only be in one orientation, this type of cloning is often called "directional," or "forced," cloning.

Reagents

- ❑ pUWL500 or pUWL501, 0.2 μg/μl (on ice)
- ❑ pUC18 or pUC19, 0.1 μg/μl (on ice)
- ❑ *Sst* I (on ice)
- ❑ *Eco*R V (on ice)
- ❑ *Hin*c II (on ice)
- ❑ 10X KGB restriction buffer (on ice)
- ❑ sterile deionized water (on ice)

- ❑ loading dye
- ❑ 1X TAE buffer
- ❑ LE agarose (low EEO) (or low-melting agarose if using the low-melt gel purification procedure in Exercise 18)
- ❑ ethidium bromide stain (1.0 μg/ml)

Supplies and Equipment

- mini ice buckets (1/group)
- 1.5-ml microcentrifuge tubes
- microcentrifuge tube opener (1/group)
- microcentrifuge tube rack (1/group)
- microcentrifuge tube float (1/group)
- 0.5–10 μl micropipetter
- 10–100 μl micropipetter
- sterile 10-μl micropipet tips
- sterile 100-μl micropipet tips
- micropipet tip discard beaker
- 250-ml Erlenmeyer flask (1 per 2 groups)
- 25-ml Erlenmeyer flask (1 per 2 groups)
- microwave oven

- top loading balance
- 37°C water bath
- 60°C water bath
- microcentrifuge
- horizontal mini gel electrophoresis chamber
- gel casting tray
- gel staining tray
- 6-tooth comb
- power source
- transilluminator
- Polaroid camera with type 667 film
- sterile razor blades
- forceps

►Procedure

A. Restriction Enzyme Digestion

Prior to coming to lab, decide whether you should use pUC18 or pUC19 as the cloning vector in this exercise (refer to your restriction map and the pUC19 map in Appendix XI). Either will work to clone the genes, but only **one** will allow expression of *lux*A and *lux*B from the *lac* promotor. In addition, you should choose whether you wish to use pUWL500 or pUWL501 as your source of insert DNA.

1. Fill your mini ice bucket with crushed ice and place the plasmids, enzymes, and reagents in it. Each group will be provided with its own set of enzymes and reagents containing slightly more than the required volume.

2. Label four sterile 1.5-ml microcentrifuge tubes A through D with a permanent marker and place them in your microcentrifuge tube rack.

3. Use the micropipetters to add the following reagents to each of the tubes. Calculate the amounts of 10X buffer and water to add to each tube to give the indicated final volumes. Tubes A and D are controls to indicate if the preparations contain any nuclease activity and to indicate what uncut plasmids look like so that you can determine if the digestion was complete. Be sure to read the Reminders following the table before proceeding.

Tube	10X buffer	pUWL____	pUC____	H₂O	*Eco*R V	*Sst* I	*Hinc* II	Final volume
A	____ μl	1 μl	—	____ μl	—	—	—	10 μl
B	____ μl	15 μl	—	____ μl	1.5 μl	1.5 μl	—	30 μl
C	____ μl	—	5 μl	____ μl	—	1 μl	1 μl	20 μl
D	____ μl	—	1 μl	____ μl	—	—	—	10 μl

Reminders:

- Always add buffer, water, and DNA to tubes prior to adding the enzymes.

- Touch the pipet tip to the side of the tube when adding each reagent. Be sure you **see** the liquid transferred to the side of the tube, especially with volumes <2 µl.

- You may use the same tip to add the same reagent to different tubes, but be sure to use a new sterile tip for each reagent.

- Check off each addition in the table as you add it. Omissions or incorrect volumes will usually prevent the restriction digestion from working.

4. After adding all ingredients, cap the tubes and vortex to mix. Pool the reagents in the bottom of the microcentrifuge tube by spinning the tubes in the microcentrifuge for 2 to 3 seconds. Be sure that the tubes are in a balanced configuration. Vortex each tube again to mix and recentrifuge to pool the digestion mix in the bottom of the tube.

5. Push all the tubes into the foam microcentrifuge tube holder and float in a 37°C water bath. Incubate for 45 to 60 minutes. At about 20 to 30 minute intervals, remove tubes B and C and vortex. Pulse the tubes in a microcentrifuge and return to the water bath.

B. Agarose Gel Electrophoresis

While the restriction digestions are proceeding, prepare and cast an agarose gel. See Exercise 4 for details on preparing, running, staining, and visualizing agarose gels.

1. Prepare one batch of **0.8% agarose** in 1X TAE buffer per two groups. If you will be using the low-melting point agarose technique to extract your DNA from the gel (Exercise 18) prepare a **1%** agarose gel from **low-melting point agarose.**

CAUTION: In the next step, you will be heating agarose to boiling. Agarose can become superheated and boil over in the microwave or when swirled. Wear gloves and eye protection when removing the agarose and swirling heated flasks.

2. Melt the agarose in a microwave oven and cool to about 60°C or place in the 60°C water bath until ready to use. Cast a gel with a **6-tooth** comb. When solidified, place the gel in the electrophoresis chamber. If you are using low-melting agarose, place the gel in the refrigerator or a cold room for 10 minutes after the gel becomes translucent. Alternately, the gel may be poured in a cold room if available.

3. After the restriction digestion has incubated for 45 to 60 minutes, remove the tubes from the water bath and place them in the microcentrifuge tube rack.

4. Label two clean tubes B' and C'. Add 20 µl (2 µg) of the digested pUWL500 or pUWL501 (tube B) to the tube labeled B'. Add 2 µl (0.05 µg) of the digested vector (tube C) and 8 µl water to the tube labeled C'. **Return the insert and vector digestions (tubes B and C) to the water bath to continue digesting.** You will run a gel on the aliquots removed from these tubes and on the uncut control plasmids (tubes A and D). A large amount of the digested pUWL500 or pUWL501 (tube B') digest will be loaded onto the gel so that you can gel purify the restriction fragment of interest for subcloning.

5. Add 1 µl of loading dye to tubes A, C', and D. Add 2 µl of dye to tube B'. **DO NOT** add loading dye to the insert and vector digestions (tubes B and C). Pulse each tube for 2 to 3 seconds in the microcentrifuge to mix the dye with the restriction digestion.

6. Remove the *Hin*d III digest of lambda DNA from your freezer box and heat the standards in the 60°C water bath for 2 to 3 minutes to melt the *cos* sites in lambda.

7. Load 10 µl of the lambda standard to wells 1 and 6. Load the entire contents of the four reaction samples in wells 2 through 5 as follows:

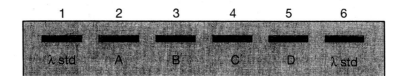

8. Electrophorese at 100–120 V until the bromphenol blue (the leading, darker dye) has migrated about three-quarters of the way to the end of the gel. You will get better separation of the desired bands if you allow the bromphenol blue to run to the end of the gel.

9. When the electrophoresis is complete, turn the voltage down to the lowest level and switch the power source off. Remove the lid to the gel box so that the gel can be removed.

C. Staining and Photodocumentation of Agarose Gel

1. Label a staining tray with your group number. Remove the tray supporting the gel from the electrophoresis box and slide the gel into the staining tray.

CAUTION: Ethidium bromide is a mutagen and suspected carcinogen. Wear gloves, eye protection, and a lab coat, and work over plastic-backed absorbent paper at all times when handling it. Carefully read precautions for handling and disposing of ethidium bromide (Appendix XII) before using ethidium bromide stain.

2. Stain the gel with ethidium bromide (1 μg/ml) for 4 to 5 minutes and rinse several times with tap or distilled water. Avoid prolonged staining to minimize the amount of ethidium intercalated into the DNA fragment you will gel purify.

3. Lower the UV blocking lid and **briefly** observe the DNA fragments produced. Because this is a preparative gel it is important to minimize the amount of UV exposure because it can result in photo-nicking of the DNA you wish to excise. Photograph your gel, working quickly to minimize photo-nicking of the DNA.

CAUTION: UV light is hazardous and can damage your eyes. *Never* look at an unshielded UV light. Always view through a UV blocking shield or wear UV blocking glasses. Some transilluminators have a switch on the lid so that the UV light switches off when the lid is opened, but many transilluminators do not have this safety feature.

4. **Do not throw your gel out!** It will be used in the next exercise to cut out and gel purify the DNA fragment that contains the *lux*A and *lux*B genes. **Immediately** after photographing your gel, remove the gel from the transilluminator and return it to your staining tray with some distilled water.

5. Tape your gel photograph in your notebook below:

6. Examine the photograph of your gel. Digested vector will appear as a single band whereas any undigested plasmid will appear as several bands. If the digestion of the vector is complete, remove the digestion from the 37°C water bath, label the tube "pUC/Sst/Hin" and place it on ice.

   It is essential that the vector be completely digested. Any undigested vector will be in the supercoiled form, which transforms much more effectively than relaxed DNA. This will result in a high percentage of transformants that lack inserts.

7. If pUWL500 was completely digested, you should not see any undigested plasmid (refer to the control lane; tube A). Instead, you should observe the predicted number of bands from your restriction map. If the digestion was complete, remove the pUWL500 or pUWL501 digestion from the 37°C water bath and place on ice.

Note: Digests can be frozen at this point and the ligation set up at a later date. If you are going to gel purify the *lux*A fragment prior to subcloning, proceed immediately to Exercise 18. The fragment must be cut out of the gel the same day the gel is run to prevent diffusion of the DNA in the gel. Gels may be stored for a maximum of 24 hours wrapped in plastic wrap in a refrigerator.

Results and Discussion

1. Consult the restriction map of pUC19 in Appendix XI. How many restriction fragments would you expect from the digest you have done with this vector? Examine the photograph of your gel. Are the number of fragments in the gel consistent with your predictions? Are there any fragments that you would not be able to see on the gel? How would you tell if both enzymes were successful in cutting the vector?

2. What fragments would you expect from the digestion of pUWL500 or pUWL501 with *Sst* I and *Eco*R V? If you have not accurately mapped these enzymes in Exercise 15, your instructor will provide a map showing their recognition sites in each of these plasmids. Which plasmid would be a better choice for purifying the fragment containing the *lux*A and *lux*B genes from an agarose gel? Why? Are the number and size of the restriction fragments resulting from this digest what you expected? Are there any fragments that you would not be able to see on the gel?

3. Does each fragment from the pUWL500 or pUWL501 digestion have one *Sst* I end and one *Eco*R V end? Which fragments can be ligated into a vector cut with *Sst* I and *Hin*c II?

4. Which restriction fragment contains *lux*A? Refer to your restriction map to determine the size of this fragment and identify it on the photograph. You will need to identify and cut this fragment out quickly in the next exercise. Are there other procedures that you could use to identify the fragment?

EXERCISE 18
Gel Purification of DNA Restriction Fragments Containing *lux*A

As you have seen in previous exercises, agarose gels are very efficient in separating restriction fragments. This separation is generally used in an analytical manner, such as in checking the efficiency of a variety of enzymatic reactions involving DNA (restriction digestion, ligation, and so on) or in restriction mapping. However, agarose gel electrophoresis can also be used in a preparative fashion, where specific DNA fragments are separated from a mixture and purified from a gel for use in subcloning, probe preparation, or restriction mapping. The advantage of this procedure is that it allows you to work with a specific piece of DNA. Gel purification can also be used to purify a size fraction of a chromosomal DNA digest as a simple and inexpensive alternative to sucrose gradient centrifugation. This allows you to simplify cloning of a gene(s) if the size of the fragment containing the gene is known. When using this procedure for subcloning a single restriction fragment, virtually all recombinants should contain the target fragment. This method can be thought of as the "slam-dunk" of cloning.

Fragment purification from agarose gels involves the electrophoretic separation of the fragments in an agarose gel, the excision of the agarose containing the fragment of interest, and the removal of the DNA from the agarose. Probably the most important factor in the recovery of usable DNA is the quality of the agarose—only high quality agarose should be used for gel purification techniques. Several suppliers have special grades of agarose that are guaranteed free of nucleases and various inhibitors of enzymatic activity. Effective separation of restriction fragments is also critical. When separating the fragments, it is important to choose an appropriate gel concentration and electrophoresis conditions to effectively separate the target fragment from other DNA. Recovery from agarose is better with higher concentrations of DNA, and it is best to have 200–500 ng of the fragment of interest. However, you should avoid excessive overloading of the gel as bands in an overloaded gel can trap and carry along DNA fragments of different sizes. This problem is magnified by running the gel at high voltages.

When excising the desired band from the ethidium bromide–stained gel, you must work quickly in order to minimize the exposure of the DNA to ultraviolet light, which can result in photo-nicking of DNA. Some transilluminators have variable power settings to allow use of a low light intensity to minimize photo-nicking. It is also important to excise the band of interest with a sterile scalpel or razor blade to reduce the chance of nuclease contamination. The minimum amount of agarose containing the DNA band should be excised as excess agarose reduces recovery of the DNA.

There are many methods to remove DNA from agarose. None of them are completely efficient and many variations exist. One common technique is electroelution, where the fragment is electroeluted out of the gel slice and then recovered. DNA is recovered by placing the gel slice in a dialysis tube that is then placed in a gel box. The fragment is electrophoresed out of the gel where it is trapped by the dialysis membrane. The DNA can then be recovered by alcohol precipitation or affinity chromatography. Commercial electroeluters are also available that allow electroelution of several gel fragments at a time. Alternatively, the DNA can be recovered from agarose by placing a piece of DEAE-cellulose membrane in a slit that is immediately ahead (toward the positive electrode) of the fragment of interest. The DNA is run into the paper by electrophoresis and subsequently recovered by eluting the DNA off the paper with a high ionic strength buffer.

Low-melting point agarose, which melts at about 65°C, is another common method of recovering DNA from agarose gels. First, a gel slice containing the fragment of interest is melted at 65°C (which will not denature the DNA). The DNA then can be recovered from this solution by either removing the agarose or removing the DNA. Agarose can be removed by phenol extraction, which collapses the agarose at the interface between the aqueous and organic phases. Alternatively, DNA can be removed by column chromatography. The molten DNA-containing agarose solution is passed through a matrix that binds DNA and is then washed free of the agarose. The DNA can then be eluted with a low ionic strength buffer. This method is awkward because you need to work at temperatures greater than 37°C. Low-melting agarose is expensive and DNA recoveries are often poor, although it has the advantage that some enzymatic reactions may be carried out directly in the melted gel.

There are a number of physical methods to remove DNA from agarose. The crush-and-soak method involves macerating the gel slice followed by soaking in a suitable buffer. The agarose is then removed by filtration or centrifugation, leaving the

DNA in solution. The freeze-and-squeeze method involves freezing the gel slice followed by squeezing the DNA-containing liquid out of the gel. The resultant solution from these procedures may then be phenol extracted to remove traces of soluble agarose, and the DNA is then ethanol precipitated.

Vogelstein and Gillespie (1979) reported a procedure that uses a chaotropic salt (such as NaI) to dissolve the agarose. The DNA is then captured on a matrix of flint glass. The glass is pelleted by centrifugation, washed free of salts and dissolved agarose, and then the DNA is eluted in a hot low ionic strength buffer. Because of its simplicity and high quality of DNA recovered, this procedure has been modified by several companies who have developed kits to purify restriction fragments, plasmids, and chromosomal DNA.

In this exercise, you will purify the gel fragment that contains the *lux*A and *lux*B genes. We will use a modification of the Vogelstein and Gillespie procedure developed by BIO 101 (GENECLEAN™) that is done in a microcentrifuge tube (BIO 101 Inc. 1989). This gel purification protocol uses a matrix that behaves like flint glass (it might be glass but this information is proprietary). The DNA will be adsorbed onto the glass in the presence of a high concentration of NaI and washed and eluted with a dilute Tris-EDTA buffer. An alternative procedure using low-melting agarose is also described. Additional procedures are also available (Sambrook, et al. 1989). The isolated fragment will be subcloned into a pUC18 or pUC19 vector in Exercise 19 and then screened by colony hybridization in Exercise 21.

PART I. DNA PURIFICATION FROM AGAROSE WITH GENECLEAN™

Reagents

- ❑ agarose gel containing the *Sst* I–*Eco*R V digest of pUWL500 or 501 (from Exercise 17)
- ❑ NaI solution (6.0 M NaI)
- ❑ Glassmilk™ solution (proprietary silica matrix in water)

- ❑ New Wash™ solution (NaCl, Tris, EDTA, in 50% ethanol) (on ice)
- ❑ sterile TE buffer (10 mM Tris, 1 mM EDTA [pH 8.0])

Supplies and Equipment

- ❑ sterile 1.5-ml microcentrifuge tubes
- ❑ sterile 2.0-ml microcentrifuge tubes
- ❑ sterile razor blade (1/group)
- ❑ blunt ended forceps (e.g., Millipore™)
- ❑ 0.5–10 μl micropipetters
- ❑ 10–100 μl micropipetters
- ❑ 100–1000 μl micropipetters

- ❑ sterile 10-μl micropipet tips
- ❑ sterile 100-μl micropipet tips
- ❑ sterile 1000-μl micropipet tips
- ❑ 55°C water bath
- ❑ electronic pan balance
- ❑ preparative transilluminator
- ❑ UV-blocking safety glasses or face shield

Procedure

A. Cutting Fragments out of Agarose Gels

 Note: The procedure is designed for agarose gels made with TAE buffer; BIO 101 markets a modified kit for use with gels made with TBE buffer.

1. Weigh an empty 2.0-ml microcentrifuge tube on an electronic balance and record the weight (to 0.01 g).

2. Examine the photograph of your gel from Exercise 17 and identify the fragment that contains the *lux*A and *lux*B genes. You will need to **quickly** identify and cut out this fragment when you reexamine the gel on the preparative transilluminator.

▼ **CAUTION:** UV light is hazardous and can damage your eyes. Prolonged exposure of uncovered skin to the UV light can cause severe sunburn. The preparative transilluminator does NOT have a switch that shuts off the lamp when the UV blocking shield is lifted. Work behind the shield, but be sure you have UV blocking goggles or a face mask on. DO NOT look at unshielded UV light.

3. Transfer your gel to a preparative transilluminator. These have a low intensity UV fluorescence to minimize photo-nicking.

4. Switch the UV lamp to the preparative mode and identify the fragment to be excised. With a sterile razor blade, carefully cut out the fragment in the smallest

slice of agarose possible. Cut by pushing the razor blade straight down rather than using a slicing motion to avoid scratching the transilluminator filter glass. Remove the fragment with flame-sterilized forceps and transfer it to the tared sterile 2.0-ml microcentrifuge tube. The agarose may be stored in a refrigerator at this point and the DNA purified at a later date.

5. Weigh the tube containing the gel slice again and calculate the weight of the gel in milligrams (mg). The weight of the gel fragment should be less than 500 mg. Assume that the weight in mg is equal to the volume in μl for subsequent calculations.

B. DNA Purification from Agarose

1. Add three volumes (3 × the volume of the gel slice) of the NaI solution to the tube containing the excised agarose gel slice and heat in a 55°C water bath for 5 minutes. Mix occasionally by inversion. Hold the tube to the light while inverting to determine if the gel slice is completely dissolved. If not, continue the incubation at 55°C until the gel slice completely disappears.

2. Vigorously vortex the tube of Glassmilk™ to resuspend the glass particles. Add 5 μl of well suspended Glassmilk™ solution to the tube and mix by inversion. Incubate the tube on ice for 5 minutes and mix occasionally by inversion.

3. Spin the tube in a microcentrifuge at 12-14,000 × g (full speed in a microcentrifuge) for **5 seconds.** Start timing when the microcentrifuge reaches full speed and do not overspin.

4. Remove the supernatant with a 100-1000 μl pipetter. Pulse the tube again in a microcentrifuge and carefully remove any remaining supernatant with a 10-100 μl pipetter.

5. Add 400 μl of the New Wash™ solution to the tube and resuspend the pellet. The pellet will be somewhat difficult to resuspend. Pipet the solution back and forth while digging the tip into the pellet.

6. Centrifuge at 12-14,000 × g for 5 seconds. Remove the New Wash™ solution from the tube with a 100-1000 μl pipetter and discard.

7. Repeat steps 5 and 6 two more times. After the third wash, pulse the tube in a centrifuge and use a 10-100 μl pipetter to remove the remainder of the New Wash™ solution.

8. Resuspend the pellet in 5 μl of TE buffer by drawing it back and forth through the pipet tip. Incubate the tube in a 55°C water bath for 3 minutes. Centrifuge at 12-14,000 × g for 30 seconds.

9. Transfer the supernatant containing the DNA to a sterile microcentrifuge tube and place on ice. A small amount of the Glassmilk™ may also transfer.

10. Repeat steps 8 and 9. Combine the second TE supernatant with the first. Store on ice or freeze until the DNA is needed in Exercise 19.

11. It is best to quantify the DNA by removing 1–2 μl of the TE solution and using one of the quantification techniques described in Exercise 9. If time does not allow this, assume a recovery of approximately 50% to estimate the DNA concentration.

PART II. GEL PURIFICATION FROM LOW-MELTING AGAROSE

Reagents

- ☐ agarose gel containing the *Sst* I-*Eco*R V digest of pUWL500 or 501 (from Exercise 17)
- ☐ "phenol ("phenol" is equilibrated with TE with 0.1% 8-hydroxyquinoline and 0.2% β-mercaptoethanol)
- ☐ "phenol:chloroform" (1:1 V/V; "chloroform" is choloroform:isoamyl alcohol [24:1])
- ☐ chloroform:isoamyl alcohol (24:1)

- ☐ 3 M sodium acetate (pH 5.2)
- ☐ 95% ethanol
- ☐ 70% ethanol
- ☐ sterile TE buffer

Supplies and Equipment

- ☐ sterile 1.5-ml microcentrifuge tubes
- ☐ sterile 2.0-µl microcentrifuge tubes
- ☐ 10–100 µl micropipetters
- ☐ 100–1000 µl micropipetters
- ☐ sterile 100-µl micropipet tips

- ☐ sterile 1000-µl micropipet tips
- ☐ microcentrifuge
- ☐ 65°C water bath
- ☐ phenol-chloroform waste beaker (in fume hood)

➡Procedure

1. Cut the desired restriction fragment from a low-melting point agarose gel and transfer to a tared sterile 2.0-ml microcentrifuge tube as described in Part IA.

2. Weigh the tube containing the gel slice again and calculate the volume of the gel. The weight of the gel in mg is equal to its volume in µl. The weight of a gel fragment should be less than 250 mg.

3. Add 3 to 5 volumes (3-5 × the volume of the gel slice) of TE buffer to the agarose fragment in the tube. The total volume of agarose and TE should not exceed 1000 µl (800 µl for a 1.5-ml tube). Close the cap and incubate in a 65°C water bath for 5 minutes to melt the agarose.

CAUTION: The next step involves the use of phenol. Phenol is toxic and can cause severe burns if it contacts your skin. Wear gloves, eye protection, and a lab coat when using it. Perform all work with phenol in the fume hood. If phenol contacts your skin, rinse immediately with cold water and notify your instructor.

4. Vortex the tube to mix and then cool to room temperature. Add an equal volume of "phenol" and vortex for 30 seconds.

5. Centrifuge for 5 minutes at 8000 × *g* at room temperature. The agarose will form a white layer between the "phenol" (bottom layer) and the aqueous phase containing the DNA (the top layer).

CAUTION: The next step involves the use of chloroform as well as "phenol."
Chloroform is toxic and is dangerous to inhale. Wear gloves, eye protection, and a lab
coat when using it. Perform all work with chloroform in the fume hood.

6. Carefully transfer the aqueous phase to a clean, sterile microcentrifuge tube, noting the exact volume transferred. Add an equal volume of "phenol:chloroform" and extract as in steps 4 and 5.

7. Transfer the aqueous phase to a third sterile microcentrifuge tube. Note the volume and extract with an equal volume of chloroform:isoamyl alcohol (24:1). **Note previous caution!**

8. Transfer the aqueous phase from this last extraction to a fourth sterile microcentrifuge tube noting the exact volume. Add 0.1 volumes of 3.0 M sodium acetate (pH 5.2) and mix. Add 2 volumes of ice-cold 95% ethanol and mix. Incubate on ice for 10 to 15 minutes to precipitate the DNA.

9. Centrifuge for 10 minutes at 12–14,000 × *g* (full speed in a microcentrifuge). Be sure your tube is balanced with a tube with an equal volume of liquid and that the hinges of the microcentrifuge tubes are pointing toward the outside of the rotor.

10. Discard the supernatant and remove the last traces with a micropipet. Wash the pellet (which will be invisible) with 1 ml of 70% ethanol. Gently rock the tube once and centrifuge for 5 minutes at 12–14,000 × *g*.

11. Decant the supernatant from the 70% ethanol wash and carefully remove the last traces with a micropipet. Air dry at your bench for 5 to 10 minutes and dissolve the pellet in 10 μl of TE buffer. Store on ice or freeze until the DNA is needed in Exercise 19.

12. It is best to quantify the DNA by removing 1–2 μl of the TE solution and using one of the quantification techniques described in Exercise 9. If time does not allow this, assume a recovery of approximately 50% to estimate the DNA concentration.

Results and Discussion

1. If you quantified the DNA you purified from the agarose gel, what was the efficiency of recovery? What factors do you think affect the efficiency of recovery? What might improve recovery?

2. Why do you think DNA binds to glass? What conditions are required for binding? Hint: What solution is the DNA in when it is bound to the glass?

3. What allows the DNA to be released from the glass? Hint: What solution is used to elute the DNA?

Literature Cited

BIO 101 Inc. 1989. The GENECLEAN™ Kit. La Jolla, CA.

Sambrook, J., E. F. Fritsch, and T. Maniatis. 1989. *Molecular cloning: A laboratory manual.* 2d ed. New York: Cold Spring Harbor Laboratory Press.

Vogelstein, B., and D. Gillespie. 1979. Preparative and analytical purification of DNA from agarose. *Proc. Natl. Acad. Sci.* 76:615–619.

EXERCISE 19
Subcloning *lux*A into a Plasmid Vector

DNA to be subcloned is ligated in much the same manner as DNA used to prepare genomic libraries (refer to Exercise 10 for a discussion of ligation reactions). Because there are far fewer combinations of recombinant DNA molecules that can form during subcloning, the insert-to-vector ratios are not as critical as with genomic cloning. In addition, restriction fragments for subcloning are often generated with two different restriction enzymes. When the vector is cut with two enzymes, a small "stuffer fragment" is generated, and the rest of the vector cannot religate back to itself. In order to reform the intact vector, the stuffer fragment must be religated to the large vector fragment. This reaction competes with the ligation of the insert DNA into the vector and if there is a molar excess of insert DNA, it is more likely that a given vector will ligate to an insert rather than to the stuffer. This greatly decreases the number of transformants that lack inserts (i.e., blue colonies when using vectors capable of α-complementation). When performing ligations for subcloning reactions, 20–100 ng of digested vector are typically ligated to an equal or slightly greater molar concentration of insert DNA in a 10–20 μl ligation reaction.

In subcloning from one plasmid into a new plasmid vector, it is important to be aware that the insert DNA is generated by digestion of the original plasmid (possibly including the same vector to be used in the subcloning). Plasmid vectors have an origin of replication and an antibiotic resistance marker gene. Thus, if the digested vector and the digested plasmid harboring the insert DNA are mixed together in a ligation reaction, there may be two vector fragments available. Sometimes this may cause confusing results, and it is advisable to determine whether the restriction fragment to be cloned needs to be purified away from the original vector.

In this exercise, we will set up ligation reactions with the pUC18 or pUC19 digested with *Hinc* II and *Sst* I and the *Eco*R V/*Sst* I restriction fragments of pUWL500 or pUWL501 from Exercise 17. We will also ligate the gel purified *Eco*R V/*Sst* I restriction fragment (from Exercise 18) to the vector, which should ensure obtaining the clone of interest. Because *Eco*R V and *Hin*c II generate restriction fragments with blunt ends, they are considered compatible enzymes. Although it is possible to ligate blunt ends, the reaction rate is slower because the substrate for DNA ligase (a 3'-OH of one fragment adjacent to a 5' phosphate of another fragment) is not held in position by complementary sticky ends. Blunt end ligations are useful, however. Because a large number of enzymes create blunt ends, numerous options to create compatible ends are available.

As we are not cloning the entire *lux* operon, the clones will not spontaneously bioluminesce. However, if the *lux*AB gene fragment was ligated in the proper orientation in the vector for expression, we should be able to identify the clone of interest by observing bioluminescence on transformation plates when an exogenous aldehyde is added (Exercise 20). We will also be able to identify the clone of interest by colony hybridization with the *lux*A probe (Exercise 21).

PART I. SETTING UP LIGATION REACTIONS

Reagents

- *Eco*R V/*Sst* I digest of pUWL500 or pUWL501 (on ice) concn. = _____ μg/μl
- purified *Eco*R V/*Sst* I fragment containing *lux*AB (on ice) concn. = _____ μg/μl
- *Hinc* II/*Sst* I digest of pUC18 or pUC19 (on ice) concn. = _____ μg/μl

- T4 DNA ligase (on ice)
- 10X ligation buffer with ATP (on ice)
- sterile deionized water (on ice)
- sterile TE buffer (on ice)

Supplies and Equipment

- mini ice buckets (1/group)
- sterile 1.5-ml microcentrifuge tubes
- microcentrifuge tube opener (1/group)
- microcentrifuge tube rack (1/group)
- 0.5–10 μl micropipetter
- 10–100 μl micropipetter
- sterile 10-μl micropipet tips

- sterile 100-μl micropipet tips
- micropipet tip discard beaker
- 37°C water bath
- 65°C water bath
- microcentrifuge
- low-temperature incubator (12–16°C)

▶Procedure

1. Fill your mini ice bucket with crushed ice and place each of the required enzymes, buffers, and DNAs in it. Each group will be provided with its own set of enzymes and reagents containing slightly more than the required volume.

2. Heat the *Eco*R V/*Sst* I digest of pUWL500 or pUWL501 (tube B from Exercise 17), the *Hinc* II/*Sst* I digest of pUC18 or pUC19 (tube C from Exercise 17), and gel purified DNA (from Exercise 18) at 65°C for 10 minutes. This inactivates most restriction enzymes. Some enzymes are not heat-inactivatable and must be destroyed by extractions with phenol and chloroform.

3. Label two sterile microcentrifuge tubes L1 and L2. You will use these to set up two ligation reactions—one with the gel purified *lux*AB fragment and one with the total pUWL500 or pUWL501 digest.

4. Combine the digested vector and the insert DNA as indicated in the following table. Based on your concentrations of vector and insert DNA preparations, calculate the required volumes of each DNA solution. Calculate the required volumes of sterile deionized water and 10X ligase buffer to make each reaction up to a final volume of 20 μl. After all additions have been made, return remaining digested vector and pUWL500 or pUWL501 DNA to your freezer box.

Tube	Digested vector	Digested pUWL____	Gel purified fragment	H₂O	10X ligase buffer	Total volume
L1	____ μl (0.1 μg)	____ μl (0.4 μg)	—	____ μl	____ μl	20 μl
L2	____ μl (0.1 μg)	—	____ μl (0.1–0.2 μg)	____ μl	____ μl	20 μl

5. After adding all ingredients, cap the tubes and vortex to mix. Pool the reagents in the bottom of the microcentrifuge tube by spinning the tubes in the microcentrifuge for 2 to 3 seconds. Be sure that the tubes are in a balanced configuration.

6. Label two sterile microcentrifuge tubes $L1/T_0$ and $L2/T_0$. Add 5 µl of the corresponding ligation mix to each tube. Spin the tubes in a microcentrifuge for 2 to 3 seconds to pool the mixtures in the bottom of the tubes. Place the tubes in your freezer box. These will be used in the next period to analyze the ligation mixes before and after ligation (by agarose gel electrophoresis).

7. Add 1 µl of T4 DNA ligase to each ligation reaction (tubes L1 and L2). Be sure you actually see a small drop of ligase transferred to the tube. Pulse the tubes in the microcentrifuge to pool the ligation mix in the bottom. Vortex the tubes and pulse again in the microcentrifuge to pool the liquid.

8. Incubate your ligation reactions at 10–12°C overnight. Ligations may also be incubated at 16°C for 1 to 4 hours.

9. Clearly label the tubes and place them in your freezer storage box.

PART II. CHECKING THE SUCCESS OF THE LIGATION REACTIONS

Reagents

☐ loading dye (room temperature)
☐ 1X TAE buffer

☐ LE agarose (low EEO)
☐ ethidium bromide stain (1.0 μg/ml)

Supplies and Equipment

☐ mini ice buckets (1 per group)
☐ sterile 1.5-ml microcentrifuge tubes
☐ microcentrifuge tube opener (1 per group)
☐ microcentrifuge tube rack (1 per group)
☐ 0.5–10 μl micropipetter
☐ 10-μl micropipet tips
☐ micropipet tip discard beaker
☐ 250-ml Erlenmeyer flask (1 per 2 groups)
☐ 25-ml Erlenmeyer flask (1 per 2 groups)
☐ microwave oven
☐ top loading balance

☐ spatula
☐ 60°C water bath
☐ microcentrifuge
☐ horizontal mini gel electrophoresis chamber
☐ gel casting tray
☐ 6-tooth comb
☐ gel staining tray
☐ power source
☐ transilluminator
☐ Polaroid camera with type 667 film

➡ Procedure

A. Agarose Gel Electrophoresis

Prior to transforming the competent *E. coli* (prepared in Exercise 11) with the recombinant DNA, we will run an agarose gel to analyze the success of the ligation reactions. This will provide an indication of the number and sizes of ligation products. See Exercise 4 for details on preparing, running, staining, and visualizing agarose gels.

1. Prepare one batch of **0.8% agarose** in 1X TAE buffer per two groups.

CAUTION: In the next step, you will be heating agarose to boiling. Agarose can become superheated and boil over in the microwave or when swirled. Wear gloves and eye protection when removing the agarose and swirling heated flasks.

2. Melt the agarose in a microwave oven and cool to about 60°C or place in the 60°C water bath until ready to use. Cast a gel with a **6-tooth** comb. When solidified, place the gel in the electrophoresis chamber.

3. While the gel is solidifying, label two microcentrifuge tubes L1/T$_{end}$ and L2/T$_{end}$. Remove the ligation tubes from the incubator (or your freezer box) and add 5 μl of each ligation mixture (L1 and L2) to the corresponding tube that you just labeled (L1/T$_{end}$ and L2/T$_{end}$). Spin in a microcentrifuge for 2 to 3 seconds to pool and mix the contents. Place the ligation reactions (L1 and L2) in your freezer box.

4. Remove and thaw the two aliquots of the ligation mixes you froze prior to adding the DNA ligase (L1/T_0 and L2/T_0). With these samples, you will be able to compare the unligated and ligated DNA for each reaction.

5. Add 1 µl of loading dye to each T_0 and T_{end} tube. Pulse the tubes for 2 to 3 seconds in the microcentrifuge to mix the dye with the ligation mixture.

6. Remove the *Hin*d III digest of lambda DNA from your freezer box. Heat the lambda standard in the 60°C water bath for 2 to 3 minutes to melt the *cos* sites in lambda.

7. Load 5 µl of the *Hin*d III digest of lambda into wells 1 and 6. Load the entire contents (6 µl) of each ligation tube to wells 2 through 5 as follows:

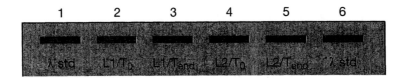

8. Electrophorese at 100–120 V until the bromphenol blue (the leading, darker dye) has migrated one-half to three-quarters of the way to the end of the gel. Good separation is not necessary as we only want to verify that the ligation reactions worked.

9. When the electrophoresis is complete, turn the voltage down to the lowest level and switch the power source off. Remove the lid to the gel box so that the gel can be removed.

B. Staining and Photodocumentation of Agarose Gel

1. Label a staining tray with your group number. Remove the tray supporting the gel from the electrophoresis box and slide the gel into the staining tray.

CAUTION: Ethidium bromide is a mutagen and suspected carcinogen. Wear gloves, eye protection, and a lab coat, and work over plastic-backed absorbent paper at all times when handling it. Carefully read precautions for handling and disposing of ethidium bromide (Appendix XII) before using ethidium bromide stain.

2. Stain the gel with ethidium bromide (1 µg/ml) for 5 to 10 minutes and rinse several times with tap or distilled water.

3. Briefly observe the gel on the transilluminator and photograph.

CAUTION: UV light is hazardous and can damage your eyes. *Never* look at an unshielded UV light. Always view through a UV blocking shield or wear UV blocking glasses. Some transilluminators have a switch on the lid so that the UV light switches off when the lid is opened, but many transilluminators do not have this safety feature.

4. Tape your gel photograph in your notebook below:

Results and Discussion

1. Examine the photograph of your gel. Did the ligations work? Can you tell if the blunt ends of the fragments were sucessfully ligated together? How many different types of recombinant DNA molecules do you think there are in each ligation?

2. Compare this photograph with the photograph of the gel from the ligations set up in Exercise 10. Does the ligated DNA look different in the two different exercises? If so, why?

3. We did not set up different insert-to-vector ratios for this ligation as we did in Exercise 10. Do you think that this is a limitation? How will you know if it is a limitation or not?

4. Were all the restriction enzymes used to generate the ligated fragments heat-inactivatable (refer to a current catalog from an enzyme supplier)? Is this a limitation? Why or why not?

EXERCISE 20

Transformation of Competent *Escherichia coli* DH5α with Subcloned DNA

The recombinant DNA molecules resulting from the ligation of the pUWL500 or pUWL501 restriction fragments into a pUC vector (Exercise 19) must be transferred into *Escherichia coli* in order to be replicated (or cloned). As was done with the recombinant DNA molecules containing genomic DNA (Exercise 12), you will transfer the ligation mixtures into competent *E. coli* by transformation. You will use the frozen competent cells prepared by the calcium chloride procedure in Exercise 11. Alternatively, fresh competent cells may be made the day prior to the lab and stored on ice. The transformed cells will then be plated on LB plates containing ampicillin and X-gal. This medium will select for cells taking up the vector or a vector containing inserts (recall that pUC plasmids contain an ampicillin resistance marker; see Appendix XI). In addition, the X-gal allows rapid identification of transformants containing inserts because pUC plasmids can be screened by α-complementation. Thus, clones containing inserts will be white, whereas colonies that have taken up only the vector will be blue (see Exercise 12).

After the transformants have grown, clones containing the *lux*A gene will be identified by colony hybridization (Exercise 21). However, it is also possible to screen *lux* subclones by light production if both the *lux*A and *lux*B genes are located on the same cloned fragment and they are expressed in the host cell. This will allow transformants to produce functional luciferase; but they will not spontaneously produce light because they lack the genes required to synthesize the aldehyde required for bioluminescence. However, if an aldehyde is supplied to the cells, they will produce light (Baldwin, et al. 1984, Engebrecht, et al. 1983), and the correct clones then can be identified in a dark room.

PART I. TRANSFORMATION OF COMPETENT CELLS

Cultures

❑ competent *E. coli* DH5α cells (on ice)

Media

❑ sterile LB broth (6 ml/group)

❑ LB plates + 50 μg/ml ampicillin + X-gal (LB/Ap⁵⁰/X-gal; 10 per group)

Reagents

❑ ligation mixtures (on ice; from Exercise 19)
❑ pGEM™-3Zf(+), 0.1 ng/μl (on ice)

❑ sterile TE buffer (on ice)

Supplies and Equipment

❑ ice chests
❑ sterile 15-ml disposable plastic culture tube (4/group)
❑ 0.5–10 μl micropipetter
❑ 100–1000 μl micropipetter
❑ sterile 10-μl micropipet tips

❑ sterile 1000-μl micropipet tips
❑ micropipet tip discard beaker
❑ 42°C water bath
❑ 37°C shaking water bath
❑ alcohol jars for flaming
❑ spreading triangles

Procedure

Note: Use careful aseptic technique throughout this procedure.

1. Remove two vials of frozen competent cells from the ultracold (−80°C) freezer and place on ice for about 15 minutes to thaw. Do not allow cells to warm and do not shake or invert vials to speed thawing. Freshly prepared (up to 48 hours old) competent cells can also be used.

2. Collect and thaw your frozen ligation mixes from Exercise 19 (tubes L1 and L2). Spin in a microcentrifuge for 2 to 3 seconds to pool the contents in the bottom of the tubes and place both tubes in your ice bucket.

3. Wrap tape completely around each of four sterile plastic 15-ml disposable culture tubes and label each with a permanent marker as follows:

   A. TE (negative control)
   B. pGEM™ (positive control)
   C. L1 (pUWL500 or pUWL 501 digest)
   D. L2 (purified *lux*A fragment)

Place tubes on ice.

4. Gently rock a thawed tube of competent *E. coli* DH5α to resuspend the cells. Using the 100–1000 μl micropipetter and a **sterile** tip, add 200 μl of the competent cells to each of the tubes labeled in step 3. Leave tubes on ice.

5. Add 10 μl of sterile TE to tube A as a negative control. Add 10 μl of the pGEM™-3Zf(+) stock (0.1 ng/μl) to tube B as a positive control. Be sure that you add the DNA **directly** into the competent cell solution.

6. Add 10 μl of each of the ligation mixes L1 and L2 to tubes C and D respectively. Be sure that you add the DNA **directly** into the competent cell solution. Mix gently and incubate tubes on ice for 20 minutes.

7. Heat-shock cells by placing them in the 42°C water bath for **exactly** 90 seconds. Return them to the rack on your bench.

8. Add 1 ml of LB broth (room temperature) to each tube and mix gently. Incubate the tubes at 37°C (preferably with gentle shaking) for 30 to 60 minutes (be sure your group number is on your tubes). Why is this incubation necessary?

9. While your tubes are incubating, collect 10 LB/Ap^{50}/X-gal plates and label them appropriately. You will spread the transformed cells on these plates as indicated in the following table:

Tube	Addition	# of plates	Volumes plated
A	TE buffer	1	100 μl
B	pGEM™-3Zf(+)	1	100 μl
C	L1 (pUWL500 digest)	4	10 μl, 30 μl, 100 μl, 200 μl*
D	L2 (gel purified *lux*AB)	4	10 μl, 30 μl, 100 μl, 200 μl*

*If the transformation efficiency of competent cells (determined in Exercise 12) is greater than 10^6/μg, plate 3, 10, 30, and 100 μl volumes

10. After the 30- to 60-minute incubation with the LB broth is complete, spread the transformed cell suspensions on the LB/Ap^{50}/X-gal plates. Plate the volumes indicated in the table in step 9 (use the spread plate technique described in Exercise 1; be sure to flame-sterilize your spreading triangle). For volumes less than 100 μl, first add 100 μl of sterile LB broth to the plates and then add the transformed cells.

11. Leave the plates upright for 5 to 10 minutes to let the fluid soak into the plates. When dry, invert and incubate for 18 to 20 hours at 37°C. **Note: Do not let the plates incubate for longer than this at 37°C.**

PART II. SCREENING FOR CLONES CONTAINING *lux*A AND *lux*B

Cultures

□ transformation plates (from Part I)

Media and Reagents

□ LB plates + 50 µg/ml ampicillin + X-gal   □ *n*-decyl aldehyde
    (LB/Ap⁵⁰/X-gal; 2/group)

Supplies and Equipment

□ cotton swabs   □ toothpick discard beaker
□ sterile toothpicks

➡Procedure

1. After 18 to 20 hours, remove your transformation plates from the 37°C incubator. Incubate for 2 to 3 hours at room temperature to allow the cells to express luciferase activity.

2. Organize your plates according to the table from Part I. First, examine your control plates. Do you have transformants on the negative control plate (A)? Do you have transformants on the positive control plate (B)? What color are the colonies on the positive control? Can you explain why?

3. Now examine the four plates prepared from each of the transformation mixes. You should have a mixture of white and blue colonies. You will use one plate from each of the two transformations for colony hybridization (Exercise 21). Choose plates with about 100 small (1 mm) colonies and place them in the refrigerator to stop their growth. Small colonies transfer and probe more uniformly, and cold plates are less likely to transfer medium to the membrane during the transfer procedure.

4. Prior to screening for bioluminescent colonies by adding exogenous aldehyde, first screen plates to see if any clones bioluminesce without the addition of aldehyde. Think about what might allow this to occur. To screen for light producing clones, go into a dark room and allow your eyes to dark adapt for 5 to 10 minutes. Mark any bioluminescent colonies with an **X** as described in Exercise 13.

5. Allow your eyes to dark adapt again and smear a small amount of *n*-decyl aldehyde on the lid of a Petri plate with a sterile swab. Reexamine the plates for bioluminescence. Clones containing *lux*A and *lux*B should produce light in 15 to 30 seconds. **Circle** any bioluminescent clones with a permanent marker.

6. Return to the lab and examine your plates to see if you have clones that are bioluminescent with the aldehyde, but not without. Pick several positive colonies with a sterile toothpick and streak on quadrants of an LB/Ap⁵⁰/X-gal plate. Incubate at 37°C for 18 to 20 hours and then at room temperature for 2 to 3 hours. These plates will be used to inoculate broths to do mini-preps on the positive clones (Exercise 22).

Results and Discussion

1. Did your controls work? If not, can you explain why?

2. Using the data from your positive control, calculate the transformation efficiency of the competent cells (see Exercise 11). If you used frozen competent cells, did the transformation efficiency change as a result of freezing (compare with the efficiency determined in Exercise 12). If your cells were freshly prepared, did the frequency change since you prepared them?

3. Did you obtain a different frequency of transformants containing inserts (based on the blue/white colony ratio) with the two different transformations (L1 and L2)? How does the blue/white ratio of the plates in this exercise compare to what you observed in the genomic cloning (Exercise 12)? If different, can you explain why?

4. Why is the addition of extracellular aldehyde required for light production in your subclone? Why is it not required for light production in clones containing pUWL500 or pUWL501?

5. Why is it important to screen your plates for bioluminescent clones **prior** to screening with aldehyde? Did you obtain any light producing clones **without** the addition of *n*-decyl aldehyde? From which ligations did these clones arise? Can you explain how this might occur (even if you did not observe any of these clones)?

6. Did you obtain any light producing clones when you added *n*-decyl aldehyde? From which ligations did these clones arise? Can you explain how this occurred? Was the frequency of light producing clones greater when the insert DNA was gel purified?

7. Which vector did you choose to do your subcloning (pUC18 or pUC19)? Compare your results with those of another group that used the other vector. Are they the same? How might the orientation of the multiple cloning site in these vectors account for any differences in the observed results?

8. Based on the restriction map you constructed in Exercises 14–16, or using information provided by your instructor, identify other restriction fragments that could be subcloned and identified by bioluminescence in the presence of *n*-decyl aldehyde. Design a flowchart showing a cloning strategy to subclone one such fragment.

9. Design a flowchart showing a cloning strategy to subclone an *Sst* I/*Xho* I fragment from the *lux*AB subclone. Hint: Consult an enzyme compatibility table in a catalog from an enzyme supplier.

Literature Cited

Baldwin, T. O., T. Berends, T. A. Bunch, T. F. Holzman, S. K. Rausch, L. Shamansky, M. L. Treat, and M. M. Ziegler. 1984. Cloning of the luciferase structural genes from *Vibrio harveyi* and expression of bioluminescence in *Escherichia coli*. *Biochemistry* 23:3663-3667.

Engebrecht, J., K. Nealson, and M. Silverman. 1983. Bacterial bioluminescence: Isolation and genetic analysis of functions from *Vibrio fischeri*. *Cell* 32:773-781.

EXERCISE 21
Colony Hybridization to Screen for
*lux*A Subclones

Screening recombinant clones by nucleic acid hybridization is a powerful detection method used to identify even nonexpressed sequences (see Exercise 13 for techniques used to screen gene libraries). The complementary nature of nucleic acid hybridization, and the ability to alter the specificity (stringency) of the reaction, provides excellent sensitivity and selectivity. With hybridization, even rare sequences can be easily found in the large and complex genetic background of the host cell.

For libraries cloned in plasmid vectors, colonies can be screened by a hybridization technique known as colony hybridization. This procedure, originally developed by Grunstein and Hogness (1975), involves the transfer of bacterial colonies to a nitrocellulose membrane. The cells on the membrane are lysed and the DNA released is subsequently denatured and immobilized to form a "DNA print" of the colonies on the plate. A nucleic acid probe can then be used to detect the presence of the target sequence in the DNA of the lysed colonies. This procedure allows simultaneous screening of thousands of colonies and is simple, reliable, and widely used. A similar procedure can also screen bacteriophage plaques on plates where a bacteriophage vector (such as lambda or M13) is used.

Colony hybridization is normally used to screen entire genomic libraries where the incidence of positive clones is low. In this exercise, we will use the procedure to screen transformants from the subcloning experiment (Exercises 17–20), although the incidence of positive clones should be so high that such a powerful screening technique is not necessary. This exercise, however, demonstrates this widely used technique without having to use a large number of plates and membranes to detect the clone of interest.

We will use several modifications of the original Grunstein and Hogness colony hybridization procedure. First, we will use nylon membranes instead of nitrocellulose because nylon is much more durable and resistant to the 0.5 M NaOH used in the lysis solution. The cells on the membrane are lysed with a solution of NaOH and

N-lauroylsarcosine similar to the alkaline lysis procedure used in some mini-prep protocols. The NaOH lyses the cells and results in the denaturation of the released DNA (it is essential to denature DNA to single-stranded molecules in order to do hybridizations). The inclusion of a mild surfactant (N-lauroylsarcosine) and heat (50°C) are modifications developed by Maas (1983). These factors aid in the lysis of the cells and denaturation of the nucleic acids. The single-stranded DNA released during lysis of each colony is immobilized on the nylon filter. The filter containing the bound DNA will then be neutralized with Tris buffer (pH 8.0) and hybridized with the oligonucleotide probe to *lux*A (see Exercise 16). This will identify clones that contain the *lux*A gene, which will then be characterized in Exercise 22.

Cultures

□ refrigerated plates from transformation (Exercise 20)

Media

□ LB plates + 50 μg/ml ampicillin + X-gal (LB/Ap⁵⁰/X-gal; 1/group)

Reagents

□ lysis solution (0.5 M NaOH, 0.1% N-lauroylsarcosine)
□ neutralization buffer (0.5 M Tris buffer [pH 8.0])
□ deionized water
□ prehybridization buffer (5X SSC, 0.5% BSA [fraction V], 1% SDS)
□ hybridization buffer (prehybridization buffer + 0.2 nM probe to *lux*A)
□ wash solution I (50°C) (0.5% SDS, 0.5% N-lauroylsarcosine, 2X SSC)
□ wash solution II (room temperature) (1X SSC, 0.5% Triton X-100)
□ alkaline phosphatase substrate buffer (containing BCIP and NBT)

Supplies and Equipment

□ 95 mm nylon membrane circles
□ glass Petri plates
□ glass spreading triangle
□ Seal-a-Meal™ heat sealer (or equivalent)
□ heat-sealable polyethylene bags (2/group)
□ Rubbermaid™ washing tray with lid (1 per group)
□ scissors
□ forceps

□ 5-ml disposable pipets (non sterile)
□ 10-ml disposable glass pipets
□ pipet aids or pipet bulbs
□ 100–1000 μl micropipetters
□ sterile 1000-μl micropipet tips
□ aluminum foil
□ 50°C shaking incubator
□ 50°C shaking water bath
□ 44°C water bath

➡ Procedure

A. Preparation of Membranes

1. Retrieve the two plates from Exercise 20 that you chose for the colony hybridization. They should have been refrigerated while the colonies were small (about 1 mm). Plates should be refrigerated (4°C) for at least 1 hour prior to use. The cooled medium will have less tendency to adhere to the nylon membrane.

2. Each group will prepare two membranes: one from a plate with cells transformed with the ligation of vector and unpurified fragments from pUWL500 or pUWL501, and the other from a plate with cells transformed with the ligation of vector and the gel purified insert. For each blot, remove a 95 mm nylon membrane with the forceps. Be sure to wear gloves when handling the membranes. Label each membrane on the tab or edge (use a pencil) with your group number and either "pUWL dig" or "Gel Pur" for the transformations with the total plasmid digest or the gel purified fragment, respectively.

3. Carefully place the membranes onto the surface of the spread plates [see Figure 21.1(a)]. Start by touching the nylon to the agar at the edge of the plate and then roll the membrane down smoothly to avoid trapping air bubbles. **Do not** move the membrane once it has made contact with the agar.

4. Gently smooth the membranes over the surface of the agar with a flame-sterilized glass spreading triangle. This will ensure contact with all of the colonies. Leave the filters in place for 3 to 5 minutes.

5. While the membranes are in contact with the surface of the agar, it is important to mark the orientation of the membranes on the plate, which will allow you to orient the hybridized membranes with the colony pattern on the plate when the positive clones are identified. Mark the orientation by placing a dot with a permanent marker on the bottom of the Petri dish directly under the three asymmetric holes near the edge of the membrane. If the membrane does not have prepunched holes, poke three holes asymetrically near the edge of the membrane with an 18-gauge needle prior to marking the plate.

6. Add 0.7 ml of the lysis solution to each of the two glass Petri dishes. Carefully peel the filter off the surface of the agar with the forceps [see Figure 21.1(b)] and then place it **colony side up** on the lysis solution in the Petri dish. There should be just enough liquid to saturate the membrane. **Do not** allow the lysis buffer to contact the colony side of the filter. Reincubate the **agar** plates used to prepare the membranes at 37°C until the hybridization of the membranes is complete. This will allow the cells remaining on the plates to grow slightly to reform colonies.

7. Carefully place the Petri dishes containing the membranes in a 50°C incubator for 15 to 20 minutes to complete the lysis and DNA denaturation. If a 50°C incubator is not available, this step may be done satisfactorily at 37°C or room temperature.

8. Remove the membranes from the Petri dishes with forceps and blot the excess fluid by placing the membranes, colony side up, on several paper towels. Do not allow the membranes to dry completely.

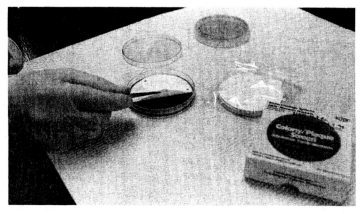

a

b

Figure 21.1
Transfer of colonies onto a nylon membrane for colony hybridization. (a) Placing the membrane onto the plate. (b) Removal of the membrane from the plate.

9. Rinse the lysis solution out of the Petri dishes with distilled water. Discard the rinse into a beaker containing disinfectant as there may be viable bacteria present. Dry the dishes with a paper towel and add 0.7 ml of the neutralization buffer (0.5 M Tris [pH 8.0]) to each dish. Place the membranes (colony side up) on the fluid and incubate at room temperature for 3 minutes.

10. Remove each membrane from the neutralization buffer and blot the membranes on paper towels to remove excess fluid. Rinse the neutralization buffer out of the Petri dishes with distilled water. Discard the rinse into the beaker containing disinfectant.

11. Place approximately 200 ml of deionized water in each of two sealable washing trays. Immerse each membrane in a separate container of deionized water. Seal the containers and incubate them for 2 to 3 minutes with gentle shaking at your bench or on a platform shaker.

12. Remove the membranes from the washing trays and place them on a paper towel (colony side up). Discard the deionized water in the discard beaker containing disinfectant.

The membranes have denatured DNA from each colony bound to the membrane in a mirror image of the colony pattern on the plates. They are now ready for hybridization with the *lux*A gene probe. If you are not going to hybridize immediately, blot the membranes dry on Whatman 3MM paper with the colony side up. When visible fluid has been absorbed, place the membrane between two pieces of 3MM paper and tape them together. Label with your group number and store in your drawer or in a desiccator. Dry membranes are stable and may be hybridized at any time.

B. Hybridization of Membranes

1. Place the membranes back to back (with the colony sides out) in a heat-sealable bag. Heat seal the bag so that it is about 1 cm larger than the membranes (see Exercise 16).

2. Cut off one corner of the bag and add 200 μl of prehybridization buffer for every square cm of membrane (assume one membrane). Carefully work out the bubbles and heat seal the bag. Submerge the bag in the 50°C shaking water bath and incubate for 15 minutes.

3. Remove the bag from the water bath and cut off a second corner. Squeeze out as much of the prehybridization buffer as possible into a sink. Roll a pipet over the bag on a flat surface to remove the remaining prehybridization buffer.

4. Add 100 μl of the hybridization buffer (containing the *lux*A probe) for every square cm of membrane. Squeeze out all of the bubbles and heat seal the bag while minimizing the loss of the hybridization buffer. Return the bag to the 50°C shaking water bath and hybridize for 30 minutes with gentle shaking.

5. Cut off a third corner of the bag and discard the hybridization buffer. Then cut the bag open on three sides, remove the membranes with the blunt ended forceps and place both membranes in a plastic washing tray.

6. Add approximately 200 ml of wash solution I that has been preheated to 50°C. The membranes should be back to back with the colony side of the membranes facing outward. Replace the lid and put the container in a 50°C incubator. Wash the membranes for 20 minutes with gentle shaking.

7. Remove the plastic container from the incubator and discard the wash solution. Add approximately 200 ml of wash solution II. Seal the container and wash the membranes at room temperature for 10 minutes with gentle agitation.

8. Transfer the washed membranes to a second heat-sealable bag (with the membranes back to back) and heat seal the bag on all sides.

9. Cut off one corner and add 8 ml of the alkaline phosphatase substrate buffer (containing BCIP and NBT). Squeeze out the bubbles, heat seal the bag, and wrap the bag in aluminum foil (the color development must be done in the dark).

10. Submerge the foil-wrapped bag in the 44°C water bath for 15 to 30 minutes or until color development is complete.

11. Stop the color development by removing the membranes from the bag and washing them thoroughly with deionized water. Blot the membranes dry on Whatman 3MM paper and store between two pieces of 3MM paper.

12. Observe the positive "DNA prints" on the membranes. Be sure that you can distinguish the blue from the X-gal cleavage (blue colonies) from the darker purple-blue dye resulting from the hybridized colonies. This will allow you to identify the colonies harboring the target piece of DNA. Invert the agar plate that the filter was lifted from and align the membrane with the colonies on the plate. Using a permanent marker, draw a circle (on the bottom of the Petri dish) around several positive colonies.

13. Divide an LB/Ap⁵⁰/X-gal plate in half by drawing a line across the bottom with a permanent marker. Use a flamed loop or sterile toothpick to pick a positive colony from the master plate and streak it on one half of the plate. Streak the other half of the plate with a second positive clone.

14. Incubate the plate for 18 to 24 hours and inoculate broths for mini-preps on the positive clones (Exercise 22). You may also pick the clone with a sterile toothpick and inoculate an LB stab for long-term storage of your clone.

Results and Discussion

1. Did you obtain clones that hybridized with the *lux*A probe? On which plate was the percentage of positive clones higher? Why?

2. Do you believe that all the colonies that hybridized with the *lux*A probe contain only the fragment of interest (the *Eco*R V/*Sst* I fragment containing the *lux*A and *lux*B genes)? How could you show conclusively that a given clone contained only this fragment?

3. How many colonies do you believe you could screen on a single Petri dish using this technique? What steps in the procedure could you optimize to screen more colonies on a single plate? If you did obtain a strong hybridization signal from an area of a plate where you could no longer resolve individual colonies, how could you still obtain the clone of interest?

4. Compare your results with results from a group that used a different vector (pUC18 or pUC19) than you. Was it possible to find the subclone of interest with either vector?

5. Is colony hybridization an effective technique to screen for the subclone of interest in this exercise? Is it required to identify your clone? Under what conditions do you think colony hybridization would be most useful?

Literature Cited

Grunstein, M., and D. S. Hogness. 1975. Colony hybridization: A method for the isolation of cloned DNAs that contain a specific gene. *Proc. Natl. Acad. Sci.* 72:3361-3365.

Maas, R. 1983. An improved colony hybridization method with significantly increased sensitivity for detection of single genes. *Plasmid* 10:296-298.

EXERCISE 22
Small-Scale Plasmid Isolations (Mini-Preps) from *lux*A Clones

After subcloning a small fragment of DNA containing a gene or genes of interest into a plasmid vector, the presence of the fragment is generally verified by performing a plasmid mini-prep. The isolated plasmid DNA can then be digested with appropriate restriction enzymes to demonstrate that the fragment of interest has been cloned into the vector. In addition, the presence of a smaller insert in the clone makes restriction mapping of insert DNA easier and more practical. The more detailed restriction map may then be used for additional subcloning (e.g., subcloning small fragments for DNA sequencing).

In the last series of exercises (Exercise 17–21), you obtained several clones that contained the *Vibrio fischeri lux*A and *lux*B genes. Clones were identified either by light production in the presence of an exogenous aldehyde (Exercise 20) or by colony hybridization with the *lux*A probe (Exercise 21). In this exercise, you will isolate plasmid DNA from these clones by a mini-prep procedure and verify that they contain the *Sst* I/*Eco*R V insert from pUWL500 or pUWL501 by restriction analysis.

In Exercise 14, we used the rapid boiling mini-prep procedure (Holmes and Quigley 1981) to isolate plasmids from bioluminescent clones. In this exercise, you will use a modification of the alkaline lysis mini-prep procedure of Birnboim and Doly (1979) (principles of this procedure were given in Exercise 14). The alkaline lysis procedure is slightly more involved than the boiling procedure but works equally as well, and either procedure can be used with *Escherichia coli*. For large plasmids, or hosts other than *E. coli*, the alkaline lysis procedure is more effective. The alkaline lysis procedure is also the basis for a number of commercially available plasmid mini-prep kits. Most of these kits couple a modification of the alkaline lysis procedure with a DNA binding step involving pulverized glass or a similar matrix to selectively purify the plasmid DNA. They are very rapid and yield higher-quality DNA than the standard alkaline lysis or rapid boiling procedures.

Following preparation of plasmid DNA, you will digest the purified plasmids with selected restriction enzymes that should then reveal the presence of a single restriction fragment equivalent in size to the fragment cloned in Exercises 17 through 21. You will need to determine which enzymes and the appropriate buffers you wish to use **prior** to coming to lab. If desired, additional restriction mapping of these clones may be done from the mini-prep DNA as part of this exercise or as a special project for Exercise 28.

Cultures

❑ overnight cultures of *E. coli* DH5α (containing recombinant plasmids)

Reagents

❑ GTE/lysozyme solution (sterile; on ice) (50 mM glucose, 25 mM Tris [pH 8.0], 10 mM EDTA, 5 mg/ml lysozyme)

❑ SDS/NaOH solution (0.2 N NaOH, 1% SDS)

❑ potassium acetate solution (pH 4.8) (on ice)

❑ TE buffer (10 mM Tris [pH 8.0], 1 mM EDTA; on ice)

❑ 95% ethanol (in freezer)

❑ RNase A (1 mg/ml; boiled to inactivate DNase; on ice)

❑ selected restriction enzymes (2/group; on ice)

❑ pUC18 or pUC19 (0.1 µg/µl; on ice)

❑ 10X restriction buffers (on ice)

❑ sterile deionized water (on ice)

❑ loading dye (room temperature)

❑ 1X TAE buffer

❑ LE agarose (low EEO)

❑ ethidium bromide stain (1.0 µg/ml)

Supplies and Equipment

❑ mini ice buckets (1/group)

❑ sterile 1.5-ml microcentrifuge tubes

❑ microcentrifuge tube opener (1/group)

❑ microcentrifuge tube rack (1/group)

❑ microcentrifuge tube float (1/group)

❑ 0.5–10 µl micropipetter

❑ 10–100 µl micropipetter

❑ 100–1000 µl micropipetter

❑ sterile 10-µl pipet tips

❑ sterile 100-µl pipet tips

❑ sterile 1000-µl pipet tips

❑ pipet tip discard beakers

❑ 37°C and 60°C water baths

❑ vortex mixer

❑ microcentrifuge

❑ 250-ml Erlenmeyer flask (1 per 2 groups)

❑ 25-ml Erlenmeyer flask (1 per 2 groups)

❑ microwave oven

❑ top loading balance

❑ spatula

❑ horizontal mini gel electrophoresis chamber

❑ gel casting tray

❑ 12-tooth comb

❑ gel staining tray

❑ power source

❑ transilluminator

❑ Polaroid camera with type 667 film

❑ 3-cycle semi-log paper

➡️Procedure

A. Alkaline Lysis Plasmid Mini-Prep

The evening before the lab you should inoculate the clones from which you wish to do mini-preps into 25 × 150-mm sterile tubes containing 5 ml of LB broth + 50 µg/ml ampicillin (see Exercise 1). Each group will mini-prep two clones, preferably one subclone identified by the aldehyde screening and one subclone identified by hybridization to the *lux*A probe. Incubate with vigorous shaking overnight at 37°C. Alternatively, it is possible to do mini-preps by scraping a small portion of bacterial growth off of a Petri plate, although yields are lower with this procedure.

1. Label two **clear** sterile microcentrifuge tubes A and B with a permanent marker. Transfer 1.5 ml of an overnight culture of a *lux*AB subclone identified by aldehyde screening to tube A. Add 1.5 ml of the overnight culture of a *lux*A subclone identified by hybridization to tube B. If doing mini-preps from cells on a plate, scrape a small glob of cells from a plate with the end of a sterile flat toothpick and resuspend in 0.5 ml of sterile saline.

2. Centrifuge at 8–10,000 × *g* (10–11,000 rpm) in a microcentrifuge at room temperature for 60 seconds. It is best not to centrifuge too long or too hard or the pellet will be difficult to resuspend. Discard the supernatant by pouring it into the discard beaker. Remove traces of supernatant with a micropipetter.

3. Vortex the tube vigorously to loosen the pellet. Add 100 µl of GTE/lysozyme and resuspend by pipetting the solution in and out several times with the micropipetter. (Note: many strains of *E. coli*, such as DH5α, lyse well in this procedure without lysozyme, and it is often omitted).

4. Add 200 µl of SDS/NaOH to each tube and mix thoroughly by gently rolling and inverting the tubes. Vigorous mixing at this point will shear the chromosomal DNA, which will prevent its precipitation in step 6.

5. Place the tubes on ice and incubate for 5 minutes with occasional gentle mixing. The suspension should clarify and become very viscous, indicating lysis of the cells.

6. Add 150 µl of ice-cold potassium acetate solution (pH 4.8), cap tubes, and mix by sharply inverting several times. You should observe a white precipitate, which is the protein, SDS, cellular debris, and high molecular weight chromosomal DNA precipitating out of solution.

7. Place the tubes on ice and incubate for 5 minutes.

8. Place the tube in a microcentrifuge in a balanced configuration with the hinges on each tube pointing toward the **center** of the centrifuge. Centrifuge the tubes at 12–14,000 × *g* (full speed in a microcentrifuge) for 3 minutes at room temperature. Wait for the other group sharing your centrifuge before beginning. Following this first centrifugation, rotate each tube 180° so that the hinges are pointing toward the **outside** of the rotor. Recentrifuge for an additional 3 minutes at 12–14,000 × *g*. This second spin will form a tighter pellet, which makes removal of the supernatant easier in the next step.

9. Label two new sterile 1.5-ml microcentrifuge tubes A and B and transfer 400 μl of each supernatant to the new tubes, being careful not to disturb the pellet. Discard the tubes with the pellets.

   Optional: You may extract the supernatants with equal volume of chloroform:isoamyl alcohol (24:1) at this point. Transfer the aqueous phase (upper) to clean tubes. This results in cleaner mini-preps and may enhance the ability of plasmids to be digested by restriction enzymes. However, this is usually not necessary.

10. Add 800 μl of 95% ethanol to each tube and mix rapidly by inversion. Incubate at room temperature for 2 to 5 minutes. What is required in addition to ethanol to precipitate DNA? Is it present in your DNA solution?

11. Centrifuge at 12–14,000 × g in a microcentrifuge for 5 minutes at room temperature. Be sure that the hinges of the microcentrifuge tubes are pointing out in the rotor. The nucleic acid precipitates are often not visible but will be along the same side of the tube as the hinge.

12. Carefully pour out the supernatants and remove the last traces with a micropipetter. You may not see a pellet—be extremely careful not to touch the tip to the side where the pellet should be.

13. Add 1 ml of 70% ethanol to each tube and wash the DNA pellets by gently rocking the tube. This washes out the salt, which will decrease the solubility of DNA and may interfere with restriction enzymes.

14. Centrifuge at 12–14,000 × g for 5 minutes at room temperature. Be sure that the tabs on the microcentrifuge tubes are pointing out in the rotor. (This centrifugation can be omitted if you have a visible pellet that stays on the wall of the tube).

15. Pour the supernatant into the discard beaker and remove the last traces with a micropipetter. It is important to remove as much as possible because excess ethanol will take a long time to evaporate. Set tubes on your bench for 5 to 10 minutes to dry (or dry pellets by carefully blowing warm air from a hair dryer over the tubes). Verify that all ethanol has been removed by sniffing the tubes for traces of alcohol odor.

16. Resuspend each pellet in 50 μl of sterile TE buffer (for lower-copy number plasmids, such as pBR322, resuspend in 20 μl). Perform restriction digestions on 2 to 4 μl of these suspensions.

B. Restriction Endonuclease Digestion

Prior to beginning this exercise, you will need to choose the restriction enzymes to use to verify that you have correctly cloned the fragment of interest. This is typically done by excising the insert (for example in Exercise 14 you removed the *lux* operon from the vector by doing a *Sal* I digest on plasmid DNA from bioluminescent clones). Additional digestions are also done to unambiguously determine that you have the clone of interest. Consult restriction maps of the vector (Appendix XI) and the insert

DNA (Exercise 15) to identify digests that will allow you to unambiguously determine if you have subcloned the fragment containing the *lux*A and *lux*B genes. You will do two digests on each mini-prep. One digest should linearize the plasmid into a single band (A1, B1) and the other should excise the insert (A2, B2). You should also select an enzyme that will linearize your vector. You may use a maximum of three enzymes (preferably two). Refer to a catalog from an enzyme supplier to select enzymes that are reasonable in cost and that will be compatible in the same buffer if doing a double digest. Fill in the table in step 3 and have it approved by your instructor before proceeding.

1. Fill your mini ice bucket with crushed ice and place each of the required enzymes, buffers, and so on, in it. Each group will be provided with its own set of enzymes and reagents containing slightly more than the required volume.

2. Using a permanent marker, label five sterile 1.5-ml microcentrifuge tubes A1, A2, B1, B2, and pUC. Place them in your microcentrifuge tube rack.

3. Complete the following table and have your digestions approved by your instructor before proceeding. Each digestion should have a final volume of 20 µl. Be sure that you record the **type** as well as **volume** of 10X buffer to use. Be sure to read the Reminders following the table before proceeding.

Tube	Buffer type	10X buffer	Mini-prep DNA	pUC DNA	H₂O	Enzymes		
A1	_____	___ µl	4 µl	—	___ µl	___ µl	___ µl	___ µl
A2	_____	___ µl	4 µl	—	___ µl	___ µl	___ µl	___ µl
B1	_____	___ µl	4 µl	—	___ µl	___ µl	___ µl	___ µl
B2	_____	___ µl	4 µl	—	___ µl	___ µl	___ µl	___ µl
pUC	_____	___ µl	—	2 µl	___ µl	___ µl	___ µl	___ µl

Reminders:

- Always add buffer, water, and DNA to tubes prior to adding the enzymes.

- Touch the pipet tip to the side of the tube when adding each reagent. Be sure you **see** the liquid transferred to the side of the tube, especially with volumes <2 µl.

- You may use the same tip to add the same reagent to different tubes, but be sure to use a new tip for each reagent.

- Check off each addition in the table as you add it. Omissions or incorrect volumes will usually prevent the restriction digestion from working.

4. After adding all ingredients, cap the tubes and vortex to mix. Pool the reagents in the bottom of the microcentrifuge tube by spinning the tubes in the microcentrifuge for 2 to 3 seconds. Be sure that the tubes are in a balanced configuration. Vortex or tap each tube to mix and recentrifuge to pool the digestion mix in the bottom of the tube.

5. Push tubes into the foam microcentrifuge tube holder and float in a 37°C water bath. Incubate for 30 to 45 minutes. If desired, you may add RNase to a final concentration of 50–100 μg/ml to each of the digesting mini-preps. Is this necessary? Why or why not? While the restriction digestions are proceeding, prepare and cast an agarose gel.

Note: Digests can be frozen at this point and analyzed at a later date.

C. Agarose Gel Electrophoresis

Each group will run an agarose gel on the digestions of its mini-preps to verify each plasmid. See Exercise 4 for details on preparing, running, staining, and visualizing agarose gels.

1. Prepare one batch of **0.8% agarose** in 1X TAE buffer per two groups.

CAUTION: In the next step, you will be heating agarose to boiling. Agarose can become superheated and boil over in the microwave or when swirled. Wear gloves and eye protection when removing the agarose and swirling heated flasks.

2. Melt the agarose in a microwave oven and cool to about 60°C or place in the 60°C water bath until ready to use. Cast a gel with a **12-tooth** comb. When solidified, place the gel in the electrophoresis chamber.

3. While the digestion is proceeding, label two tubes A/UC and B/UC. Add 2 μl of the uncut plasmid (from the mini-preps) to each of these tubes. Add 7 μl of TE buffer and 1 μl of loading dye to each tube and pulse for 2 to 3 seconds in a microcentrifuge to pool the ingredients. These will serve as uncut controls for each restriction digestion.

4. After the restriction digestion has incubated for 30 to 45 minutes, remove the tubes from the water bath and place them in the microcentrifuge tube rack.

5. Add 2 μl of loading dye to each tube. Pulse each tube for 2 to 3 seconds in the microcentrifuge to mix the dye with the restriction digestion.

6. Remove the *Hin*d III digest of lambda DNA from your freezer box and heat the standards in the 60°C water bath for 2 to 3 minutes to melt the *cos* sites in lambda.

7. Load 5 μl of the *Hin*d III digest of lambda as a size standard into wells 2 and 10. Load 10 μl of the digested and undigested plasmids into the wells in the gel as

follows (UC refers to uncut plasmid; indicate the enzymes you used below each digest):

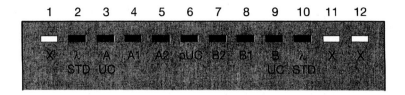

8. Electrophorese at 100 V until the bromphenol blue (the leading, darker dye) has migrated about three-quarters of the way to the end of the gel.

9. When the electrophoresis is complete, turn the voltage down to the lowest level and switch the power source off. Remove the lid to the gel box so that the gel can be removed.

D. Staining and Photodocumentation of Agarose Gel

1. Label a staining tray with your group number. Remove the tray supporting the gel from the electrophoresis box and slide the gel into the staining tray.

CAUTION: Ethidium bromide is a mutagen and suspected carcinogen. Wear gloves, eye protection, and a lab coat, and work over plastic-backed absorbent paper at all times when handling it. Carefully read precautions for handling and disposing of ethidium bromide (Appendix XII) before using ethidium bromide stain.

2. Stain the gel with ethidium bromide (1 μg/ml) for 5 to 10 minutes and rinse several times with tap or distilled water.

3. Briefly observe the gel on the transilluminator and photograph.

CAUTION: UV light is hazardous and can damage your eyes. Never look at an unshielded UV light. Always view through a UV blocking shield or wear UV blocking glasses. Some transilluminators have a switch on the lid so that the UV light switches off when the lid is opened, but many transilluminators do not have this safety feature.

4. Tape your gel photograph in your notebook below:

Results and Discussion

1. Carefully examine the photograph of your agarose gel. What is the rationale for loading the lanes in the order indicated? Measure the distance migrated by each of the bands of the lambda DNA digest and plot on 3-cycle semi-log paper to make a standard curve. Measure the distance migrated by each of the restriction fragments and calculate their size based on the standard curve.

2. Do both of the subclones that you analyzed contain the expected insert? Are the two subclones the same? By what criteria?

3. Do you observe a fragment in your digestions that is the same size as the pUC18 or pUC19 DNA cut once?

4. Did you identify any restriction endonuclease recognition sites in the insert DNA that could be used to confirm that the clone contains the correct insert? Suggest additional restriction digests to further confirm your clone.

5. Can the lack of a restriction endonuclease site be used to confirm that a subclone is correct? How could this be used to confirm the clones you have identified by colony hybridizations or bioluminescence in the presence of *n*-decyl aldehyde?

6. Can you propose other methods of confirming that the subclone contains the correct fragment?

Literature Cited

Birnboim, H. C., and J. Doly. 1979. A rapid alkaline extraction procedure for screening recombinant plasmid DNA. *Nucleic Acids Res.* 7:1513.

Holmes, D. S., and M. Quigley. 1981. A rapid boiling method for the preparation of bacterial plasmids. *Anal. Biochem.* 114:193.

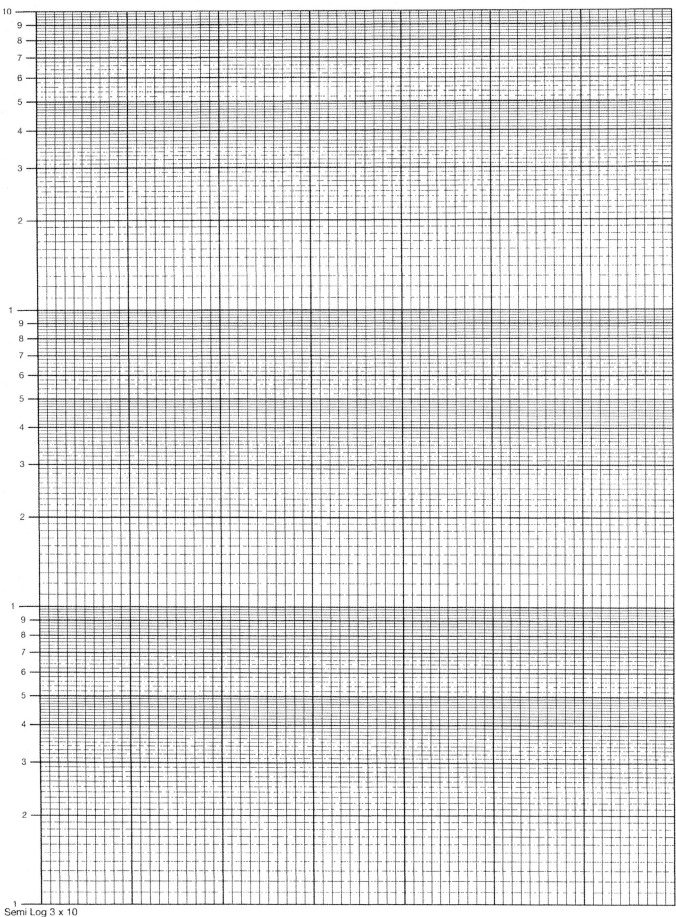

Semi Log 3 x 10

VI
Advanced Techniques

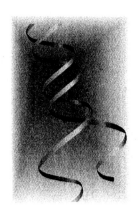

PART SIX

EXERCISE 23

Amplification of *lux*A from Natural Isolates by the Polymerase Chain Reaction (PCR)

The field of molecular biology has been driven by the development of theories and procedures that revolutionize the way scientists think about and address problems. DNA cloning, Southern blotting, and DNA sequencing are but a few examples. The polymerase chain reaction (PCR), developed at the Cetus Corporation by Kary Mullis (Saiki, et al. 1985; K. B. Mullis, U.S. patent 4,683,195; K. B. Mullis, U.S. patent 4,683,202), is having a revolutionary impact on how scientists from all the life sciences address problems. Based on the significance of his contribution, Kary Mullis was awarded the 1993 Nobel Prize in Chemistry for the development of the PCR.

PCR is an *in vitro* method for the enzymatic synthesis of specific pieces of DNA. The procedure is simple in concept, yet elegant in overall design. The process requires two oligonucleotide primers that hybridize to opposite strands of the target DNA and flank the region to be amplified (see Figure 23.1). A suitable DNA polymerase, the four deoxyribonucleoside triphosphates (dNTPs), and Mg^{2+} are also required. Amplification results from successive cycles of denaturation and oligonucleotide priming of the target DNA followed by the enzymatic synthesis of the strand of DNA that is complementary to the target DNA.

To perform a PCR amplification, a mixture containing the target DNA, oligonucleotide primers, dNTPs, and a heat-stable DNA polymerase is heated to 90–95°C to denature the strands of the target DNA (Saiki, et al. 1988). The solution is cooled to a temperature that allows the primers to anneal to their complementary sequence on the target DNA and provide the 3'-OH required for DNA synthesis. The DNA polymerase then synthesizes a new DNA strand complementary to the target. The polymerase used (*Taq* polymerase) is isolated from the thermophilic bacterium *Thermus aquaticus*. This thermophilic enzyme can tolerate the high temperature denaturation step and is stable throughout all of the reaction cycles. The newly synthesized strand initially extends a variable distance beyond the other primer-binding site on the complementary strand (see Figure 23.1), creating additional primer-binding sites on the newly synthesized strand. This thermal cycling scheme is repeated numerous times with the DNA synthesized during the previous cycles serving as a template for each subsequent cycle. With each cycle the strands are denatured at high temperature, the primers annealed, and then extended by the polymerase. The result is a doubling of the target DNA present with each cycle. This exponential accumulation of DNA sequences can theoretically produce over a millionfold amplification of the target in 20 cycles. In practice, the amplification efficiency is less than 100%, and 30 to 40 cycles are commonly done.

The PCR is potentially capable of producing detectable quantities of product from a single molecule of template DNA. In most PCR procedures, a wide range of template DNA concentrations will result in the desired amplification. However, the PCR is an example of where more is not necessarily better, and often the specificity and yield of a PCR amplification can be improved by decreasing the amount of template DNA as well as primers, enzyme, and dNTPs in the reaction mix.

New applications of the PCR are being developed constantly. This procedure can be used to amplify and clone extremely rare sequences from the large genomes of eukaryotes, screen phage and plasmid libraries, produce templates for DNA sequencing, produce large quantities of probe DNA, diagnose infectious and genetic diseases, amplify minute quantities of DNA at crime sites for forensic analysis, and create fingerprint-like patterns of amplified DNA for the identification of individuals and paternity testing. The applications are virtually endless.

In this exercise, you will amplify the *lux*A gene from the luminescent bacteria you isolated from fresh seafood or seawater in Exercise 3. The template DNA for this exercise will be prepared from a single colony of your natural bioluminescent isolates by a boiling lysis method. The purity of the template DNA is less important in PCRs than in other molecular biology procedures involving DNA; in this protocol you will only do a short centrifugation of the lysate to remove large cell debris from the template. The PCR of this template will amplify an approximately 700 bp *lux*A fragment. The products of these PCRs will be probed with the *lux*A probe in Exercise 24. The PCR products may also be used for cloning the *lux*A gene from the various natural isolates or used as a template for DNA sequencing.

The components used in this exercise are supplied by Perkin Elmer and include *Taq* polymerase, 10X reaction buffer, dNTPs, control template DNA, and control primers. Positive and negative controls are extremely valuable when setting up experimental PCRs. In addition to the reactions designed to amplify the *lux*A gene from your natural isolate, each group will set up one positive control reaction with control DNA and primers provided in the Perkin Elmer *GeneAmp*™ kit. The positive control

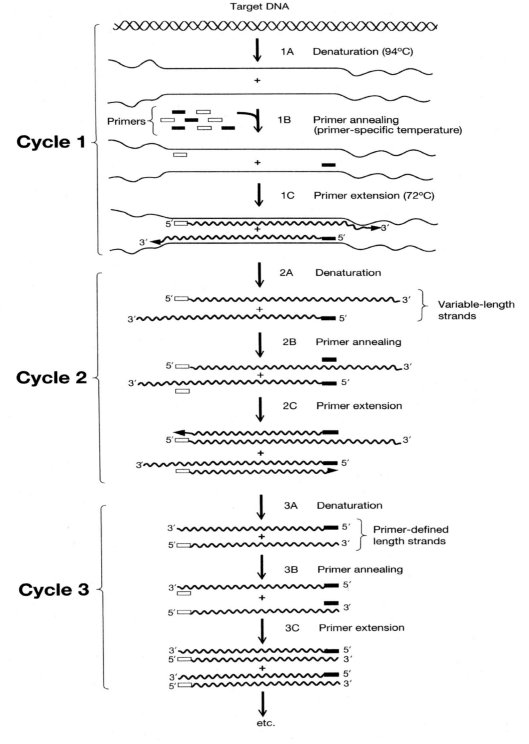

Figure 23.1
The polymerase chain reaction. Note that in the first cycle, the primers are extended beyond the target region to be amplified (defined by the location of the two primers). In subsequent cycles, the primers are extended only to the binding site of the second primer, resulting in an amplified product defined by the two primers. To simplify the diagram, additional cycles involving the original and variable-length strands have been omitted. They persist in the reaction, but variable-length strands increase at an arithmetic rate, while the primer-defined length strands increase exponentially. Thus, after 20 to 30 cycles, virtually all the product consists of the primer-defined length strands.

reaction uses bacteriophage lambda DNA and primers constructed to amplify a 500 bp fragment of the lambda genome. We will also perform additional controls omitting primers and template DNA, and we will confirm the specificity of the primers in a control using nontarget template DNA.

Cultures

- ❏ 2 different bioluminescent isolates (from Exercise 3; on plates with 1-2 mm colonies)
- ❏ *E. coli* DH5α (on LB plate with 1-2 mm colonies)

Reagents

- ❏ control template (lambda DNA; 0.1 µg/ml)
- ❏ lambda control primer 1 (4 µM) 5'-GATGAGTTCGTGTCCGTACAACTGG-3'
- ❏ lambda control primer 2 (4 µM) 5'-GGTTATCGAAATCAGCCACAGCGCC-3'
- ❏ nucleotide mix (dATP, dCTP, dGTP, dTTP; 1.25 mM each)
- ❏ *lux* 7 primer (2.8 µM) 5'-AAAAGGATCCTCAGAACCGTTTGCTTCAAAACC-3'
- ❏ *lux* 8 primer (2.8 µM) 5'-ACACAAGCTTCTACTGGATCAAATGTCAAAAGGACG-3'
- ❏ *Taq* DNA polymerase (5 units/µl)
- ❏ 10X *Taq* polymerase buffer (0.1 M Tris [pH 8.3], 0.5 M KCl, 1.5 mM $MgCl_2$, 0.1% gelatin)
- ❏ 25 mM $MgCl_2$
- ❏ sterile mineral oil (for thermal cyclers without heated lids)
- ❏ BRL 1 kb ladder (0.05 µg/µl)
- ❏ 1X TAE buffer
- ❏ ethidium bromide stain (1.0 µg/ml)
- ❏ sterile deionized water
- ❏ loading dye (room temperature)
- ❏ LE agarose (low EEO)

Supplies and Equipment

- ❏ 0.5–10 µl micropipetter
- ❏ 10–100 µl micropipetter
- ❏ sterile 10-µl micropipet tips
- ❏ sterile 100-µl micropipet tips
- ❏ micropipet tip discard beaker
- ❏ sterile thin-walled PCR tubes (0.2 or 0.6 ml)
- ❏ rotor adapters for 0.2 or 0.6 ml PCR tubes
- ❏ sterile 1.5-ml microcentrifuge tubes
- ❏ microcentrifuge tube opener
- ❏ microcentrifuge tube rack
- ❏ boilable microcentrifuge tube float
- ❏ boiling water bath
- ❏ mini ice buckets
- ❏ 250-ml Erlenmeyer flask (1 per 2 groups)
- ❏ 25-ml Erlenmeyer flask (1 per 2 groups)
- ❏ microwave oven
- ❏ autoclave gloves
- ❏ top loading balance
- ❏ 60°C water bath
- ❏ thermal cycler
- ❏ microcentrifuge
- ❏ gel casting tray with 6-tooth comb
- ❏ gel staining tray
- ❏ horizontal mini gel electrophoresis chamber
- ❏ power source
- ❏ transilluminator
- ❏ Polaroid camera with type 667 film

➡Procedure

A. Preparation of Template DNA from Natural Bioluminescent Isolates

Each group will prepare template DNA from two natural bioluminescent isolates (Exercise 3). If you do not have two isolates, you may use *Vibrio fischeri* or *Photobacterium phosphorium* for the amplification. In addition, some groups will be assigned a control with nontarget DNA and should prepare template DNA from *E. coli* DH5α.

1. Label two 1.5-ml sterile microcentrifuge tubes A and B for your bioluminescent isolates. If you have been assigned *E. coli* as a nontarget DNA control (control D5 in Part B), label a third tube D5. Include your group number on all tubes and indicate the strain designation of each bioluminescent isolate below:

   Tube A: _____

   Tube B: _____

   Tube D5: *E. coli* DH5α

2. Add 15 μl of sterile deionized water into each labeled tube. With the flat end of a sterile toothpick, scrape one well-isolated 1–2 mm colony from the Petri plate streaked with your bioluminescent isolate (or *E. coli*). Resuspend the colony in the appropriate tube containing the 15 μl of water by twirling the toothpick in the water. Repeat with your second isolate.

CAUTION: In the next step, you will be using a boiling water bath. Boiling water is hazardous in the lab, especially when elevated. Use caution when adding and removing tubes from the water bath and when working in the vicinity of boiling water.

3. Place the tubes of resuspended cells in a boilable floating microcentrifuge rack and float in a boiling water bath for 12 minutes.

4. Carefully remove your tubes from the boiling water bath and centrifuge at 12–14,000 × g (full speed in a microcentrifuge) for 10 seconds.

5. Label two new sterile microcentrifuge tubes A-1/100 and B-1/100 (groups doing control D5 also label a tube D5-1/100). Prepare a ⅟₁₀₀ dilution of each boiled supernatant in the new tubes with deionized water. Aliquots of these dilutions will serve as the template DNA in the PCRs for amplification of the *lux*A gene from your isolates.

B. Preparation of the PCR Amplifications

1. Label a sterile 1.5-ml microcentrifuge tube MM and add your group number. Prepare the following master mix (sufficient for four PCRs with slight excess) as indicated in the following table. The master mix contains the reagents common to all reactions you will set up at a 1.25X concentration. When the primers and template DNA are added to the mix (steps 4 and 5), all components will be at a 1X concentration. Preparing a master mix saves time in setting up individual reactions, ensures that all components are present in the final PCRs at identical concentrations, and minimizes the amount of expensive enzyme used.

1.25 × Master Mix for Four Reactions

Component	Volume	Final concentration in reaction
deionized water	111.0 µl	—
10X *Taq* pol buffer	22.5 µl	1X
25 mM MgCl$_2$*	9.0 µl	2.5 mM
dNTP mix†	36.0 µl	200 µM (each dNTP)
Taq polymerase (5 units/µl)	1.5 µl	≈ 1.5 units/50 µl rxn
	Total: 180 µl	

*The *Taq* polymerase buffer contains MgCl$_2$ (to provide a final concentration of 1 5 mM), but **additional** MgCl$_2$ is added to raise the Mg^{2+} concentration to 2 5 mM This is the empirically determined optimum for amplification with the *lux*A-specific primers. The Mg^{2+} concentration should be optimized for each PCR as too little or too much MgCl$_2$ can reduce the efficiency of a PCR or cause amplification of nontarget sequences.
†A freshly prepared mixture of dNTPs is made from stock solutions of each nucleotide to give a mix with each dNTP at 1 25 mM

2. After all the additions have been made, set a 10–100 µl micropipetter to 100 µl and pipet the solution in and out several times to mix the *Taq* polymerase into the master mix. The enzyme is supplied in 50% glycerol and will settle to the bottom of the tube if not mixed. Store the master mix on ice until it is dispensed into the PCR tubes.

3. Label three thin-walled PCR tubes A through C and add your group number. Label a fourth tube with D1, D2, D3, D4, or D5 to correspond with your assigned negative control (see step 5). Place tubes on ice.

4. Add the following reagents to the labeled PCR tubes in the order listed. Tubes A and B are PCRs designed to amplify *lux*A from your isolates. Use the ¹⁄₁₀₀ dilutions of the boiled cells of your bioluminescent isolates for A and B templates. Tube C is a positive control reaction with the bacteriophage lambda (λ) DNA and primers provided in the Perkin Elmer *GeneAmp*™ kit. Each group will also set up one additional assigned negative control in step 5. Check off each addition as you add the various components. Keep the reactions on ice until they are placed in the thermal cycler.

Tube	Master mix	*lux* 7 primer	*lux* 8 primer	Isolate A template (¹⁄₁₀₀)	Isolate B template (¹⁄₁₀₀)	λ 1 primer	λ 2 primer	λ Control template
A	40 µl	2.5 µl	2.5 µl	5 µl	—	—	—	—
B	40 µl	2.5 µl	2.5 µl	—	5 µl	—	—	—
C	40 µl	—	—	—	—	2.5 µl	2.5 µl	5 µl

5. Tube D will contain one of five negative control reactions assigned to your group by your instructor. The first four negative controls contain the same additions as tube C in step 4 with one or more reagents omitted. The last control (D5) contains template DNA prepared from *E. coli* DH5α to demonstrate the specificity of the control primers. Check off each addition as you add the various components. Keep the reactions on ice until they are placed in the thermal cycler.

Tube	Master mix	λ 1 primer	λ 2 primer	λ Control template	H₂O	DH5α template (¹⁄₁₀₀)
D1	40 μl	2.5 μl	2.5 μl	—	5 μl	—
D2	40 μl	—	2.5 μl	5 μl	2.5 μl	—
D3	40 μl	2.5 μl	—	5 μl	2.5 μl	—
D4	40 μl	—	—	5 μl	5 μl	—
D5	40 μl	2.5 μl	2.5 μl	—	—	5 μl

6. Cap all the PCR tubes and vortex to mix. Pulse the tubes for 2 to 3 seconds in a microcentrifuge to pool the contents in the bottom. Use special rotor adapters as the small PCR tubes do not fit in the standard microcentrifuge rotor.

7. If you are using a thermal cycler with a heated lid, the tubes are ready to go into the thermal cycler. If not, overlay the reaction mixture with 50 μl of sterile mineral oil and cap the tubes. Return tubes to your ice bucket until all groups are ready to load the thermal cycler. When each group's tubes are in the thermal cycler, we will begin the amplification.

Part C. Amplification of *lux*A from Natural Bioluminescent Isolates

1. Your instructor will supervise the loading of the thermal cycler and demonstrate how the cycle times and temperatures are programmed. The first step will be a 5-minute 95°C incubation to denature the target DNA and to inactivate proteases that may be present in the impure template DNA. The machine will then run 30 amplification cycles. The first 5 cycles will be run under the following conditions:

> denaturation: 94°C   30 seconds
> annealing: 37°C   60 seconds
> primer extension: 72°C   60 seconds

The next 25 cycles will be the same except that the annealing temperature will be raised to 55°C to increase the stringency of the reaction.

2. After the amplification is complete (approximately 2.5 to 4 hours), place your tubes on ice or in the freezer until they are analyzed. Many thermal cyclers may be programmed to hold the tubes at 4°C until they are removed.

3. If you added oil to your PCR tubes, you will need to separate the reaction mix from the oil as follows:

   a. Transfer the total volume of each tube (including oil) to a piece of clean Parafilm™ with a micropipetter set at 100 µl. Be careful not to allow air to enter the reaction mix. Label the Parafilm™ below each drop of liquid with a permanent marker. The oil will spread over the surface of the Parafilm™ but the aqueous PCR mixture will not. This results in an egg-like drop on the Parafilm™ with the oil resembling the egg white and the reaction the yolk.

   b. Label four new sterile 1.5-ml microcentrifuge tubes A through D and add your group number. Reset your micropipetter to 5 µl less than the volume of the aqueous PCR mixture and carefully suck up the "yolk" containing the aqueous reaction. Be careful not to draw up any oil in the pipet tip. Wipe the pipet tip with a clean Kimwipe™ and transfer the oil-free reaction to the corresponding labeled 1.5-ml microcentrifuge tube. Store in your freezer box until analyzed.

D. Agarose Gel Electrophoresis

1. Each group will run an agarose gel to analyze its PCRs. Prepare one batch of **1.2%** **agarose** in 1X TAE buffer per two groups. We will use a slightly higher agarose concentration to give better resolution of the small PCR products.

CAUTION: In the next step, you will be heating agarose to boiling. Agarose can become superheated and boil over in the microwave or when swirled. Wear gloves and eye protection when removing the agarose and swirling heated flasks.

2. Melt the agarose in a microwave oven and cool to about 60°C or place in the 60°C water bath until ready to use. Cast a gel with a **6-tooth** comb. When solidified, place the gel in the electrophoresis chamber.

3. Label a set of microcentrifuge tubes A' through D'. Add 20 µl of your PCR to the new corresponding tubes. Store the remainder of the PCR in your freezer box. Add 2 µl of loading dye to each tube and pulse in the microcentrifuge for 2 to 3 seconds to mix the dye with the PCR.

4. Load 5 µl of the BRL 1 kb ladder into lanes 1 and 6. Load the entire contents of each tube into the wells in the gel as follows:

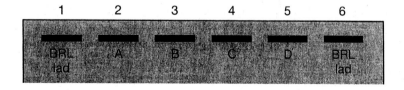

5. Electrophorese at 100 V until the bromphenol blue (the leading, darker dye) has migrated about three-quarters of the way to the end of the gel.

6. When the electrophoresis is complete, turn the voltage down to the lowest level and switch the power source off. Remove the lid to the gel box so that the gel can be removed.

E. Staining and Photodocumentation of Agarose Gel

1. Label a staining tray with your group number. Remove the tray supporting the gel from the electrophoresis box and slide the gel into the staining tray.

CAUTION: Ethidium bromide is a mutagen and suspected carcinogen. Wear gloves, eye protection, and a lab coat, and work over plastic-backed absorbent paper at all times when handling it. Carefully read precautions for handling and disposing of ethidium bromide (Appendix XII) before using ethidium bromide stain.

2. Stain the gel with ethidium bromide (1 μg/ml) for 5 to 10 minutes and rinse several times with tap or distilled water.

3. Briefly observe the gel on the transilluminator and photograph. **Do not discard your gel!** Carefully remove it from the transilluminator and return it to the gel staining tray. You will use this gel to perform a Southern blot in the next exercise.

CAUTION: UV light is hazardous and can damage your eyes. *Never* look at an unshielded UV light. Always view through a UV blocking shield or wear UV blocking glasses. Some transilluminators have a switch on the lid so that the UV light switches off when the lid is opened, but many transilluminators do not have this safety feature.

4. Tape your gel photograph in your notebook below:

Results and Discussion

1. Examine the photograph of your gel and see if you can identify the template DNA (bacteriophage lambda) in the positive control reaction (tube C containing lambda DNA). Refer to the size of lambda (Appendix XI) to determine where to look on the gel. Recall that the lower limit of detection of DNA in an ethidium bromide stained gel is about 2 ng and that one-fifth of the amplification product was run on the gel. **Should** you be able to see the template?

2. Examine the photograph for the presence of the PCR product in your positive control (tube C). Using the BRL ladder size standards (see Exercise 15), prepare a standard curve and estimate the size of the product and record the results in the table in question 4. Can you estimate approximately how much product was produced?

3. Examine the lane corresponding to your assigned negative control (tube D). Are there any bands present? Record your results in the table in question 4. What was the purpose of each control?

4. Examine the lanes corresponding to the PCRs of the DNA from your bioluminescent isolates as a template (tubes A and B). In which reactions did significant amplification occur? Is there only one band in each lane? What could account for the presence of additional bands from these reactions? What parameters could be altered in an attempt to produce only the desired product? Using the size standards, estimate the size of the product(s) observed and record your data in the following table.

PCR reaction	Product observed	Product size(s)
Tube A		
Tube B		
Tube C		
Tube D___		

5. Examine the region of the gel below your PCR product(s) for additional ethidium bromide staining material. What could account for this?

6. Assuming each of the cells in the 2-mm bioluminescent colony that you chose had only one copy of their genome, and that the volume of each cell is 2 μm^3 (roughly equivalent to *E. coli*), calculate the number of copies of template you added to your PCR tubes. You may make any assumptions you wish about the shape of the colony.

   The PCR has been used to amplify DNA from a single cell. Is it surprising that your reactions worked?

7. The *E. coli* genome is approximately 5×10^6 bp, and the atomic mass of a base pair of DNA is approximately 660 daltons. Based on these assumptions, calculate the mass of DNA added to your PCR tubes. Express your answer with the correct number of significant figures. Hint: 1 atomic mass unit = 1.66×10^{-24} g. How does the mass of template compare to that in the lambda control reaction?

8. How many of the assumptions stated above are probably invalid? State your reasons. After you have examined the results of your experiments individually, record your results on the board for class discussion.

Literature Cited

Saiki, R. K., S. Scharf, F. Faloona, K. B. Mullis, G. T. Horn, H. A. Erlich, and N. Arnheim. 1985. Enzymatic amplification of β-globin genomic sequences and restriction site analysis for diagnosis of sickle-cell anemia. *Science* 230:1350–1354.

Saiki, R. K., D. K. Gelfand, S. Stoffel, S. J. Scharf, R. Higuchi, G. T. Horn, K. B. Mullis, and H. A. Erlich. 1988. Primer directed enzymatic amplification of DNA with a thermostable DNA polymerase. *Science* 239:487–491.

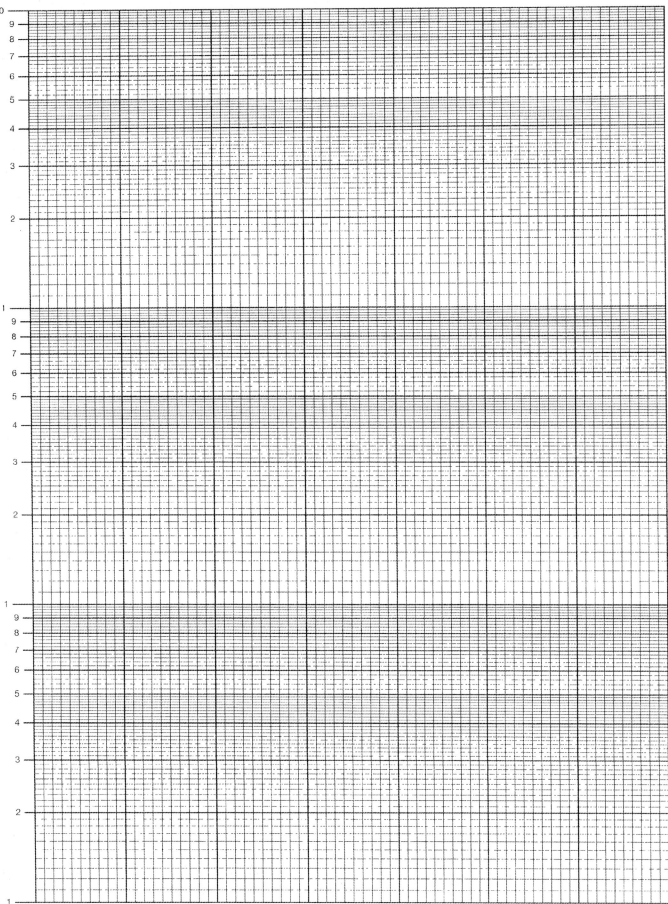

Semi Log 3 x 10

EXERCISE 24
Southern Blotting and Hybridization of PCR Products

In Exercise 16 you performed a Southern blot on a gel containing various restriction digestions of pUWL500 or pUWL501. The transfer membrane was then hybridized with an oligonucleotide probe complementary to *lux*A to identify the location of the *lux*A gene on the restriction map generated in Exercise 15. These exercises demonstrate the power of Southern blotting coupled with nucleic acid hybridization, although there are numerous other applications of these techniques. One such application is the identification of a sequence of interest in PCR products separated on agarose gels. Southern blotting of gels containing PCR products followed by hybridization of the resultant membranes can verify that the sequence of interest was amplified. The combination of PCR, Southern blotting, and hybridization techniques has numerous applications in the diagnosis of genetic and infectious diseases, DNA fingerprinting, and so on.

In this exercise, you will blot the PCR products that you separated on an agarose gel in Exercise 23 and probe the transferred products with the *lux*A gene probe. The probe was designed to be complementary to a portion of the *lux*A gene of the *Vibrio fischeri* strain used in this manual (MJ1), but it is not necessarily complementary to the *lux*A genes of all bioluminescent bacteria. Because there is only a slight possibility that any of the environmental isolates obtained in Exercise 3 are the same strain of *V. fischeri* used in this manual, the probe may or may not anneal to your PCR product(s). As you conduct this experiment, you should think about the factors that determine whether the probe will hybridize to the PCR products of a given isolate. Also consider what a positive hybridization with your PCR product(s) tells you about the nature of your isolate.

PART I. SOUTHERN BLOTTING OF AN AGAROSE GEL CONTAINING PCR PRODUCTS

Reagents

- gel from Exercise 23
- 0.4 M NaOH
- 5X SSC

- deionized water
- pUWL500 (10 ng/μl freshly prepared in 0.4 M NaOH)

Supplies and Equipment

- 0.5–10 μl micropipetter
- sterile 10-μl micropipet tips
- disposable gloves
- Rubbermaid™ washing trays (2/group)
- glass baking dish
- gel support platform
- 5-ml glass pipet or glass rod
- Whatman 3MM paper wicks (cut to exact width of gel and 25 cm long; 2/group)
- nylon membrane (cut to exact size of gel; 1/group)

- Whatman 3MM paper blotter (cut to exact size of gel; 4/group)
- paper towel blotters (cut to exact size of gel; 5 cm high stack/group)
- masonite board (cut to exact size of gel; 1/group)
- platform shaker
- 10 × 12-cm 3MM paper for drying nylon after blotting (4/group)

Procedure

Note: Always wear gloves when handling nylon membranes. Oils from your skin will prevent proper wetting of the membrane and subsequent transfer of the DNA.

1. Set up a Southern blot of your gel containing the PCR products as described in Exercise 16. Transfer a minimum of 1 hour (you may allow the transfer to proceed overnight).

2. After the transfer is complete, remove and discard the paper towels and 3MM blotters. Be sure you are wearing gloves, and be careful not to disturb the nylon membrane. Remove the nylon and gel as a single unit and lay them (gel side up) on a clean, dry piece of 3MM paper.

3. Mark the position of the wells with a black fine-point permanent marker or soft pencil. Peel off and discard the gel.

4. With the membrane DNA side up, add 2 μl of the 10 ng/μl alkaline solution of pUWL500 to the lower (opposite the wells) left-hand corner of the membrane as indicated in the following diagram:

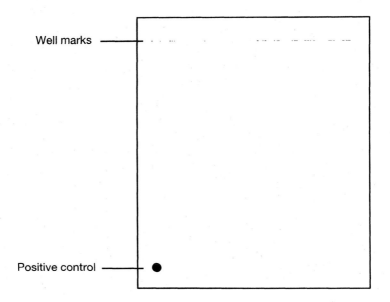

Like the transferred DNA, the denatured pUWL500 will irreversibly bind to the nylon membrane. Because pUWL500 contains the *V. fischeri lux* operon (and hence the probe target sequence), this will serve as a positive control. By placing the control DNA under one of the DNA standard lanes, it will not interfere with the hybridization of the PCR products in the adjacent lanes.

5. Float the nylon, DNA side down, on 5X SSC in a washing tray. Gently agitate at room temperature for 5 minutes.

6. Remove the nylon membrane from the SSC, then place the membrane with the DNA side up on 3MM paper to blot dry. When visible fluid has been absorbed, place the membrane between two pieces of 3MM paper and tape them together. Label with your group number and store in your drawer or in a desiccator. Dry membranes are stable and may be hybridized at any time.

PART II. HYBRIDIZATION OF TRANSFER MEMBRANE WITH *lux*A PROBE

Reagents

☐ prehybridization buffer (5X SSC, 0.5% BSA [fraction V], 1% SDS)
☐ hybridization buffer (prehybridization buffer + 0.2 nM probe to *lux*A)
☐ Probe sequence: 5' GATGGAACTCTATAATGAAATTGC 3'
☐ wash solution I (50°C) (0.5% SDS, 0.5% N-lauroylsarcosine, 2X SSC)
☐ wash solution II (room temperature) (1X SSC, 0.5% Triton X-100)
☐ alkaline phosphatase substrate buffer (containing BCIP and NBT)
☐ deionized water

Supplies and Equipment

☐ dried nylon membrane from Southern blot
☐ Seal-a-Meal™ heat sealer (or equivalent)
☐ heat-sealable polyethylene bags (2/group)
☐ Rubbermaid™ washing tray with lid (1/group)
☐ scissors
☐ forceps

☐ 5-ml disposable glass pipets (nonsterile)
☐ pipet aids or pipet bulbs
☐ aluminum foil
☐ 50°C shaking water bath
☐ 50°C shaking incubator
☐ 44°C water bath

➡ Procedure

1. Remove the membrane from between the two sheets of 3MM paper. Remember to wear gloves when handling the membrane. Hybridize with the *lux*A probe as described in Exercise 16. By using the same wash buffers and wash temperatures, you will use the same **stringency** as in the previous hybridizations.

2. Transfer the washed, hybridized membrane to a second heat-sealable bag and heat seal so that the bag is approximately 1 cm larger than the membrane on all sides.

3. Add 100 μl of the alkaline phosphatase substrate buffer (containing NBT and BCIP) for every square cm of membrane. Squeeze out air bubbles, heat seal the bag, and wrap the bag in aluminum foil (the color development must be done in the dark).

4. Submerge the foil-wrapped bag in the 44°C water bath and incubate for 15 to 60 minutes.

5. Check your membrane for color development periodically. Check to see if the positive control spot turns purple. This indicates that the color development is working. If you also see purple-blue bands corresponding to the PCR products in the gel, stop the color development by removing the membrane from the bag and rinsing it thoroughly in deionized water. If you do not see any bands turning blue, continue the color development for 1 to 2 hours before rinsing.

6. Place the membrane on 3MM paper to dry. You may store the developed membrane between two pieces of Whatman 3MM paper.

Results and Discussion

1. Examine your hybridized membrane to see if the positive control spot gave a signal. Why is it important to have a positive control on this membrane?

2. Examine each lane of your membrane for the presence of DNA fragments identified by the probe. Compare these bands with those on the photograph of the gel (from Exercise 23). Did all of your PCR products hybridize with the *lux*A probe? If so, what does this result tell you about your environmental isolate? Does a positive signal indicate that your isolate was *V. fischeri* MJ1? Why or why not?

   If one of your PCR products did not hybridize to the probe, what does this tell you about your environmental isolate?

3. If you had more than one PCR product on your gel, examine the membrane to see if both, one, or none of the bands hybridized with the probe. Explain your result.

4. Are all of the purple-blue bands on the nylon membrane of the same intensity (examine the membranes of other groups as well as your own)? If not, how might you explain variations in the intensity? Does the intensity of the hybridization signal correspond to the intensity of ethidium fluorescence of the DNA fragments separated on your gel?

5. Are there bands that developed on the hybridized membrane that are not visible on the photograph of the agarose gel? If so, what does this suggest about the relative sensitivity of DNA hybridization compared to ethidium bromide fluorescence?

6. The hybridization performed in this exercise was done at high stringency. How might altering the stringency of the hybridization change the results you observed? When might there be an advantage to performing a lower stringency hybridization?

7. If you identified one or more of your PCR products as containing the *lux*A sequence, can you envision any further use for the remaining PCR product? If so, you may wish to consider further analysis of these products as a special project for Exercise 28.

EXERCISE 25
DNA Sequencing of *lux* Genes from Plasmid Templates

Determination of the base sequence of nucleic acids is one of the most powerful techniques used in molecular biology. The ability to determine the sequence of DNA has led to major advances in the understanding of the basic processes of life. Through comparative sequence studies, regulation of gene expression has become better understood, as have virtually all biological processes involving nucleic acids. The examination of sequence data has also allowed researchers to address evolution and phylogeny in a rational and quantitative fashion.

DNA may be sequenced by either the chain scission method of Maxam and Gilbert (1977) or, more commonly, by the chain termination method of Sanger, et al. (1977). The Sanger procedure can use a variety of DNA templates, including denatured double-stranded plasmid DNA, denatured linear fragments, or single-stranded DNA from bacteriophage M13 or phagemid clones.

The Sanger procedure involves synthesis of complementary copies of a single-stranded template DNA. Synthesis of the DNA strand is initiated by an oligonucleotide primer designed to anneal just before the region of DNA to be sequenced, and this primer is extended by a DNA polymerase. A variety of DNA polymerases are used in DNA sequencing, including the Klenow fragment of DNA polymerase I, reverse transcriptase, *Taq* polymerase (from *Thermus aquaticus*), Vent™ polymerase (from a thermophilic deep-sea archaebacterium), or bacteriophage T7 DNA polymerase. DNA synthesis occurs in the presence of 2',3'-dideoxyribonucleotides (dideoxys or ddNTPs; see Figure 25.1), which are analogs of 2'-deoxyribonucleoside triphosphates lacking the 3'-OH required for DNA synthesis. Thus, the random incorporation of a dideoxy into an elongating DNA molecule prevents further synthesis of that chain because the lack of the 3'-OH group prevents the formation of the phosphodiester bond to the next base. This results in a nested set of variable-length strands, each ending in a specific base (corresponding to the dideoxy used) and allows the sequence to be determined (see following text).

218

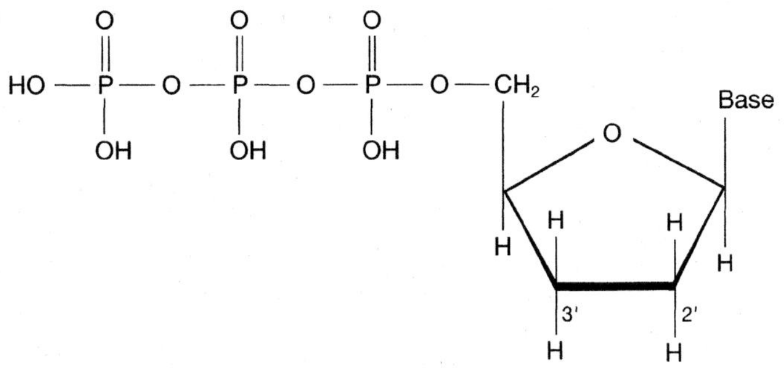

Figure 25.1
Chemical structure of a chain-terminating 2', 3'-dideoxyribonucleotide. Note the absence of the 3'-OH required for addition of the next nucleotide.

In order to detect the formation of newly synthesized DNA complementary to the DNA template, most DNA sequencing procedures incorporate an easily identifiable reporter molecule into the synthesized strands. The reporter molecule may be either a radioactive molecule or a fluorescent dye. Until recently, radioactivity (generally ^{32}P, ^{33}P, or ^{35}S) has been the most widely used reporter. Radioactivity may be added to an oligonucleotide primer by transferring a single radioactive phosphate to the 5' end with a kinase (**end labeling**). Alternatively, several radioactive nucleotides may be incorporated as the primer is extended a short distance by the DNA polymerase (**extension/labeling**). In this exercise, you will do radioactive sequencing by extension/labeling with an ^{35}S-labeled nucleoside triphosphate ([α-^{35}S]dATP; see Figure 25.2).

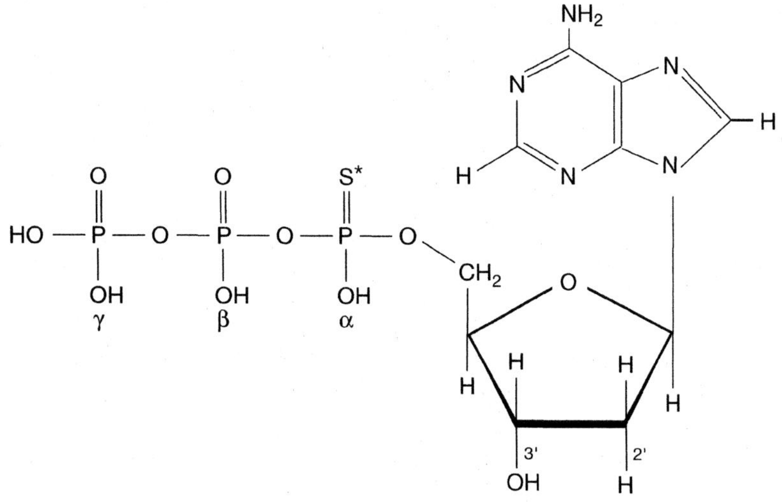

Figure 25.2
Radioactive nucleotide ([α-^{35}S]dATP) used in DNA sequencing by extension labeling. Note the location of ^{35}S in the nucleotide.

When sequencing is performed using either end-labeled primers or extension/labeling, the reaction is split into four separate reaction tubes, with each tube containing only one of the four dideoxyribonucleotides that randomly terminate DNA synthesis (see Figure 25.3). Under these conditions, DNA synthesis in each of the four reaction tubes results in the production of a random set of radiolabeled DNA fragments produced when the polymerase inserts a molecule of the ddNTP present in that reaction (ddG, ddA, ddT, or ddC). Each of the four reaction tubes also contains all four of the deoxyribonucleotides (dNTPs) in molar excess to the ddNTP present so that only a small fraction of the DNA molecules being synthesized will terminate at a given nucleotide (G, A, T, or C). Thus, one of the reaction tubes will contain variable-length fragments ending only in ddG, another containing fragments ending only in ddA, and so on. Prior to being loaded onto the polyacrylamide gel, the newly synthe-sized DNA in each of the four reaction tubes is denatured, separating it from the tem-plate. Following electrophoretic separation, the gel is exposed to film, and the relative position of images of the fragments in the four lanes is used to determine the sequence (see Figure 25.3).

The majority of nucleic acid sequencing today is automated and uses nonradioac-tive fluorescent reporter groups to detect the newly synthesized DNA. In this version of the original Sanger method, each dideoxy is labeled with a different fluorescent reporter. Once the fluorescent dideoxys are incorporated into the newly synthesized DNA, they are detected by a laser detector. Because the laser detection method dis-criminates between fragments ending in the four different labeled dideoxys, the entire reaction can run in a single tube and be loaded into a single lane on a gel. X-ray film-dependent detection is not required because the laser detector determines the order in which the DNA fragments ending in the different ddNTPs (labeled with fluorescent reporters) pass a fixed point in the gel.

Figure 25.3 (opposite)
Sanger dideoxy DNA sequencing from a double-stranded plasmid template using extension labeling. Prior to beginning the sequencing reactions, the double-stranded template is dena-tured with alkali. The sequencing reactions are done in three steps: (i) primer annealing; (ii) primer extension in the presence of radioactive nucleotides; and (iii) termination reactions. In the termination reactions, the extension-labeling mix is divided into four tubes, each contain-ing a different dideoxynucleotide. These are incubated to synthesize a nested set of radioactive DNA fragments. An aliquot of each of the four tubes is denatured, then loaded onto a poly-acrylamide gel and separated electrophoretically. The box at the bottom represents an autoradi-ograph of the gel, indicating the position of radioactive fragments. The lane showing the smallest fragment (i.e., **A**) indicates the base that the fragment ends in and is the **first** base in the sequence. A fragment that is one base longer will be the next band above the band just read. The lane containing this band (**A** again) indicates the base this fragment ends in and hence the **next** base in the newly synthesized DNA. In a similar manner, the remaining sequence is read moving up the autoradiograph in the 5'→3' direction.

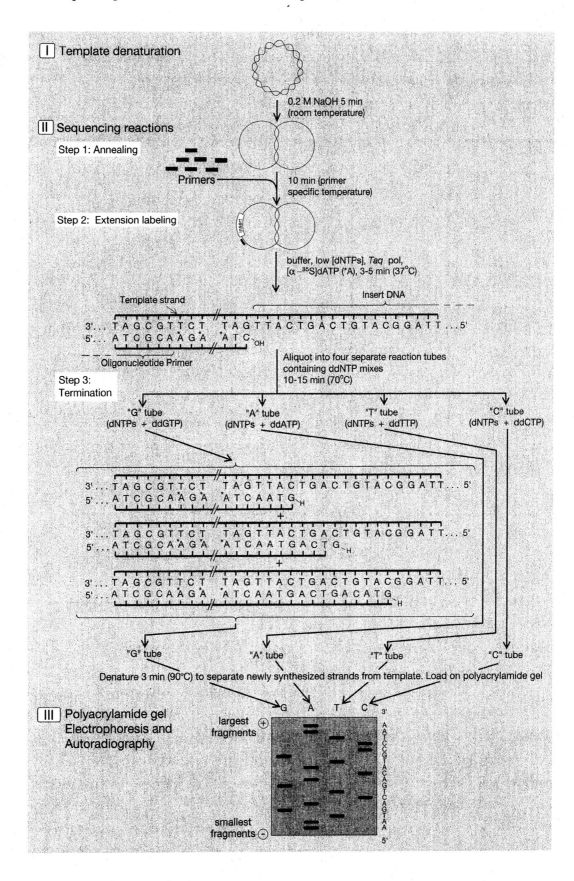

The TaqTrack™ dideoxy sequencing protocol (from Promega Corporation) we will use in this exercise uses a plasmid template, *Taq* DNA polymerase, and extension labeling with [α-³⁵S]dATP. In order to produce the single-stranded DNA template required for the polymerase, the plasmid strands are separated (denatured) with alkali to allow annealing of the oligonucleotide sequencing primer. Other sequencing methods use high temperature to denature the template or start with single-stranded template generated from M13 or phagemid clones.

Although the TaqTrack™ procedure can be used with end-labeled primers, we will use extension labeling of the newly synthesized DNA with ³⁵S. The procedure may be divided into three basic steps: (i) **annealing**; (ii) **extension/labeling**; and (iii) **termination** (see Figure 25.3). In the annealing step, a short oligonucleotide primer hybridizes to the single-stranded plasmid template. A "universal" primer is commonly used to hybridize to a region flanking the multiple cloning site of pUC-derived cloning vectors. In pGEM™-3Zf(+), there are two sets of primer binding sites flanking the multiple cloning site (see Figure 25.4). One set corresponds to the bacteriophage SP6 or T7 promoters, and the other set corresponds to the pUC/M13 forward and reverse primer binding sites that flank the two phage promoters. The use of universal primers allows determination of the sequence of any cloned insert DNA between the primer binding sites. In each DNA annealing reaction, one of these primers anneals to the single-stranded template—this template:primer complex serves as the substrate for the DNA polymerase. The stringency of this hybridization (annealing) is controlled by temperature and must be optimized for each primer.

In the extension/labeling reaction, the primer is extended a short distance (40–50 bp; but **not** extended into the cloned DNA to be sequenced) by using limiting concentrations of the four nucleotides (dNTPs) and an excess of a single radiolabeled dNTP ([α-³⁵S]dATP; see Figure 25.3). By limiting the concentration of nucleotides, the reaction time, and the reaction temperature (37°C, which is well below the optimum for *Taq* polymerase), the radiolabel is incorporated uniformly into the beginning of each newly synthesized DNA molecule.

In the final step, termination, the growing radiolabeled DNA chains are extended in the presence of all four dNTPs and a limiting amount of one of each of the four dideoxy NTPs (ddNTPs). This step is done by dividing the extension/labeling reaction into four separate tubes, each containing a nucleotide mixture with a low concentration of one dideoxyribonucleotide added. The reaction is carried out at 70°C (the optimum for the thermostable *Taq* polymerase), which allows rapid incorporation of dNTPs (>60/second). Each of the four reactions will generate a nested set of terminated nucleotide strands up to several hundred bases in length that will contain frag-

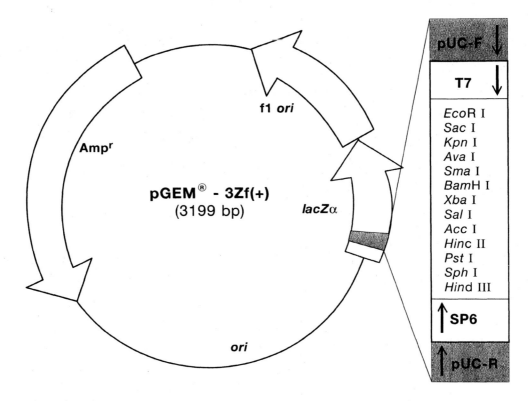

Figure 25.4
Location of the universal sequencing primer binding sites in pGEM™-3Zf(+). Modified from
Promega Corporation 1996, with permission.

ments with a common origin (the primer and a radioactive label) but end with a different ddNTP (ddG, ddA, ddT, or ddC).

The four separate sequencing reactions contain a set of radioactively labeled
DNA fragments of varying lengths with a common 5' end. The reactions are loaded
into four adjacent lanes on a polyacrylamide gel that contains a denaturant (either
urea or formamide) to prevent intrastrand base pairing of the single-stranded DNA
fragments. Separation of single-stranded DNA fragments by denaturing polyacrylamide
gel electrophoresis allows much better resolution than agarose, and the sequencing
gels have the ability to resolve DNA fragments that vary in size by only a single base.
Further information on the use of polyacrylamide gel electrophoresis can be found in
Appendix IV.

After electrophoresis, the gel is fixed to remove the urea, dried, and exposed to
X-ray film (**autoradiography**). When the X-ray film is developed, the position of the
radioactively labeled DNA fragments separated on the gel is visualized. The relative
position of the radiolabeled DNA fragments on the autoradiograph is subsequently
used to determine the DNA sequence (see Figure 25.3). Several factors affect the
amount of DNA sequence information that can be determined from a sequencing gel,

including the percentage of acrylamide and bis-acrylamide used and the length of the gel. Under routine conditions, between 300–400 bases of sequence can be determined from a set of sequencing reactions on a single gel.

Nucleic acid sequencing is an exciting technology that has become somewhat routine, although it requires diligence and care to maximize the amount of information derived from a given template molecule. Additionally, sequencing can be hazardous because it requires the use of toxic acrylamide and may use radioisotopes (several systems now use nonradioactive reporters). In this exercise, we will be using ^{35}S-labeled dATP because of its lower emission energy compared to ^{32}P (see Appendix XIII). ^{35}S does not present a significant **external** radiation hazard as the radiation emitted is barely able to penetrate the outer layer of the skin (^{35}S does pose a significant health hazard if internalized). The low energy of this isotope also results in autoradiographs with sharp bands due to its short penetration distance in air and most solids.

In this exercise, some groups will sequence the insert DNA in pUWL506, a 0.7 kb *lux* subclone containing a portion of the *lux*A gene. This clone was constructed using pGEM™-3Zf(+) as a vector, which allows use of the pUC/M13 forward and reverse sequencing primers (see Figure 25.4). Other groups will sequence subcloned DNA from different regions of the *lux* operon as assigned by your instructor. Each group will set up two sets of sequencing reactions, each set using different primers (the forward and reverse primers) that bind to sites on opposite ends of the cloned DNA. Thus, the two reactions will allow sequence to be determined from opposite ends of the insert DNA. Because you will be using potentially hazardous radioactivity, you must be thoroughly familiar with the precautions required for safe handling, detection, and disposal of isotopes (see Appendix XII).

A. Preparing a Sequencing Gel

 Note: Sequencing apparatuses vary widely in size. This protocol is designed for pouring 20 × 40 cm gels and must be modified for larger gels. See Appendix XV for additional recipes and protocols for preparing sequencing gels.

Reagents

- ❏ Acryl-a-Mix™ acrylamide solution (6% acrylamide:bis [19:1], 8 M urea, 1X TBE buffer, 0.05% TEMED; 37.5 ml/group in squeeze bottle with pouring spout)
- ❏ 10% ammonium persulfate (APS)
- ❏ Dawn™ dish-washing liquid (or equivalent)
- ❏ Pam™ nonstick cooking spray (aerosol can)
- ❏ Kodak Photoflow™
- ❏ 95% ethanol (in squirt bottle)

Supplies and Equipment

- ❏ acrylamide waste container
- ❏ 100–1000 µl micropipetter
- ❏ sterile 1000-µl micropipet tips
- ❏ peg rack
- ❏ sequencing gel clamps
- ❏ plastic wrap

- ❏ assembled gel sandwich containing:
  short (20 cm × 38 cm) sequencing plate
  long (20 cm × 40 cm) sequencing plate
  sequencing gel spacers
  sharkstooth sequencing comb
- ❏ plastic-backed absorbent paper
- ❏ Kimwipes

➡️Procedure

1. Due to variations in the thickness of the plates, spacers, and combs, a set of sequencing plates has been prechecked and assembled so that the fit of the components is within tolerances that will produce a usable gel. Examine the assembled sequencing gel plate apparatus at your station, but do not disassemble. Note the number and position of the clamps holding the spacers and sharkstooth comb in place between the gel plates. Note th\ the comb has been installed with the teeth pointing away from the gel. When \ur gel has been poured, you will install the comb in the same orientation.

2. Mark the outside of both plates near the sh 'stooth comb with a small piece of tape so that you may reassemble them in the \ ne orientation after cleaning.

3. Unclamp the comb from the top of the plates, \d, as you remove it, notice how precisely it fits between the glass sequencing plate Place the comb on the plastic-backed absorbent paper. The fit of the comb is ve. \ important as a loose fit will allow samples to leak into adjoining wells when loa\ \ If the comb is too tight, it is difficult to install.

4. Unclamp the sequencing plates. Remove the spacers and \ \ce them on the plastic-backed absorbent paper along the sides of the plates. K\ \ spacers in the same orientation as they were in the assembled plates.

5. Transport the glass sequencing plates and the peg rack to \ sink with distilled water for cleaning. Scrub the short and long plates with D\ \™ dish-washing detergent and a moistened paper towel. Never use a razor blade or abrasive cleaner. Avoid nicking or scratching the plates; rough treatment of the plates may cause pits that will retain radioactivity and cause the polyacrylamide to stick to the glass during separation of the plates.

6. Rinse each plate thoroughly with tap water followed by three distilled water rinses. Place the plates in a peg rack and air dry. Larger plates may require more than one peg rack. A rinse with 95% ethanol will speed the drying process.

7. Return the clean plates to your bench. Lay the plates flat on the plastic-backed absorbent paper at your bench with the marked side of each plate (the outside) down. Wearing gloves to avoid the transfer of oils from your skin to the plates, coat the gel side (opposite of your marks) of both plates with 95% ethanol and

polish with Kimwipes. The plates are sufficiently clean when a dry Kimwipe slides evenly and smoothly over the surface of the dry plate. Because the gels are so thin (0.4 mm), small particles of dirt or oil can cause air bubbles to be trapped in the gel during the pouring.

8. Apply two **brief** sprays of Pam™ (use the smallest amount possible) to two evenly spaced areas on the gel side of the **short plate**. Spread the cooking oil over the surface of the plate with Kimwipes and then polish 3 to 4 times with fresh Kimwipes until only a nearly invisible oil film remains evenly distributed over the plate. This treatment helps the gel release from the short plate when the plates are separated.

9. Apply four drops of Kodak Photoflow™ evenly distributed down the center of the gel side of the **long gel plate**. Spread and polish as described in step 8 (you will feel a slight drag as you polish). This step will make the gel adhere to the long glass plate when the plates are separated.

10. Place the long plate, treated side up, flat on the peg rack centered over the plastic-backed absorbent paper. Lay the spacers flat on the plate along the long edges in their original orientation.

11. Place the short plate on top of the spacers with the far end of the short plate even with the far end of the long plate. The spacers should be even with the **bottom** (far end) and outer edges of both plates.

12. Clamp the edges of the plates together along the sides with four binder clamps per side. The assembly should now be in the same configuration as it was when you disassembled it (minus the comb; see Figure 25.5).

Figure 25.5
A set of clamped sequencing gel plates ready to pour.

13. The clamped gel plate sandwich should now be laying flat on the peg rack with the short plate on top and with the uneven ends of the plates facing you (this will be the top of the gel). Recheck the fit of the comb to confirm that it will fit snugly into the space between the top of the gel plates. Remove the comb and place it on the absorbent paper in front of the gel plates.

CAUTION: In the next step, you will be working with acrylamide, which is a potent neurotoxin and should be handled with extreme care. Always wear gloves, a lab coat, and eye protection when working with acrylamide solutions. Work over plastic-backed absorbent paper and place all acrylamide waste in the proper receptacle.

14. Your instructor will demonstrate the pouring of a sequencing gel by capillary action. When you return to your bench, add 200 µl of 10% ammonium persulfate to the bottle containing 37.5 ml of premixed acrylamide. **Gently** swirl to mix. Avoid mixing air into the solution, which will inhibit the polymerization and may cause bubbles in the gel.

15. Open the pouring spout on the plastic squeeze bottle of acrylamide solution, invert, and place the spout next to one spacer on the long glass plate, but touching the short plate:

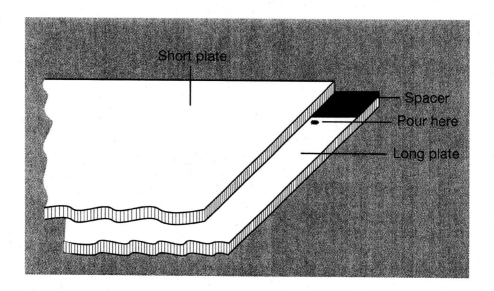

16. Immediately pour the gel by squeezing the acrylamide solution out of the bottle so that it flows into the gel plates (see Figure 25.6). Leave the plates flat while pouring; the acrylamide solution should flow evenly between the plates by capillary action. As the solution flows into the space between the plates, slide the pouring spout to the center of the plates while continuing to dispense acrylamide solution. Keep an excess of acrylamide mix at the comb end of the gel plates or bubbles will enter the gel plates when inserting the comb and ruin the gel. Squeeze the bottle with even pressure and do not allow air to reenter the bottle through the pour spout. Tap firmly with your fingers on the top of the gel plates near the leading edge of the acrylamide to aid the solution in flowing evenly into the plates. When the gel plates are full, do not discard the excess acrylamide solution in the squeeze bottle—set the bottle on the plastic-backed absorbent paper next to your gel.

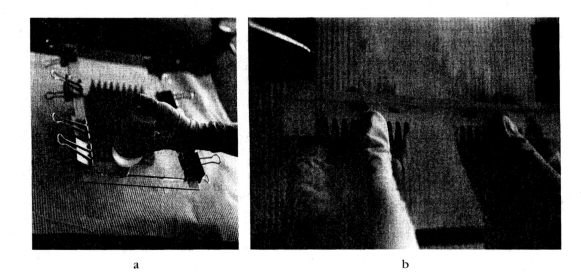

a   b

Figure 25.6
Casting a sequencing gel. (a) Pouring the gel by capillary action. (b) Installing the sharkstooth comb. Note that the comb is installed **backwards** with the teeth pointing out.

17. Immediately after pouring your gel, install the sharkstooth comb **backwards** (flat side in) between the plates until the leading edge of the slots in the comb are even with the top of the short plate (the points of the teeth will be pointing away from the gel and should be approximately even with the edge of the long plate). Hold the comb in place by placing two additional clamps over the comb while the gel is polymerizing (you will remove and reinstall the comb with the teeth pointing in after the gel has polymerized). Remove your acrylamide contaminated gloves and place them on the plastic-backed absorbent paper.

18. After 15 to 30 minutes, check the remaining acrylamide solution in the squeeze bottle to be sure that it has polymerized. If polymerized (which indicates your gel will be polymerized), remove the clamps on the top and bottom of the gel only. Be careful not to disturb the comb.

19. Place plastic wrap over both ends of the gel to prevent drying and promptly replace the clamps. The gel will be ready to run in 1 hour but may be stored for 1 to 2 days at room temperature if a moistened paper towel is inserted under the plastic wrap to keep the gel from drying out.

B. Preparing DNA Sequencing Reactions

Reagents

- template DNA (pUWL506 or other plasmid containing cloned *lux* DNA) (0.25 µg/µl)
- dideoxy termination mixes containing:
    ddATP (and all 4 dNTPs)
    ddCTP (and all 4 dNTPs)
    ddGTP (and all 4 dNTPs)
    ddTTP (and all 4 dNTPs)
- pUC/M13 forward primer (10 µg/ml, 24mer, CGCCAGGGTTTTCCCAGTCACGAC)
- pUC/M13 reverse primer (10 µg/ml, 22mer, TCACACAGGAAACAGCTATGAC)
- 2 M NaOH, 2 mM EDTA
- 2 M ammonium acetate, pH 4.6

- ice-cold 95% ethanol
- ice-cold 70% ethanol
- sterile deionized water
- extension/labeling nucleotide mix (contains all four dNTPs)
- [α-^{35}S]dATP, specific activity approximately 1000 Ci/mmol
- *Taq* DNA polymerase (5 U/µl)
- stop solution/loading dye (95% formamide, 10 mM NaOH, 0.05% bromphenol blue, 0.05% xylene cyanol)
- Count Off™ solution in plastic squirt bottle

Supplies and Equipment

- sterile 1.5-ml microcentrifuge tubes
- sterile 0.6-ml microcentrifuge tubes
- microcentrifuge tube opener
- microcentrifuge tube rack
- mini ice bucket
- 0.5–10 µl micropipetter
- 10–100 µl micropipetter

- sterile 10-µl micropipet tips
- sterile 100-µl micropipet tips
- radioactive tip and tube discard beaker
- 70°C heat block
- 37°C heat block
- thin window Geiger counter (optional)

➡Procedure

Notes:

- It is important that you have all tubes labeled and materials organized before beginning this exercise because the reactions proceed rapidly. Your success in this exercise will be related to your understanding of the procedure prior to coming to the lab. You should prepare a detailed flowchart prior to beginning the lab. The timing of the extension/labeling reaction is critical, and the actual sequencing reactions are finished in approximately 10 minutes.

- Because you will be working with radioactivity, you **must** wear gloves, a lab coat, and proper shoes (no sandals). Perform all manipulations with radioactive materials over absorbent plastic-backed paper. Thoroughly study the precautions for handling and disposal of radioactive materials (see Appendix XIII) prior to beginning the lab.

- Each group will set up two sets of sequencing reactions with the same DNA template (one using the pUC/M13 forward primer and one using the pUC/M13 reverse primer). One partner should set up one set of reactions and the other partner the second set.

Assigned plasmid: _____

1. Fill your mini ice bucket with crushed ice and place each of the required enzymes, buffer, and DNAs in it. Each group will be provided with its own set of enzymes and reagents containing slightly more than the required volume.

2. Label four **0.6-ml** microcentrifuge tubes with your group number and GF, AF, TF, or CF. Label a second set of tubes GR, AR, TR, and CR. R or F refers to the sequencing primers to be used in the reaction (i.e., forward or reverse) and the letters A, T, G, and C refer to the dideoxy present in the dNTP mix.

3. To the tubes labeled AF and AR, add 1.0 µl of the ddATP termination mix. Add the solution in a single drop in the bottom of the tube. In a similar manner, add 1.0 µl of the ddCTP, ddGTP, and ddTTP termination mixes to the remaining corresponding tubes. Cap the tubes to prevent evaporation and then place on ice.

4. Label one **1.5-ml** microcentrifuge with your **group number** and **F** (for the forward sequencing reaction) and a second tube with your **group number** and **R** (for the reverse reaction). Add 18 µl (approximately 4.5 µg or 2 pmol) of your assigned template DNA to each tube.

5. Add 2 µl of 2 M NaOH/2 mM EDTA directly into the template DNA solution in each tube. Set the micropipetter to 10 µl and mix by gently drawing the solution back and forth 3 to 4 times in the pipet tip. Incubate for 5 minutes at room temperature. What does this reagent do to your template DNA?

6. Neutralize by adding 2 µl of 2 M ammonium acetate, pH 4.6, directly to the mixture in each tube and mix as above. Add 75 µl of cold 95% ethanol, mix, and place the tubes on ice for 5 minutes. What is occurring in this step?

7. Spin the samples for 10 minutes at 12–14,000 × *g* (full speed in a microcentrifuge). Be sure that the hinges on the tubes are facing out from the center of the rotor.

8. Promptly remove the supernatant with a 10–100 µl micropipetter, taking care not to disturb the pellet (which will be invisible along the side of the tube below the hinge). You may leave a trace of ethanol in the tube.

9. Wash the pellet to remove excess salt by slowly adding 200 µl of cold 70% ethanol. Add the ethanol to the side of the tube opposite of the pellet while holding the tube in a nearly horizontal position (to avoid disturbing the pellet). Centrifuge in a microcentrifuge for one minute at 12–14,000 × *g*.

10. Remove all traces of the supernatant with a 10–100 µl micropipetter, taking care not to disturb the pellet (which is invisible). Set the tubes with lids open on your bench for 10 to 20 minutes to dry. You may speed this process by using a hair dryer on the low setting to blow warm air over the open tubes.

11. Add reagents to the tubes containing the dried DNA pellet (____-F and ____-R) as indicated in the following table. First, calculate the volume of 5X buffer and sterile water to add to give a final volume of 25 µl. The tube labeled ____-F should receive only the **forward** primer and ____-R should receive only the **reverse**

primer. After all additions have been made, mix by gently drawing the solution back and forth 4 to 5 times with a micropipetter set at 20 μl.

Tube	H₂O	Extension/ labeling mix	5X buffer	Forward primer	Reverse primer	Total volume
____-F	___ μl	2 μl	____ μl	2 μl*	—	25 μl
____-R	___ μl	2 μl	____ μl	—	2 μl*	25 μl

*Approximately 2 pmol

12. Anneal the primers by placing both tubes in a 37°C heat block for at least 10 to 30 minutes. The kinetics of primer annealing at a given temperature and concentration are dependent on primer length (and other factors such as G-C content) and the temperature and time may vary depending on the primer used. The 24mer forward and 22mer reverse primers used here anneal in 10 minutes.

CAUTION: The balance of this procedure involves the use of a significant amount of radioactivity. Read and understand the information in Appendix XIII prior to the lab. THINK! You must wear gloves, safety glasses, and a lab coat at all times. All work must be performed over plastic-backed absorbent paper. Discard all contaminated tips and tubes in the beaker labeled "RADIOACTIVE DISCARD." If you think you have contaminated yourself or your work area with radioactivity, do not touch anything. Immediately notify your instructor, who will assist you in decontamination.

13. Add 2 μl of [α-³⁵S]dATP (approximately 20 μCi) to the annealed primer/template mixture in the tube labeled ____-F (you will add the labeled nucleotide to the ____-R tube separately; see step 19). Set the pipetter to 10 μl and mix by gently pipetting the reaction mixture in and out 3 to 4 times. Discard the tip in the **radioactive tip discard beaker.**

14. Add 0.8 μl of sequencing grade *Taq* polymerase (5 U/μl) to the tube labeled ____-F. Set the pipetter to 10 μl and pipet in and out 3 to 4 times to mix. Discard the tip in the **radioactive tip discard beaker.**

15. Incubate at 37°C for 5 minutes. **Do not extend this incubation.** If you wish to read sequence close to the primer, this incubation can be shortened to approximately 3 minutes.

16. Immediately add 6 μl of the extension/labeling reaction mixture from your 1.5-ml microcentrifuge tube labeled ____-F to the first of the four 0.6-ml tubes containing the ddNTP mix (GF). Place the remaining ____-F extension/labeling mixture in the 37°C heat block with the lid open. Be sure you add the extension/labeling mixture directly to the 1 μl of ddGTP solution at the bottom of the tube. Mix by pipetting in and out once and discard the tip in the **radioactive tip discard beaker.** Place the dideoxy reaction in the 70°C heat block and note the time this incubation begins.

17. Open the cap of the second 0.6-ml tube containing a ddNTP mix (AF). Repeat step 16 for this tube and for each of the remaining tubes containing the ddNTPs (TF and CF). All four dideoxy reactions should be set up within 1 to 2 minutes to prevent the extension/labeling reaction from synthesizing new DNA into the sequence you wish to read.

18. Discard the empty radioactive extension/labeling reaction tube (____-F) in the **radioactive tip discard beaker.**

19. Repeat steps 13–18 with the reverse primed template (____-R tube) DNA and the dideoxy tubes labeled GR, AR, TR, and CR.

20. Incubate each set of four tubes at 70°C for 10 to 15 minutes to allow the termination reactions to occur.

21. After the incubation, remove the tubes from the 70°C heat block (in the order they were placed in the block) and add 4 µl of stop solution to each tube. Mix by pipetting in and out 3 to 4 times, cap, and store on ice until you are ready to load the gel.

    Examine the recipe for the stop solution. What ingredients are common with loading dye used in agarose gel electrophoresis? What ingredients are specific for DNA sequencing and what is their function?

Note: The samples may be stored at −20°C in a freezer licensed for radioactive storage if the gel is to be run at a later date.

C. Loading and Running the Sequencing Gel

Reagents

- DNA sequencing reactions from Part B
- stop solution
- 1X TBE buffer
- 0.5X TBE buffer
- 1X TBE + 1 M sodium acetate
- Count Off solution (in plastic squirt bottle)
- deionized water (in plastic squirt bottle)
- Lubriseal™ stopcock grease

Supplies and Equipment

- 0.5–10 µl micropipetter
- 10-µl sequencing micropipet tips
- radioactive tip and tube discard beaker
- 90°C heat block
- DNA sequencing electrophoresis apparatus
- high voltage electrophoresis power source (2000–4000 volts)
- aluminum cooling plate (if not part of sequencing apparatus)
- metal spatula with thin, flat blade
- Pasteur pipet tipped with plastic tubing
- stick-on sequencing gel thermometers
- thin-window Geiger counter (optional)
- Kimwipes

Procedure

Note: This procedure may vary slightly depending on the type of sequencing apparatus used.

1. Carefully unclamp your polymerized 6% acrylamide gel. Be careful not to bump the comb while you are handling the gel. Wipe the crystallized urea from the exposed surfaces and edges of the gel plates with a moistened paper towel. Failure to remove the urea may allow the buffer to travel down the edges of the plates during electrophoresis.

2. With the gel laying on the absorbent paper, **carefully** loosen the teeth of the sharkstooth comb from the crystallized urea with a metal spatula with a thin, flat blade. Do not push the spatula in too far under the comb or it will begin to separate the gel plates. Carefully remove the comb by pulling gently and evenly on both sides. Rinse the comb with deionized water and dry with a Kimwipe.

3. Rinse the large well created by the comb with deionized water from a squirt bottle. Blot the excess water off the gel plates with a Kimwipe.

4. Reinstall the sharkstooth comb with the teeth pointing toward the gel matrix. Using even, gentle pressure, push the comb between the plates until the teeth have just entered the gel (<0.5 mm). This will be demonstrated by your instruc-tor. Do not push the comb too far into the gel matrix or your sequencing bands will be bowed and difficult to read.

5. Place foam pads coated with Lubriseal™ stopcock grease on the spacers protruding above the top corners of the short plate (they should span the height difference between the short and tall plates).

6. Lay the sequencing electrophoresis apparatus horizontally on the bench and clamp the gel sandwich (with the foam pads affixed) onto the apparatus. The foam pads should seal the gap between the tall glass plate and the upper buffer chamber of the sequencing electrophoresis apparatus.

   If your electrophoresis apparatus does not have a built-in ceramic or aluminum cooling plate, clamp an aluminum cooling plate onto the front of the gel sandwich. Place the cooling plate approximately 1 cm above the plexiglass blocks in the lower buffer chamber so that it will not contact the buffer.

7. Place the electrophoresis apparatus in the vertical position and recheck the position of the foam pads. Press down on the pads to be sure they have formed a seal against the top edge of the short glass plate. The pads frequently shift up as the plates are clamped into place, causing the buffer in the top chamber to leak out.

Note: The distance travelled by any DNA fragment in a gel is inversely proportional to the log of its size. Thus, the bands in the top of the gel will be compressed and the bands in the lower regions of the gel will be more separated than is required for accurate reading. When using 1X TBE in both the top and bottom buffer tanks, the gel will provide excellent resolution of small DNA fragments. Thus, 1X TBE is a useful buffer system if you need to determine the DNA sequence close to the primer. However, you may increase the total amount of readable sequence from a sequencing gel with a gradient buffer system. By increasing the concentration of ions in the lower buffer relative to the upper buffer, smaller DNA fragments near the bottom of the gel will be compressed, allowing more readable sequence near the top of the gel. To run a gradient gel, use 0.5X TBE in the top buffer chamber and 1X TBE containing 1 M sodium acetate in the bottom chamber.

8. Add 1X TBE (or 0.5X TBE for gradient gels) to the upper buffer chamber until the buffer flows over the top edge of the short glass plate and just contacts the gel.

9. Check for leaks by examining the edges of the plates for accumulating buffer. Wipe the edges of the glass plate just below the foam blocks with a dry Kimwipe to determine if a good seal has been achieved. Check the Kimwipe for traces of moisture.

10. If the seal around the foam pads has held, fill the lower buffer chamber with 1X TBE (or 1X TBE containing 1M sodium acetate for gradient gels) until the buffer level is 0.5 cm above the bottom edge of the glass plates.

11. Flush the wells created by the sharkstooth comb by squirting buffer from the upper buffer chamber into each well with a plastic-tipped Pasteur pipet. Be sure to remove any bubbles trapped in the sample wells, as air will prevent the samples from evenly entering the gel. The plastic tip on the Pasteur pipet prevents the tip from breaking, which could lead to glass fragments falling into the sample wells and blocking entry of the samples.

12. It is important to load the gel in a pattern that will allow you to recognize the front and back sides of the autoradiograph after the film is developed. One way to do this is to load the two left most lanes with the same sample (i.e., load the G lane twice). Alternately, you can asymmetrically pattern the loading of the samples by skipping lanes between sets such that the orientation of the resulting autoradiograph will be unambiguous.

    After choosing a method to allow you to orient your autoradiograph, use a permanent marker to label the lanes (formed by the sharkstooth comb) on the glass plate in the order G, A, T, C. Do not load samples in the outside lanes if possible because the current runs differently near the outer edges of the gel, resulting in a sequence that is difficult to read. Label your group number and the template and primer used just below each set of four reactions. Record this information in your notebook, as the marks on the plate will be washed off before the autoradiograph is developed.

Note: Nongradient sequencing gels are commonly preheated prior to loading samples by running the unloaded gel until it reaches the recommended temperature (see step 18). This minimizes formation of intrastrand secondary structure in the single-stranded DNA when the reactions are loaded on the gel. However, *Vibrio fischeri* DNA has a low G + C content and this heating step can generally be eliminated when sequencing DNA from this organism. The gradient gels used in this exercise do not require preheating because they generate heat rapidly in the upper portion of the gel.

13. Prior to loading your sequencing reactions, practice loading one or more unused lanes of the gel. Attach a gel loading tip to your 0.5–10 µl micropipetter (these are special tips flattened at the end to allow them to slide between the glass plates). Withdraw 2 µl of stop solution and examine the tip so you will be familiar with what 2 µl looks like in these tips when you load your samples.

a

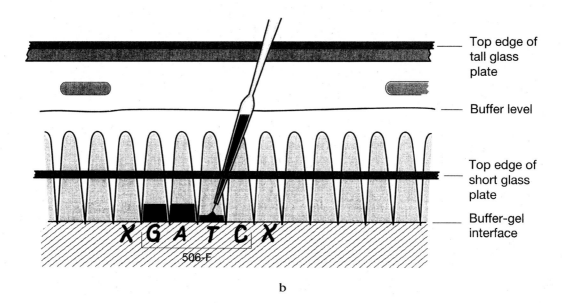

b

Figure 25.7
Loading a sequencing gel. (a) Using a micropipetter to load a sample. (b) An enlargement showing the location of the gel loading tip when loading. The location of the tip is illustrated in the diagram—a rear view as seen through the upper buffer chamber.

To load the sample, point the tip at an angle toward the gel and insert the flat tip between the glass plates in a well formed by the sharkstooth comb [see Figure 25.7(a)]. Add the sample slowly with the tip 3–4 mm above the gel/buffer interface to prevent piercing the gel [see Figure 25.7(b)]. Practice in the unused lanes until you are proficient loading the samples without poking the gel or trapping bubbles in the sample wells. If you are having difficulty, ask your instructor for assistance. Practicing loading also allows you to determine if the samples will leak into adjoining lanes when loaded.

CAUTION: The balance of this procedure involves the use of a significant amount of radioactivity. Read and understand the information in Appendix XIII prior to the lab. THINK! You must wear gloves, safety glasses, and a lab coat at all times. All work must be performed over plastic-backed absorbent paper. Discard all contaminated tips and tubes in the beaker labeled "RADIOACTIVE DISCARD." If you believe you have contaminated yourself or your work area with radioactivity, do not touch anything. Immediately notify your instructor, who will assist you in decontamination.

14. Transfer the first set of four sequencing reactions to the 90°C heat block. Heat the sequencing reactions for 3 to 5 minutes immediately before loading onto the gel. Why is this important?

15. While the samples are heating, flush the four wells to be loaded with a plastic tipped Pasteur pipet. This needs to be done immediately prior to loading the samples to remove urea, which diffuses from the gel and interferes with band resolution.

16. Load 2 µl of each radioactive sample into the adjacent labeled lanes. You may use the same pipet tip for each set of four reactions, but change tips between templates/primers or if there is reaction mix left in the tip that won't discharge.

17. When all groups sharing one gel have loaded their samples, close the buffer chamber lids and plug the electrophoresis leads into the high voltage power source.

18. Attach a stick-on thermometer to the glass plate (or aluminum cooling plate if the cooling plate is not built into the electrophoresis chamber). This will allow you to monitor the temperature of the gel during electrophoresis.

Note: The gels will be run at a voltage that will maintain a gel temperature of 45–50°C (approximately 1600–2000 volts for most sequencing gels). Most power supplies used for DNA sequencing can be run using constant volts, constant amps, or constant watts. Running a gel with either constant volts or amps will cause the temperature of the gel to vary because the resistance of the gel changes as it heats. After the gel has heated to 45–50°C, the temperature of the gel may be held constant by switching the power source to constant watts. The wattage required to maintain temperature must be empirically determined for a given electrophoresis system and will vary depending on ambient temperature, efficiency of the cooling plate, height of the electrophoresis unit, and so on.

19. Turn on the power source and set the run parameters. Adjust the output as needed during electrophoresis to maintain 45–50°C.

20. Run the 6% **gradient gels** until the bromophenol blue dye reaches the bottom of the gel and the xylene cyanol dye is 12–14 cm from the bottom (approximately 1.5–2.5 hours). Run the **nongradient gels** until the bromphenol blue dye just runs off the gel before turning off the power (approximately 1 to 2 hours). On 6% polyacrylamide gels, the bromophenol blue and xylene cyanol dyes comigrate with 26 and 106 bp DNA fragments, respectively. Note that the bromophenol blue dye does not migrate off the bottom of the gradient gel. Why?

21. When the gels have run long enough for the dyes to migrate to the desired position in the gel, turn off the power source.

CAUTION: Always disconnect a lead from the power source before touching the electrophoresis apparatus. Power sources have capacitors that may allow current to run through the gel for some period after the power supply is turned off.

22. Remove the radioactive buffers from the electrophoresis apparatus. There are a number of methods to do this. Many electrophoresis units have valves in the buffer chambers that can be connected to a hose so that the chambers can be drained by gravity. Electrophoresis units without drain valves are best drained by vacuum aspiration. Regardless of the method used, the buffer must be transferred into a container labeled "Liquid Radioactive Waste." The lower buffer will be more radioactive than the top buffer.

23. Unclamp the radioactive gel sandwich from the electrophoresis apparatus and place it flat on a bench top covered with plastic-backed absorbent paper. Arrange the gel on the absorbent paper with the comb side facing toward you and the long plate on the bottom.

24. Rinse the buffer chambers of the sequencing gel apparatus 8 to 10 times in a sink licensed for low-level radioactive waste disposal.

D. Disassembly and Drying of Sequencing Gels

Reagents

- [ ] gel fixation solution (5% methanol, 5% acetic acid)
- [ ] Dawn™ dish-washing liquid (or equivalent)
- [ ] Count Off™ solution in plastic squirt bottle

Supplies and Equipment

- [ ] peg rack
- [ ] plastic wrap
- [ ] metal spatula with thin, flat blade
- [ ] gel fixing tray
- [ ] Whatman 3MM paper (cut 1 cm larger than the gel dimensions; 3 sheets/gel)
- [ ] gel dryer
- [ ] vacuum pump
- [ ] dry ice-ethanol trap
- [ ] soda lime trap
- [ ] plastic-backed absorbent paper
- [ ] X-ray cassette (film holder)
- [ ] solid radioactive waste disposal container
- [ ] liquid radioactive waste disposal container
- [ ] thin-window Geiger counter (optional)

➡ Procedure

Because of the relatively low energy of the beta particles released from ^{35}S, you will need to dry the gel to reduce its thickness. This allows the radioactivity to escape the gel matrix and expose the X-ray film efficiently. Reducing the thickness of the gel also makes the bands on the autoradiograph sharper because the radionucleotide incorporated in the DNA fragments on the gel is physically close to the film during exposure. The urea in the gel matrix quenches the beta particles released from ^{35}S decay events. Therefore, you must remove the urea and fix the gel prior to drying.

CAUTION: The gel in this procedure will be radioactive. Read and understand the information in Appendix XIII prior to the lab. THINK! You must wear gloves, safety glasses, and a lab coat at all times. All work must be performed over plastic-backed absorbent paper. If you believe you have contaminated yourself or your work area with radioactivity, do not touch anything. Immediately notify your instructor, who will assist you in decontamination.

1. Carefully remove the sharkstooth comb from your gel and place it on the absorbent paper. Slide a thin spatula into the space between the glass plates on one edge of the space left by the comb and **gently** pry the plates apart. The gel will usually stick to the long (Photoflow™ treated) plate and readily release from the short (Pam™ treated) plate. If sections of the gel remain bound to both plates, ask your instructor for assistance.

2. Carefully submerge the gel (attached to the long glass plate) in a tray containing gel fixation solution. Soak the gel for 15 to 20 minutes to allow urea to diffuse out of the gel.

3. Carefully remove the plate with the attached gel and tilt slightly to let the excess fixative drain into the tray. Place the plate/gel on plastic-backed absorbent paper.

4. Wet a single paper towel so that it is evenly moistened but not dripping. Unfold the towel and place it on the surface of the gel. Carefully remove the paper towel by lifting from the edge to avoid separating the gel from the glass plate. Repeat with all areas of the gel until excess fixation solution is blotted from the gel surface.

5. To remove the gel from the long plate, lay a piece of dry Whatman 3MM paper (cut 1 cm larger than the gel) on the surface of the gel. Gently press on the entire surface of the 3MM paper to stick the paper to the gel. Do not reposition the paper once it has contacted the gel.

6. Lift a corner of the paper (the gel should remain attached to the paper) and pull the paper and the gel together off the glass plate (see Figure 25.8). Place the gel (gel side up) on 2 additional sheets of Whatman 3MM paper (cut 1 cm larger than the gel).

7. Fold back the flexible plastic cover to the gel dryer and remove the mylar sheet. Place the entire stack of Whatman 3MM paper with the gel up on the gel dryer. Cover the gel with a single sheet of plastic wrap. Do not reposition the plastic wrap once it has contacted the surface of the gel. The plastic wrap prevents the gel from sticking to the mylar sheet during drying and will be removed after the gel is dry. Lay the mylar sheet on the gel and fold the flexible plastic cover of the dryer over the mylar sheet.

Figure 25.8
Removing a gel from the glass plate.

CAUTION: Gel drying systems require a trap system to prevent liquid and vapors (particularly acetic acid) from corroding the vacuum pump and fouling the oil (see Figure 25.9). The vacuum pump can be seriously damaged after a single use if water, acetic acid, or methanol are allowed into the crankcase. Never run a vacuum pump with an open valve in the system so that it cannot pull a vacuum. The oil in the crankcase will trap air, which can also damage the pump.

8. Connect the gel dryer to the pump with the appropriate traps (see Figure 25.9). Refer to the manufacturer's specifications for the drying system you are using. The cold trap will trap vapors and the soda lime trap traps acid vapors. The 3-way stopcock between the traps and the pump allows you to direct the vacuum to the dryer.

9. When all the connections are made, close the stopcock valve so the pump pulls a vacuum against the closed valve. Turn on the vacuum pump and let it run for 2 to 3 minutes.

10. Turn the stopcock to apply vacuum to the gel dryer. The flexible plastic cover on the gel dryer should promptly seal.

11. Set the thermostat on the gel dryer to 80°C. The gel should dry in 1 to 4 hours depending on the size of the gel and the strength of the pump.

Note: While the gel is drying, rinse the comb, spacers, and glass sequencing plates 8 to 10 times in a sink licensed for low-level radioactive waste disposal. They may now be washed as nonradioactive.

12. Once the gel is dried, turn the 3-way stopcock valve to release the vacuum from the gel dryer (but not the pump; the pump should still pull a vacuum against the closed valve). Let the vacuum pump run, pulling a vacuum against the stopcock

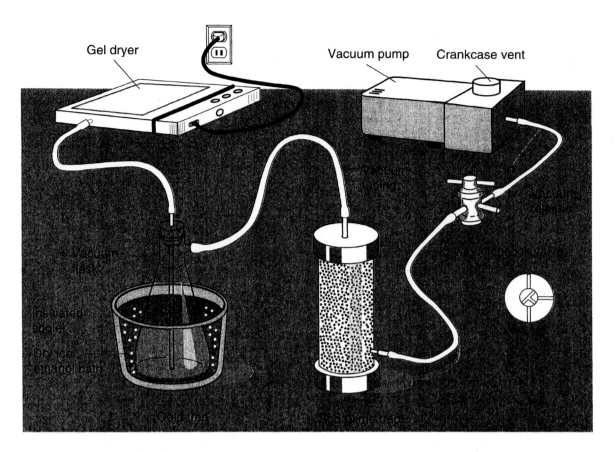

Figure 25.9
Components of a gel drying system. The enlargement of the 3-way stopcock shows the closed position.

valve for 10 minutes with the crankcase vent on the pump open. This allows any acetic acid, water, or methanol not captured in the traps to be released from the crankcase. After 10 minutes, close the crankcase vent and turn the vacuum pump off. Immediately vent the vacuum from the system by opening the stopcock valve so pump oil will not be sucked out of the crankcase and into the soda lime trap.

13. Wearing gloves, remove the gel from the gel dryer and immediately place the single sheet of Whatman 3MM paper containing the dried gel in an open X-ray film cassette (or paper film holder) with the gel side up. **Caution: The gel is still radioactive.** Discard the remaining two sheets of Whatman 3MM paper in the solid radioactive waste container.

14. Remove the plastic wrap from the gel and discard in the solid radioactive waste container. If the plastic wrap is left on the gel, it will absorb most of the radiation, preventing an image from forming on the X-ray film.

15. If available, scan the surface of the gel with a thin-window Geiger counter. The counting efficiency of a Geiger counter with ^{35}S is poor, but you should be able to detect radioactivity at the top of the gel (where the larger DNA fragments are)

if your reactions worked properly. Where would the radioactivity be if the reaction chemistry failed, for example, due to inactive DNA polymerase?

The gel is now ready for autoradiography.

E. Autoradiography of Sequencing Gels

Reagents

□ developer (Kodak GBX)
□ fixer (Kodak GBX)

Supplies and Equipment

□ X-ray film cassette containing   □ wire X-ray film holder
  dried sequencing gel   □ developing tanks
□ X-ray film (Kodak X-omat AR)

 Procedure

> **CAUTION:** You will be using undeveloped X-ray film in the next step. Never turn on a light other than a safelight or open the door to the darkroom with the X-ray film out of the light tight package. The briefest exposure to light can ruin the entire package of film.

1. Transfer your dried gel in the X-ray film cassette to a photographic darkroom. Turn off the lights and turn on the safelight. Using only the illumination of a safelight, open the box of X-ray film and remove one sheet. Lay the X-ray film over the dried gel and seal the X-ray film cassette. Place the unused film back in the light-tight box **before** you turn the lights back on or open the door.

2. Expose the film for 16 to 48 hours. Transfer the film cassette to a darkroom, turn off the room light, and turn on the safelight. Wearing gloves, remove the film from the cassette and clip the film into a wire film holder.

3. Turn on the rinse water to the developing tanks.

4. Submerge the film in the developer tank for 20 to 30 seconds. Hold the film up to the safelight to check the intensity of the developing bands.

5. Place the film back into the developer if necessary until it is properly developed. Maximum development will occur within 2 to 3 minutes.

6. Rinse the film for 20 to 30 seconds in the rinse tank and place it in the fixer solution for 2 to 4 minutes. You may now turn on the lights (if **no** other undeveloped film is out).

7. Rinse the film in the rinse tank for 10 to 20 minutes to remove the fixer. Hang the developed film to air-dry.

8. Examine your autoradiograph for the presence of bands. Guidelines for reading DNA sequence are given in Exercise 26.

F. Monitoring for Radioactive Contamination

Reagents

☐ scintillation cocktail

Supplies and Equipment

☐ paper disks (for wipe tests)
☐ 20-ml scintillation vials

☐ solid radioactive waste disposal container
☐ liquid scintillation counter

➤ Procedure

Federal regulations require monitoring of work areas when radioisotopes are used. Contamination from high energy isotopes such as ^{32}P can be readily monitored using a Geiger counter. Geiger counters are inefficient in detecting the low-energy emissions from ^{35}S, so wipe tests are used to examine work areas and equipment for contamination. Wipe tests are done by wiping a defined area of a benchtop, floor, faucet, piece of equipment, and so on, with an absorbent paper disk and then quantifying the radioactivity on the disk by liquid scintillation counting. The purpose of Part F is to monitor radioactive contamination of equipment and work areas following the completion of this exercise.

1. After your gel has been placed on film, roll up and discard all plastic-backed absorbent paper in the solid radioactive waste container.

2. Perform wipe tests on the following items. For items with large surface areas, wipe a 10 cm² area. For each wipe test, place the paper disk in a scintillation vial and label the lid of the vial to identify what was wiped.

   a. the 0.5–10 μl micropipetter used for the radioactive portions of this exercise; wipe the entire outside surface

   b. the upper surfaces of the heat blocks used for DNA sequencing

   c. the pipet boxes used for the radioactive portions of the exercise; wipe the entire box

   d. the inner surface of the flexible plastic cover of the gel dryer

   e. the inner gel support grate in the gel dryer

   f. the vacuum pump; swab exposed surfaces

   Your radiation safety officer will specify additional items to be monitored. Typically, areas on the benchtops and floor are tested as well as faucets, sinks, light switches, and doorknobs.

3. Add 10 ml of scintillation cocktail to each of the labeled vials.

4. Add 10 ml of scintillation cocktail to an empty scintillation vial marked "Background" to determine the background radiation.

5. Count all vials in a liquid scintillation counter (use a channel set for counting ^{14}C as ^{35}S has an almost identical emission energy).

6. Compare the test counts with the background to determine if any contamination occurred. Report results to the Radiation Safety Officer. Contaminated items or areas must be decontaminated and retested until free of radioactivity.

Results and Discussion

1. Why was it necessary to heat the samples to 90°C prior to loading the gel? What is the function of the urea in the gel? Why is it necessary to run the gels at a voltage that heats the gels to 45–50°C? Some gel recipes call for the addition of formamide (a small molecule that is very efficient in forming hydrogen bonds with DNA) to the gel. The stop solution used in the exercise contains formamide. Speculate as to the function of this compound in DNA sequencing gels.

2. What are the advantages of using a thermostable DNA polymerase in sequencing reactions? Are these enzymes more useful for sequencing DNA from some sources than others? Why or why not?

3. Some DNA polymerases used in DNA sequencing work in the presence of formamide (see question 1). Based on your understanding of the enzymology of DNA synthesis, what is the function of the formamide if added to the sequencing reactions?

4. Under certain circumstances, it is useful to add helix destablizing protein (also known as single-stranded binding protein) to the DNA sequencing reactions. What is the function of this protein in DNA replication *in vivo*? What function may it serve in DNA sequencing reactions and under what circumstances might it be particularly useful?

5. Several DNA polymerases will incorporate base analogs into the DNA that base pair weakly with the four normal bases found on the DNA template. Examples include 7-deaza deoxyguanosine triphosphate and deoxyinosine triphosphate. Both compounds are dGTP analogs that pair weakly with dCTP on the template strand. Under what circumstances might it be appropriate to substitute one of these base analogs for dGTP in DNA sequencing reactions? Why?

6. Many scientists performing manual DNA sequencing prefer to end label the primers rather than use extension labeling (done in this exercise) to incorporate radioactivity into the newly synthesized DNA fragments. When using an end labeling procedure, a separate reaction involving polynucleotide kinase is used prior to the actual DNA sequencing reactions to transfer a single radioactive $^{32}PO_4$ from ^{32}P-labeled ATP to the 5' end of the primer. At which position must the ^{32}P be incorporated into the nucleoside triphosphate to be used as a substrate by the polynucleotide kinase in an end labeling reaction (refer to Figure 25.2)? Which procedure (end labeling or extension labeling) do you believe would be better to

use when determining the DNA sequence of a region very near the primer binding site on the template? Why?

7. In general, 300–400 bases of DNA sequence can be read from a single set of DNA sequencing reactions using a given primer. If you have cloned a piece of DNA larger than 400 bases, you cannot completely determine the sequence of one strand from one reaction using a universal sequencing primer. Can you think of a method that could be used to determine the entire sequence of a 2 kb insert cloned into pGEM™-3Zf(+)? Briefly describe one method (there are several currently in use) and explain how it could be used to determine the entire sequence of a large insert. You may wish to draw a brief flowchart as part of your answer.

8. Cycle DNA sequencing combines the reaction chemistry of DNA sequencing with the polymerase chain reaction (PCR). Using your knowledge of both reactions, draw a flowchart that illustrates how the PCR could be used with Sanger dideoxy sequencing to generate a nested set of DNA fragments that could be resolved on a gel to yield an autoradiograph similar to the one generated in this exercise. What components would be used in the reaction tubes? Cite as many potential advantages of this procedure as you can.

Literature Cited

Maxam, A. M., and W. Gilbert. 1977. A new method for sequencing DNA. *Proc. Natl. Acad. Sci.* 74:560–564.

Sanger, F. S., S. Nicklen, and A. R. Coulson. 1977. DNA sequencing with chain terminating inhibitors. *Proc. Natl. Acad. Sci.* 74:5463–5467.

EXERCISE 26
Computer Analysis of DNA Sequences Using the World Wide Web

The development of nucleic acid and protein sequencing techniques resulted in tremendous advancements in the biological sciences. However, researchers soon discovered that they were faced with a problem: massive amounts of data. The ever-growing amount of data needed to be stored, accessed, and manipulated in order to be of use in answering biological questions. A solution to this problem was first proposed by Devereux, et al. (1984), who described methods using the power of computers to store and manipulate biological sequence data. This publication has become one of the most cited papers in print.

The introduction of this paper states: "The rapid advances in the field of molecular genetics and DNA sequencing have made it imperative for many laboratories to use computers to analyze and manage sequence data. UWGCG (The University of Wisconsin Genetics Computer Group, GCG) was founded when it became clear to several faculty members at the University of Wisconsin that there was no set of sequence analysis programs that could be used together as a coherent system and be modified easily in response to new ideas. With intermural support a computer group was organized to build a strong foundation of software upon which future programs in molecular genetics could be based."

What started as a group of researchers working together to solve a common problem has grown into a major industry dedicated to keeping pace with the needs of the scientific community. GCG is now one of a number of privately owned companies that develop and market sequence analysis software. These sequence analysis packages are essentially collections of complex algorithms (programs) that can be used to manipulate and analyze DNA, RNA, and protein sequences. Versions of these software packages are available for VAX/VMS, Unix, IBM-PC, Macintosh, and other common operating platforms.

Sequence analysis software programs allow researchers to enter, annotate, store, and retrieve sequence information as well as offer a wide variety of analyses of sequence data. Years of software development by academia, government agencies, and private industry have led to powerful algorithms that allow researchers to search any unknown biological sequence against vast databases containing all known biological sequence information. Additional programs allow one to compare, align, or overlap sequence information. Users may also translate DNA sequence into protein sequence; back-translate protein sequence into nucleic acid sequence; rapidly locate restriction endonuclease recognition sites, probe binding sites, transcriptional terminators, potential coding regions in any reading frame, or intron splicing sites; examine RNA, DNA, or protein structure; and submit new sequence information to databases accessible to scientists around the world. All these tasks may be done with the click of a mouse.

Extensive databases now store biological sequence information in powerful computers that are accessible to all individuals using the World Wide Web (WWW). Standard formats allow researchers to submit new sequence information electronically and to retrieve information from the databases for further manipulation on their own computers. Some of the databases store DNA sequences (GenBank and EMBL), whereas others store protein sequences (SwissProt and PIR-Protein). Additional databases store ribosomal RNA sequences that can be used in phylogenetic analyses or vector sequences that are useful in cloning projects. For some applications, databases and software on the WWW eliminate the need to purchase expensive sequence analysis software.

Driven by the vast resources and computer power required to sequence and manipulate the human genome, DNA sequencing and computer analysis of sequence information has become increasingly automated. Machines called digitizers can read sequence information directly from autoradiographs, and these scanners are linked to computers with software that automatically enters the derived sequence in a format that easily allows further manipulation. Modern automated sequencers use nonradioactive, fluorescently labeled dideoxyribonucleotides in sequencing machines that allow laser detectors to read DNA sequence from four sequencing reactions in a single lane of a polyacrylamide gel (see Exercise 25). The sequence data is then automatically entered into a computer. Many countries have large sequencing facilities that rapidly generate and analyze vast amounts of DNA sequence information in several ongoing genomic sequencing projects.

This exercise is not intended to do justice to the power and flexibility of software used to manipulate and analyze biological sequence information. Rather, it is intended only to provide the first-time user a glimpse of the versatility of sequence analysis software in computational biology. Without question, this powerful and ever-changing research tool will continue to have a major impact on how researchers approach biological questions.

In this exercise, you will read the DNA sequence from autoradiographs generated in the previous exercise. If your DNA sequencing efforts were unsuccessful, or if you did not complete Exercise 25, your instructor will provide you with sequence information read from a relevant autoradiograph. You will then connect to remote databases via the WWW and access powerful software that will allow you to enter your sequence (referred to as your **query sequence**). The software will then search through the millions of bases of DNA sequence in the databases to determine if your sequence is similar to other known sequences.

Supplies and Equipment

- ☐ autoradiograph of DNA sequencing gel (from Exercise 25) or DNA sequence provided by your instructor
- ☐ light box (or large sheet of white paper)
- ☐ computer with access to the WWW and connected to a printer

➡ Procedure

A. Reading the DNA Sequence from an Autoradiograph

1. Determine the proper orientation of your autoradiograph. The portion of the film corresponding to the top of the gel will contain bands that are closely spaced in comparison to those at the bottom of the gel. Locate the pattern you used to distinguish the front from the back of the autoradiograph (Exercise 25, Part C). Orient the autoradiograph so you are examining the front and the lanes read G, A, T, C.

2. With a fine point, permanent marker, label each set of sequencing reactions at the top of the autoradiograph (above the region of readable sequence). First label each set of four lanes with the DNA template and primer used in your sequencing reactions (i.e., pUWL506/pUC-forward). Then label the individual lanes of each set of reactions with the dideoxy used in the reactions. If you loaded the samples in the order specified in Exercise 25, the order should be G, A, T, C.

3. Read the sequence in each set of reactions by starting from the bottom of the autoradiograph and recording the order of the DNA bands from bottom to top (see Figure 25.3). To do this, locate the smallest (lowest) resolvable band in your set of four sequencing reactions. The lane this band is in (G, A, T, or C) indicates that it is a DNA fragment that **ends** in that base. This is your first base of readable sequence. A fragment that is **one** base longer will be the next band above the band just read. The lane this band is in will indicate which base that fragment ends in. Because this fragment is one base longer than the first fragment, the lane indicates the **next** base in the newly synthesized DNA. Thus, the sequence is read

from bottom to top in the 5'→3' direction. In a similar manner, continue to read the DNA sequence by moving up the autoradiograph.

You may wish to use a small piece of paper with "G A T C" written on the upper edge and spaced to match the lanes in the autoradiograph. This will help you to keep track of your place as you read and transcribe the sequence onto paper. Read until you are unable to clearly determine the order of the bands. In most cases, you should be able to read 200–300 bases of sequence from a single reaction. Each member of the group should read the sequence independently and compare his or her sequence to reach consensus. If your sequences do not agree, carefully reexamine the relevant region of the autoradiograph.

4. Consult Appendix XI and examine the DNA sequence of the multiple cloning site of pGEM™-3Zf(+). Use this information to determine the exact position where the vector sequence ends and the insert DNA sequence begins. Use the information in Appendix XI to determine which strand of DNA the sequence corresponds to. Will it be the same for both the forward and reverse primers?

B. Entering DNA Sequence and Searching Databases

Note:
This exercise requires basic familiarity with the use of the World Wide Web. Updates on sequence analysis software and database search tools using the WWW are available at the Home Page maintained by the authors of this manual. Due to the rapid rate that available software and Web sites change, we advise you to check the University of Wisconsin-La Crosse Molecular Biology Home Page for updates prior to beginning this exercise (see step 2).

1. Bring your recorded DNA sequence (with the vector and insert DNA sequences clearly marked) and this manual to a computer connected to the WWW. Enter your DNA sequence from 5' to 3' using any standard word processing program. Insert a space every 10 bases to provide reference points when searching for specific sequences, such as the junction between the vector and insert sequence, probe binding sites, and so on. Record any other pertinent information about this sequence in a separate place in your word processing file (date, template, primer, position of vector-insert junction, and so on).

2. Load the WWW browser software on your computer (Netscape Navigator™, Omniweb™, etc.). You will use this to connect to a WWW-linked gateway server that allows remote access to biological sequence databases. A partial listing of these servers is maintained at many institutions, including the WWW Molecular Biology Home Page at the UW–La Crosse. You may access this information by connecting to:

http://www.uwlax.edu/MoBio

Examine the **Molecular Biology Links** section of the Home Page. This will provide you with several links to database servers suitable to perform this exercise.

This section of the Home Page also provides links to a wide variety of servers with information relevant to molecular biology.

3. Connect to one of the recommended servers. Use the help function to familiarize yourself with the capabilities of the server. The server will allow access to database search engines such as the **BLAST** (Basic Local Alignment Search Tool) sequence homology program (Altschul, et al. 1990). BLAST contains several programs with different functions. Blast<u>n</u> compares the sequence of your DNA (the **query sequence**) to all DNA sequences in several databases simultaneously. Blast<u>x</u> translates your query DNA sequence into protein sequence in all six possible reading frames. It will then search databases containing the sequences of all known proteins (and hypothetical proteins identified by translation of nucleic acid sequence) for sequence similarity to your sequence at the amino acid level.

4. Connect to BLAST so that you can search the databases to determine what sequence(s) most closely resemble the DNA sequence you read from your autoradiograph. Use the on-line help functions to read more about the power and utility of the BLAST programs. The help functions will also be useful in interpreting your results.

5. Enter 50 bases of your group's consensus DNA sequence into the window designed to accept user query sequences. Be sure to enter DNA sequence from the insert only. Most modern operating systems will allow you to copy the sequence information from your word processing text file to the query window of the BLAST server. You may also type in your sequence.

6. Perform a Blastn search to compare your query sequence to the nucleic acid database. Print out the results of your search.

7. Examine the results of your Blastn search and address the following questions:

   a. What sequences in the database most closely resemble your DNA sequence? How closely do they match your query sequence?

   b. Did you (or the researchers that entered the sequence in the database) make any mistakes in reading the autoradiograph?

   c. Do you believe that all of the sequences identified in your search of the databases have a biologically significant reason for matching your query sequence?

   d. How many of the sequences identified by the Blastn program do you believe are relevant to the DNA sequence you are investigating?

8. Return to the BLAST menu and resubmit your query sequence using the Blastx program. Print the results of this search.

9. Examine the results of your Blastx search and address the following questions:

   a. What protein sequences in the databases most closely resemble the translation of your DNA sequence? How closely does the best protein sequence match the translation of your sequence?

   b. Did your Blastx search identify any sequences that are different from those in the output of the Blastn search?

10. Reexamine the results of your Blastn search. Retrieve the DNA sequence file from the database that is **most similar** to your sequence. To do this, return to the screen showing the output of your search and click the mouse on the name of the sequence. You may use the **accession number** assigned to the sequence to retrieve the sequence file. Accession numbers are assigned to each sequence deposited into a database (i.e., X06758). Print out the file that you have retrieved.

11. Similarly, retrieve the protein sequence file corresponding to the best match identified in the search you performed using the Blastx program.

12. Carefully examine the information in the **headers** of the DNA and protein sequence files you retrieved (headers are text in the printout that precedes the actual DNA or protein sequence). If you were conducting research on bioluminescence, what information present in the header would you find useful in learning more about the relationship of your query sequence to similar DNA or protein sequences in the databases?

C. Accessing Additional Information Based on the Results of DNA Sequence Similarity Analysis

1. Return to the main menu of the database server you are using. Most databases may also be searched using author names, accession numbers, the names of genes or proteins, and key words that are relevant to the sequence. Scroll through the main menu and use the help functions to learn how to conduct searches using queries other than DNA sequence.

2. Use the information from the headers of the DNA and protein sequence files you retrieved to identify key words that could be used to conduct additional searches of the databases to retrieve information relevant to bioluminescence. Conduct several searches of the databases using key words of interest. Print any interesting information you retrieve for class discussion.

3. Return to the main menu of the server you are using. Most molecular biology servers are also linked to **Prosite**, a database of protein sites and patterns (protein **motifs**). Check the headers from the sequence files you retrieved for Prosite accession numbers.

4. Access the Prosite database and view the files related to the accession numbers from your sequence file headers. What information can you find about bioluminescence from a Prosite search? Print a copy of any relevant information for class discussion.

Note: If your institution has one of the commercially available sequence analysis software programs, you may perform additional analyses of your generated DNA sequence. One such function is to find the recognition sites for a variety of restriction endonucleases, which is very useful when working with a given sequence of DNA (see Exercise 15). To date, this capability is not available through the WWW, but if additional options for sequence analysis become available, they will be described on the UW–La Crosse Molecular Biology Home Page.

Results and Discussion

1. How do a Blastn and a Blastx search differ? Which do you believe would be a better tool to identify sequences in databases that are biologically relevant to a given DNA query sequence? Why?

2. The Prosite information on bacterial luciferase enzymes notes that there is extensive similarity between the LuxA and LuxB proteins. What is the relationship of these two proteins in the biological production of light?

3. Using the LuxA protein sequence as a query, use the Blastp program to search the protein databases for sequences similar to the LuxA sequence. Carefully examine the output. Do any of the sequences identified in the search correspond to LuxB protein sequences? What might your analysis indicate about the evolution of the luciferase enzyme in *Vibrio fischeri*? How might you use the computer to determine if the sequence relationship between the LuxA and LuxB proteins is common in diverse luciferase enzymes?

4. Many insects, such as fireflies, are bioluminescent. The enzyme responsible for bioluminescence in these insects is also termed luciferase. Using the skills you have acquired in this exercise, conduct database searches to determine the relationship between the procaryotic and eucaryotic bioluminescence systems. Do the procaryotic and eucaryotic luciferase enzymes catalyze the same reaction?

5. Conduct a BLAST search using the sequence of the *lux*A probe used in Exercises 16, 21, and 24 as a query sequence. Compare the results of this search with the data that you obtained by searching the databases with the DNA sequence from your autoradiograph. Did the computer find sequences with similarity to the probe sequence that had been identified in previous searches? Are the sequences a perfect match in all cases? Would you expect all bioluminescent bacteria to harbor genes that would bind the probe? Does this information support the results you obtained in Exercise 24?

Literature Cited

Altschul, S. F., W. Gish, W. Miller, E. W. Myers, and D. J. Lipman. 1990. Basic local alignment search tool. *J. Mol. Biol.* 215:403-410.

Devereux, J., P. Haeberli, and O. Smithies. 1984. A comprehensive set of sequence analysis programs for the VAX. *Nucleic Acids Res.* 12:387-395.

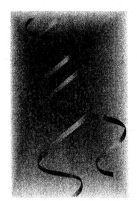

EXERCISE 27
Mapping the *Vibrio fischeri* Genome by Pulsed Field Gel Electrophoresis

Recent developments in molecular biology have allowed researchers to begin to physically characterize the entire genomes of simple organisms, such as bacteria and yeast, and those of some higher organisms, such as the plant *Arabidopsis thaliana* and the nematode *Caenorhabditis elegans.* The ultimate goal of this analysis is to obtain the nucleotide sequence of the entire genome of these organisms. These projects are viewed as model systems in the effort to sequence the entire 3-billion base pair (bp) human genome. The overall approach in these studies is initially to map the cleavage sites of rare cutting restriction endonucleases on the chromosome(s) and then to use the restriction map to break the DNA into progressively smaller fragments that can be ordered and sequenced. To accomplish the daunting task of sequencing the entire human genome, many techniques have been developed to manipulate and analyze DNA fragments ranging in size from several hundred base pairs to several megabases (Mb; 1 Mb = 10^6 bp). Most genomic mapping strategies currently in use rely on newly developed electrophoretic methods capable of separating DNA fragments from 20 kilobases (kb) to several Mb in length.

Conventional agarose gel electrophoresis (such as that used in this course) is generally only able to resolve DNA fragments of 20 kb or smaller. Although agarose gel electrophoresis is usually adequate to resolve restriction fragments generated in most applications in molecular biology, this technique is unable to separate the much larger DNA fragments required in whole genome analysis. This limitation of conventional agarose electrophoresis is a function of the shape of DNA molecules in solution (or in a gel) and their movement through a gel matrix when exposed to an electrical field.

DNA fragments in solution exist as a random coil. When subjected to the constant electric field used in conventional agarose gel electrophoresis, DNA molecules migrate through the gel matrix in a process called "reptation." This movement (and hence the name) is analogous to a snake moving through grass. When induced to move toward something (like lunch), the snake uncoils, the head selects the path through the vegetation, and the elongated body follows. Similarly, when DNA in an agarose gel is subjected to an electric field, it elongates (uncoils) and moves through the gel toward the positive electrode in a linear form. Small DNA fragments migrate more quickly because they uncoil more completely, have a smaller cross sectional area, and encounter less resistance than large DNA fragments from the agarose matrix during movement through the gel. The upper size limit of DNA fragments resolvable in conventional electrophoresis is approximately 40 kb. Separation of fragments of this size requires the use of fragile gels made with low concentrations of agarose. The lack of resolution of the larger molecules is because they do not completely uncoil and molecules larger than 40 kb have similar cross sectional areas. Thus, in conventional agarose gel electrophoresis with a constant electrical field, all DNA fragments above 20–40 kb appear as a single, unresolved band (recall in Exercise 4 the similar migration of the undigested 48 kb lambda DNA and the 23 kb restriction fragment of lambda).

A solution to this problem was developed in the 1980s by Schwartz and Cantor (1984), who devised an electrophoresis system capable of separating yeast chromosomes **several hundred kb** in length. This new system, termed **pulsed field gel electrophoresis (PFGE),** functioned by periodically altering the orientation of the electric field applied to the gel (see Figure 27.1), which resulted in greatly enhanced resolution of large DNA fragments. Periodic switching of the orientation of the electric field causes the DNA molecules to briefly reassume their random coil configuration when the current is reoriented, and then to elongate and realign with the direction of the new field. Time-lapsed photographs of DNA moving through agarose gels and computer modeling studies demonstrated that the time required for reorientation

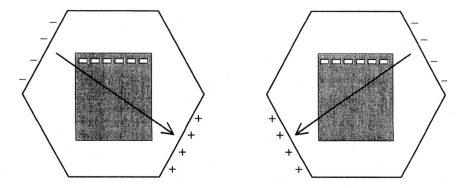

Figure 27.1
Electric field reorientation in a pulsed field electrophoresis chamber (modified from Bio-Rad, with permission).

of DNA molecules after the direction of the electric field has been switched largely depends on the size of the DNA molecule. As the direction of the electric fields are alternated, the larger molecules spend more time reorienting in the new field than smaller DNA molecules, and thus large DNA molecules move more slowly through the gel than smaller DNA molecules (see Figure 27.2). DNA molecules move through a PFGE gel a straight line that is actually the result of the sum of many short zig-zag steps taken by the molecule.

The term "pulsed field" essentially refers to any electrophoretic process using more than one electric field in which the fields are systematically switched. The switch interval, or pulse time, refers to the amount of time that a given field is active, and the angle between the alternating fields is defined as the reorientation angle. Since the development of the pulsed field gel electrophoresis system by Schwartz and Cantor, several variations in the configuration of the hardware used in PFGE has led

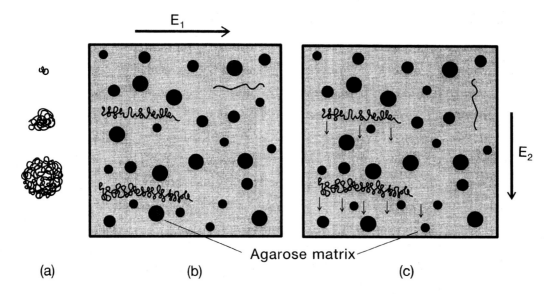

Figure 27.2
Schematic illustration of DNA separation in conventional and pulsed field gel electrophoresis. (a) DNA of various sizes exists as a random coil in the absence of an external electric field. The DNA molecules diagrammed represent fragments of 5, 50, and 500 kb (from top to bottom). (b) Separation of DNA molecules in conventional gel electrophoresis with a static electrical field (E_1). Note that each DNA fragment is aligned in the field, but the 50 and 500 kb molecules have essentially the same cross sectional area in the gel and migrate at the same rate. (c) In a pulsed field gel, the first electric field [E_1 in (b)] is turned off and a second field (E_2) is activated in a new direction. Shortly after the new field is applied, the smallest fragment has realigned in the direction of the new field, while the large molecules have not reoriented. The 500 kb fragment will take longer to reorient and begin migration in the new direction than the 50 kb fragment. Thus, after many repeated reorientations of the electrical field, the smaller fragment will migrate farther than the larger one (modified from Birren, B., and E. Lai. 1993. *Pulsed field electrophoresis: A practical guide,* p. 2. San Diego, CA: Academic Press. Reprinted by permission).

to some complex nomenclature. Most of the variations involve the electrode geometry, the method used to change the direction of the electric field, or the electric circuitry. A summary of the variations of PFGE are presented in Table 27.1.

TABLE 27.1 Acronyms for Pulsed Field Gel Electrophoresis Systems

Acronym	Electrophoresis system	Date
PFGE	Pulsed field gradient gel electrophoresis	1984
OFAGE	Orthogonal field alternation gel electrophoresis	1984
TAFE	Transverse alternating field electrophoresis	1986
FIGE	Field inversion gel electrophoresis	1986
CHEF	Contour clamped homogeneous electric field	1986
RGE	Rotating gel electrophoresis	1987
	Crossed-field gel electrophoresis	1987
Rotaphor	Rotating electrodes gel electrophoresis	
Waltzer	Crossed-field gel electrophoresis	1989
PACE	Programmable autonomously controlled electrodes	1988
ZIFE	Zero integrated field electrophoresis	1990
ST/RIDE	Simultaneous tangential/rectangular inversion decussate electrophoresis	1991

From Birren, B., and E. Lai. 1993 *Pulsed field electrophoresis: A practical guide*, p 5 San Diego, CA: Academic Press. Reprinted by permission

The components of a PFGE system include an electrophoresis chamber configured to control the run temperature, a power supply, and a switching device that changes the direction of the electric fields (see Figure 27.3). Most electrophoresis chambers are connected to a chiller and pump that circulates cooled (typically 14°C) buffer over the surface of a gel. Temperature control is important in PFGE as the temperature during electrophoresis can significantly affect the migration of DNA within the gel. Cool temperatures also reduce the run times (by allowing the gels to be run with higher voltage), which minimizes breakdown of the buffer associated with the long run times required for the separation of large DNA molecules. The electrical requirements for PFGE are generally less than 300 V at 500 mA or less, but precise control of the strength and orientation of the electric field is required for good separation. A range of 1.5–10 V/cm is commonly used and varies depending on the size of the DNA fragments to be resolved on the gel. The switching device controls the switch interval. Switch intervals required for resolution of DNA fragments range from 0.1 seconds (for 5–10 kb fragments) to 30 to 40 minutes (for 2–10 Mb fragments).

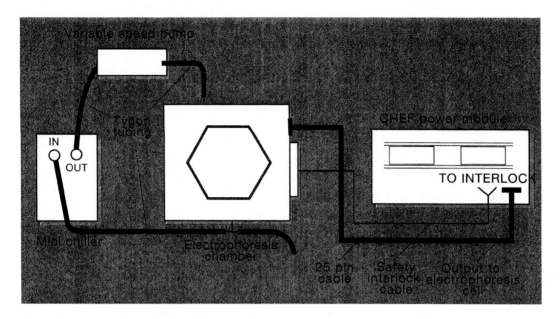

Figure 27.3
Configuration of a pulsed field gel electrophoresis system (modified from Bio-Rad with permission).

Another problem associated with separation of such large DNA fragments is that shear forces associated with the pipetting and organic extraction steps in traditional DNA purification protocols readily break large DNA molecules into fragments less than 100 kb. To avoid shearing, DNA used for genomic mapping is generally purified from whole cells embedded in blocks of agarose (plugs). The cells are lysed by enzymatic digestion and/or the use of detergents while embedded in agarose. This protects the DNA released upon lysis from shear forces that can break the DNA. Following lysis, cellular components other than DNA (RNA and protein) are digested enzymatically, and then the plugs are incubated in a buffer to allow the digestion products to diffuse out of the gel. The high molecular weight genomic DNA is too large to diffuse out of the agarose and remains embedded in the gel. Following equilibration in a suitable restriction endonuclease buffer, the DNA can be digested while embedded in the agarose with restriction endonucleases that cut infrequently. The conditions required to liberate intact cellular DNA from different species and tissue types vary widely, but numerous procedures are available for DNA isolation from different sources.

In this exercise, you will prepare intact chromosomal DNA from *Vibrio fischeri* MJ1 using the Bio-Rad Bacterial Genomic DNA Plug Kit. You will then digest the DNA with selected restriction enzymes that are predicted to cut the genome infrequently and generate fragments suitable for resolution by PFGE. The resulting fragments and size standards will be separated by electrophoresis on the Bio-Rad Clamped Homogeneous Electric Field (CHEF™) PFGE electrophoresis system. Following staining with ethidium bromide, the size of the restriction fragments will be determined, which will allow you to calculate the genome size of *V. fischeri*. Although not part of this exercise, mapping of individual restriction sites and specific genes can be done by Southern blotting and hybridization.

PART I. PREPARATION OF INTACT GENOMIC DNA FROM *Vibrio Fischeri*

Cultures

- [ ] *Vibrio fischeri* MJ1 broth culture (OD_{600} 1.5–2.0; 5 ml per group)

Media

- [ ] sterile GVM broth

Reagents

- [ ] Reagents from CHEF™ Bacterial Genomic DNA Plug Kit:
    cell suspension buffer (10 mM Tris [pH 7.2], 20 mM NaCl, 50 mM EDTA; on ice)
    2% CleanCut™ agarose (melted and cooled to 50°C)
    proteinase K (25 mg/ml)
    proteinase K reaction buffer (100 mM EDTA, [pH 8.0], 0.2% sodium deoxycholate, 1% sodium lauryl sarcosine; on ice)
    lysozyme (25 mg/ml)
    lysozyme buffer (10 mM Tris, [pH 7.2], 50 mM NaCl, 0.2% sodium deoxycholate, 0.5% sodium lauryl sarcosine; on ice)
    1X wash buffer (20 mM Tris [pH 8.0], 50 mM EDTA)
    0.1X wash buffer
- [ ] sterile deionized water
- [ ] 100 mM phenyl-methylsulfonyl fluoride (PMSF) in isopropanol

Supplies and Equipment

- [ ] CHEF™ plug molds
- [ ] mini ice buckets
- [ ] sterile 2.0-ml microcentrifuge tubes
- [ ] sterile 1.5-ml microcentrifuge tubes
- [ ] 13 × 100-mm tubes (disposable)
- [ ] microcentrifuge tube opener (1/group)
- [ ] microcentrifuge tube rack (1/group)
- [ ] microcentrifuge tube float (1/group)
- [ ] 0.5–10 µl micropipetter
- [ ] 10–100 µl micropipetter
- [ ] 100–1000 µl micropipetter
- [ ] sterile 10-µl micropipet tips
- [ ] sterile 100-µl micropipet tips
- [ ] sterile 100-µl micropipet tips heated to 55°C (in insulated cooler)
- [ ] sterile 1000-µl micropipet tips
- [ ] micropipet tip discard beaker
- [ ] PMSF discard beaker (in fume hood)
- [ ] 50°C water bath
- [ ] 37°C water bath
- [ ] microcentrifuge
- [ ] spectrophotometer
- [ ] disposable 1-cm cuvettes
- [ ] liquid discard beaker with disinfectant

▶Procedure

A. Period 1

1. Fill your mini ice bucket with crushed ice and place the cell suspension buffer, lysozyme buffer, and lysozyme stock solutions in it. Each group will be provided with its own set of tubes containing slightly more enzyme and reagents than the required volume.

2. Prepare a ⅕ dilution of the *V. fischeri* culture in a 13 × 100-mm tube. Check the optical density of this dilution with a spectrophotometer using a ⅕ dilution of sterile GVM broth as a blank. Correct for the dilution to determine the optical density of your *V. fischeri* culture.

3. If the absorbence of the culture is not between 1.5 and 2.0, dilute the culture to an OD_{600} of 1.5. This will provide the proper cell density for the DNA isolation protocol.

4. Label a sterile 1.5-ml microcentrifuge tube with your group number. Add 100 µl of a 1.5 OD_{600} culture to the labeled tube.

5. Centrifuge the tube in a microcentrifuge at 8–10,000 × *g* for 1 minute to pellet the cells. Carefully remove the supernatant with a micropipetter.

6. Add 100 µl of ice-cold cell suspension buffer. Resuspend the cell pellet by carefully drawing the cell suspension up and down with a micropipetter. Do not allow air bubbles to enter the cell suspension.

7. Place the tube in a foam microcentrifuge tube float and incubate in a 50°C water bath for 1 to 5 minutes to equilibrate the temperature.

8. Add 4 µl of lysozyme solution (25 mg/ml) to the cell suspension. Set the micropipetter to 10 µl and pipet in and out 3 to 4 times to mix. Again, be careful to avoid forming air bubbles in the cell suspension. Return the tube to the 50°C water bath to maintain temperature. Proceed **immediately** to the next step.

9. Add 100 µl of molten 2% CleanCut™ agarose (precooled to 50°C) to the cell suspension and pipet in and out 3 to 4 times to mix. Return the tube containing the cell suspension–agarose mix to the 50°C water bath to keep the mixture from solidifying.

10. Place a pipet tip preheated to 55°C (from a pipet tip box in an insulated cooler) on your 10–100 µl micropipetter. Preheating the tips and working rapidly minimizes the risk of the agarose solidifying in the pipet tip in the next step.

 CAUTION: Avoid introducing nucleases from your skin by wearing gloves when handling plug molds.

11. Immediately add 100 µl of the cell suspension–agarose mixture to each of two plug molds. Be careful to prevent air bubbles from being transferred into the plug molds.

12. Place the plug molds containing the cell suspension–agarose mixture on ice for 10 minutes to allow the agarose to harden.

13. While the agarose plug is solidifying, label a sterile 2.0-ml microcentrifuge tube with your group number. Add 1 ml of lysozyme buffer and 40 µl of lysozyme (25 mg/ml) to the labeled 2.0-ml tube. Gently mix the contents by inversion.

14. Peel back the tape from the base of the plug mold, and, using the tab attached to the tape at the base of the plug mold, push both plugs into the 2.0-ml tube containing the lysozyme solution.

15. Place the labeled tube containing the agarose plugs in a foam float and incubate in a 37°C water bath for 1 hour.

16. Using a 100–1000 μl micropipetter, slowly remove the lysis buffer from the tube, taking care not to damage the agarose plugs. Add 1.8 ml of 1X wash buffer and mix by gently inverting the microcentrifuge tube. Remove and discard the buffer, taking care not to damage the agarose plugs.

17. Add 1.0 ml of proteinase K buffer to the tube, and check to see that the plugs are suspended in the solution. Note that the plugs at this point are translucent.

18. Add 40 μl of proteinase K (25 mg/ml) to the tube. Incubate at 50°C overnight (16 to 24 hours). This step may be extended up to 48 hours.

B. Period 2.

19. Examine the plugs in your tube after the proteinase K digestion. Have they changed in appearance? What do you think accounts for this change? Remove the proteinase K solution as described in step 16. Add 1.8 ml of 1X wash buffer to the tube containing the plugs and incubate for 1 hour at room temperature with gentle shaking.

20. Remove the 1X wash buffer with a micropipetter, leaving the plugs in the tube. Add 1.8 ml of fresh 1X wash buffer and transfer your tube to a fume hood.

 CAUTION: In the next step, you will be working with PMSF, which is a neurotoxin. Wear gloves, a lab coat, and eye protection when working with this compound, and work in a fume hood.

21. Add 20 μl of PMSF solution to the tube containing the agarose plugs. Discard the tip in the beaker in the fume hood labeled "PMSF Discard."

22. Mix the tube containing the agarose plugs by gentle inversion. Be sure that the plugs are in solution and not stuck on the sides of the tube above the solution. Incubate for 30 to 60 minutes at room temperature with gentle shaking.

23. In the fume hood, remove the PMSF solution from the tube containing the plugs with a micropipetter. Discard the PMSF solution and the micropipetter tip in the beaker in the fume hood labeled "PMSF Discard."

24. Wash the plug for 1 hour in 1.8 ml of 1X wash buffer with gentle shaking at room temperature. Remove the 1X wash buffer from the tube containing the agarose plug with a micropipetter as before.

25. Repeat the wash (step 24) with 1.8 ml of fresh 1X wash buffer.

26. Add 1 ml of **0.1X wash buffer** to the plugs and store the sample at 4°C. The lower concentration of wash buffer will equilibrate the sample at a concentration of EDTA that will not inhibit restriction endonuclease activity.

Note: The plugs at this point are stable for 3 to 6 months at 4°C if stored in **1X** wash buffer. However, they must be equilibrated in 0.1X wash buffer prior to digestion with restriction endonucleases.

PART II. RESTRICTION DIGESTS OF GENOMIC DNA EMBEDDED IN AGAROSE

Reagents

- ❑ 10X restriction buffer (appropriate for your assigned enzyme)
- ❑ agarose plugs (in 0.1X wash buffer)
- ❑ restriction endonucleases (assigned by instructor)
- ❑ 1X wash buffer (20 mM Tris [pH 8.0], 50 mM EDTA)
- ❑ sterile deionized water

Supplies and Equipment

- ❑ mini ice buckets
- ❑ sterile 2.0-ml microcentrifuge tubes
- ❑ sterile 1.5-ml microcentrifuge tubes
- ❑ microcentrifuge tube opener (1/group)
- ❑ microcentrifuge tube rack (1/group)
- ❑ microcentrifuge tube float (1/group)
- ❑ 0.5–10 µl micropipetter
- ❑ 10–100 µl micropipetter
- ❑ 100–1000 µl micropipetter

- ❑ sterile 10-µl micropipet tips
- ❑ sterile 100-µl micropipet tips
- ❑ sterile 1000-µl micropipet tips
- ❑ micropipet tip discard beaker
- ❑ sterile flat toothpicks
- ❑ sterile Petri dishes (1/group)
- ❑ sterile razor blades (1/group)
- ❑ 37°C water bath
- ❑ microcentrifuge

➤Procedure

Due to variations in the G + C content of DNA from different organisms, it is possible to predict which restriction endonucleases will cleave the genome of a given organism infrequently and provide restriction fragments suitable for resolution by pulsed field gel electrophoresis. Most catalogs from suppliers of restriction enzymes contain a section describing enzymes that are useful in cleaving the genome of representative organisms for PFGE. The catalogs also describe appropriate conditions for the digestion of DNA embedded in agarose plugs. The effectiveness of digestion with a given restriction endonuclease, will be a function of many factors, including the size of the enzyme, its activity over the length of the digestion, the percent of agarose in which the DNA is embedded, the type and purity of the DNA substrate, and the incubation temperature.

In this portion of the exercise, your instructor will assign different restriction endonucleases to some students for digestion of their DNA. Other students will prepare control reactions to assure that the DNA samples are free of nuclease contamination.

1. Using sterile deionized water and 10X restriction endonuclease buffer appropriate for your assigned enzyme, prepare 1 ml of 1X restriction endonuclease buffer in a sterile 1.5-ml microcentrifuge tube. Check your calculations with your instructor. Each student should make their own 1X buffer.

2. Label a 2.0-ml microcentrifuge tube with your group number and your assigned enzyme. Transfer 0.5 ml of your 1X restriction endonuclease buffer to the tube.

3. Retrieve your group's agarose plugs from the refrigerator. Pour the 0.1X wash buffer containing the agarose plugs into a sterile Petri dish. If the plug remains in the tube, it may be carefully removed with a sterile toothpick.

4. With a sterile razor blade, cut a slice of the agarose plug that will fit into the wells of a pulsed field gel. Your instructor will specify the dimensions of the gel slice as this will depend on the choice of comb used when casting the gel. For 15-well combs, each lane will require approximately one-third of a full sized plug cut along the narrow axis (1–2 mm × 5 mm).

5. With the corner of your sterile razor blade, transfer the piece of the agarose plug to the sterile 2.0-ml microcentrifuge tube containing the 0.5 ml of 1X restriction endonuclease buffer. Make sure that the slice of the plug is suspended in the liquid and not stuck to the side of the tube. Incubate the samples for 30 to 60 minutes at room temperature to allow the buffer to equilibrate within the agarose plug.

6. Using the corner of a sterile razor blade, transfer the remainder of the plugs from the Petri dish to a sterile 2.0-ml microcentrifuge tube containing 1 ml of 1X wash buffer. Label the tube "PFGE plugs" and add your group number. Return the plugs to the refrigerator in the rack labeled "PFGE plugs in 1X wash buffer."

7. After the 30 to 60-minute incubation in step 5, remove the 1X restriction endonuclease buffer with a micropipetter, leaving the agarose slice in the microcentrifuge tube. Add 130 μl of fresh 1X restriction endonuclease buffer to the tube. Make sure that the agarose slice is suspended in the buffer and not stuck on the side of the tube.

8. Consult the appropriate section in the catalog or applications guide from the supplier of your assigned restriction enzyme describing the digestion conditions for use with samples for PFGE. Add the amount (units) of your assigned restriction endonuclease reported in the catalog to give complete digestion of a standard sample in an overnight incubation. Note that most enzymes give complete cutting in an overnight digestion with less than the number of units supplied in 1 μl of the enzyme. If this is the case, add 1 μl of your assigned restriction endonuclease. However, you may need to add more than 1 μl to achieve complete cutting with some enzymes. Set the 10–100 μl micropipetter to 50 μl and pipet in and out 3 to 4 times to mix. Be careful not to damage the agarose slice.

9. Incubate the samples on ice for 20 minutes to allow the restriction endonuclease to diffuse into the agarose slice. This incubation will allow the enzyme to diffuse into close contact with the DNA before it is activated by increasing the temperature.

10. Incubate the digests overnight (for 16 to 20 hours) at a temperature that gives maximum activity of your assigned enzyme (refer to the catalog from the enzyme supplier). For most enzymes, the incubation temperature is 37°C, but some thermophilic enzymes require higher temperatures and others cut best at room temperature.

11. Following digestion with the restriction enzyme, your plugs are ready for loading into the pulsed field gel. The digested plugs may also be stored at 4°C for 2 to 3 weeks prior to loading the gel by removing the restriction enzyme buffer and adding 1 ml of 1X wash buffer. However, to prevent diffusion of DNA fragments smaller than 100 kb out of the gel during storage, remove the 1X wash buffer after 1 hour at 4°C and store the plugs without buffer at 4°C.

PART III. PULSED FIELD GEL ELECTROPHORESIS OF DIGESTED GENOMIC DNA

Reagents

- [ ] *Saccharomyces cerevisiae* DNA standards (225 kb-2.2 Mb)
- [ ] *Hansenula wingei* DNA standards (1-3.1 Mb)
- [ ] lambda ladder DNA standards (48.5 kb-975 kb)
- [ ] 0.5X TBE buffer
- [ ] 1X TAE buffer
- [ ] molecular biology certified agarose
- [ ] pulsed field certified agarose
- [ ] ethidium bromide stain (1.0 µg/ml)

Supplies and Equipment

- [ ] 250-ml Erlenmeyer flask
- [ ] 25-ml Erlenmeyer flask
- [ ] metal spatula with thin, flat blade
- [ ] microwave oven
- [ ] top loading balance
- [ ] Bio-Rad CHEF™ PFGE system
- [ ] gel casting tray and platform

- [ ] small bubble level
- [ ] 15-tooth pulsed field gel combs
- [ ] 60°C water bath
- [ ] gel staining tray
- [ ] transilluminator
- [ ] Polaroid MP4 camera with type 667 film

➡ Procedure

A. Casting the Pulsed Field Agarose Gel

CAUTION: In the next step, you will be heating agarose to boiling. Agarose can become superheated and boil over in the microwave or when swirled. Wear gloves and eye protection when removing the agarose and swirling heated flasks.

Note: Due to the long run times required to separate DNA by PFGE and the expense of the equipment used, several groups or the entire class will load their samples on a single gel.

1. Prepare 110 ml of 1% agarose in the appropriate buffer for each gel (100 ml will be used to pour each gel and the remainder used to seal the wells after the plugs have been inserted in the gel). Various grades of agarose are tailored for separating different sized DNA fragments. We will use Bio-Rad Pulsed Field Certified™ agarose in 1X TAE buffer to resolve large DNA fragments (500 kb-2 Mb) and Molecular Biology Certified™ agarose in 0.5X TBE buffer to resolve smaller DNA fragments (10 kb-2 Mb). Melt the agarose in a microwave and cool the molten agarose to 60°C before casting the gels.

2. Because the gels are large, it is important that they be cast on a perfectly level surface. Use the small bubble level to determine if the gel casting tray is level. If not, adjust as needed. Assemble the casting tray and slide the platform into the casting stand as demonstrated by your instructor.

3. Install the 15-well comb in the casting stand and adjust the comb so that the bottom of the teeth are approximately 2 mm above the surface of the platform.

4. Pour approximately 100 ml of the molten agarose into the casting stand to give a thickness of 5–6 mm. Place the remaining 10 ml of molten agarose in a 60°C water bath. Allow the gel to solidify for 30 minutes at room temperature.

5. Prepare two liters of 0.5X TBE for the Molecular Biology Certified™ agarose gel (for the small fragments) and two liters of 1X TAE for the Pulsed Field Certified™ agarose gel (for the larger fragments). Place the appropriate buffer in the electrophoresis chamber of a Bio-Rad CHEF™ PFGE system.

6. Turn on the pump and adjust the flow to a setting of 70. Remove the bubbles from the system by elevating the hoses as the pump is running.

7. Turn on the chiller and adjust the set temperature to 14°C. Circulate the buffer for about 30 minutes or until the temperature in the chamber has reached 14°C.

8. Carefully remove the comb and comb holder after the gel has hardened. Leave the gel in the casting stand; it is now ready for loading.

B. Loading Agarose Plug Samples into a Pulsed Field Gel

1. Using two flame-sterilized thin spatulas, place your digested agarose plugs onto the front walls of the well as demonstrated by your instructor (see Figure 27.4).

2. Load the DNA size standards on each gel. Be sure to record the position of each sample in the gel in your notebook.

Figure 27.4
Loading agarose plugs into a pulsed field gel.

3. When all of the samples have been loaded, seal the samples into the wells with molten agarose. Add molten agarose (saved from Part A, step 4) drop-wise with a Pasteur pipet to the sample wells until they are completely filled and sealed. Allow the agarose to harden for approximately 10 minutes at room temperature.

4. Remove both end gates from the gel casting stand. Slide the platform (containing the gel) out of the casting stand and place the platform with the gel attached in the electrophoresis chamber of the CHEF™ system. Do not remove the gel from the platform, or the current created by the flowing buffer in the electrophoresis chamber will move the gel during the run.

5. Adjust the buffer height so that it covers the gel by about 2 mm.

6. Enter the run parameters for each gel as follows:

Parameter	Large fragment gel (1X TAE buffer)	Small fragment gel (0.5X TBE buffer)
Field strength	6 V/cm	6 V/cm
Run time	24 hr	16 hr
Initial switch time	10 sec	1.5 sec
Final switch time	80 sec	20.5 sec

7. Start the run and note the bubbles emitted from the electrodes as the gel runs. This will allow you to observe the current transfer to different electrodes as the fields are switched.

C. Staining and Photodocumentation of Pulsed Field Gels

1. After the electrophoresis is complete (the CHEF system will shut the current off automatically), remove the platform supporting the gel from the electrophoresis box and slide the gel into a staining tray.

2. Shut off the chiller and drain the buffer from the electrophoresis chamber.

3. Flush the system by circulating 2 liters of distilled water in the electrophoresis chamber. Drain the water from the system and shut off the pump and drive module.

 CAUTION: In the next step, you will be working with ethidium bromide, which is a mutagen and suspected carcinogen. Wear gloves, eye protection, and a lab coat, and work over absorbent plastic-backed paper at all times. Carefully read precautions for handling and disposing of ethidium bromide (Appendix XII) before using ethidium bromide stain.

4. At the staining station, flood the gel with ethidium bromide solution (1 μg/ml) and stain for 20 to 30 minutes.

5. Pour the ethidium bromide solution back into the storage bottle using the funnel provided. Rinse the gel 2 to 3 times in distilled water.

6. Destain for 1 to 3 hours to reduce the background fluorescence in the gels to a level suitable for photography.

7. Using a large plastic kitchen spatula, carefully lift the gel from the tray and slide it onto a large transilluminator. Be careful to avoid trapping bubbles under the gel.

CAUTION: UV light is hazardous and can damage your eyes. *Never* look at an unshielded UV light. Always view through a UV blocking shield or wear UV blocking glasses. Most large transilluminators do not have UV blocking lids to protect you from exposure.

8. Lower the UV blocking lid, turn on the transilluminator, and briefly observe the DNA fragments present in each lane. Turn off the transilluminator.

9. If you plan on doing a Southern transfer and hybridization of the DNA separated in the gels, place a clear plastic ruler along the side of the gel with the "zero" on the ruler aligned with the wells. Adjust the camera so that the gel fills the field of view, and focus on the ruler with the room light on.

10. Turn off the room light and turn on the transilluminator. Photograph the gel and tape your gel photograph in your notebook below:

Results and Discussion

1. Using the photographs of gels run with size standards in Figure 27.5, identify the fragments on your gel corresponding to each of the size standards. Are all the fragments in each standard lane resolved on your gel? Plot the mobility of the standards on each gel using 3-cycle semi-log paper.

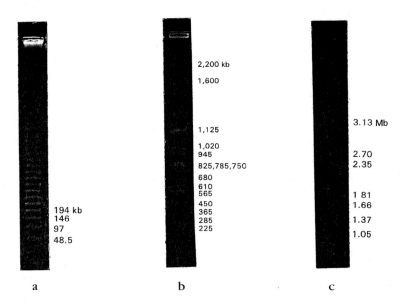

a   b   c

Figure 27.5
Separation of size standards using pulsed field electrophoresis. (a) Lambda ladder standard. (b) *S. cerevisiae* standard. (c) *H. wingei* standard. Modified from Bio-Rad, with permission.

2. Examine the lane corresponding to your assigned digest. Determine the size of each fragment in your lane.

3. Examine the other lanes in the gel. How many fragments resulting from a genomic digest do you believe would be ideal for initial mapping of a bacterial genome? Did each enzyme produce fragments useful for pulsed field mapping of the *V. fischeri* genome? Which of the enzymes do you believe cleaved the genome too frequently to be used in initial mapping studies? Which of the enzymes do you believe cleaved the genome too infrequently?

4. Calculate the size of the *V. fischeri* genome. Hint: Can you calculate the size of the genome by looking at any single digest? Do you need to look at the pictures of both gels to do this?

5. What is the average size of the DNA fragments generated by digestion of the *V. fischeri* genome with *Sal* I? Is this what you would have predicted by purely statistical considerations (refer to Exercise 8)? How might this have affected the efficiency of cloning the *lux* genes?

6. What would be the next step in creating a restriction of the *V. fischeri* genome? What would need to be done to map the site of each of the enzymes used in this exercise?

7. After having constructed a genomic restriction map, how would you determine the location of specific **genes** on the map?

Literature Cited

Bio-Rad Corporation. 1992. CHEF™ Pulsed field electrophoresis.

Birren, B., and E. Lai. 1993. *Pulsed field electrophoresis: A practical guide.* San Diego, CA: Academic Press.

Fonstein, M., and R. Haselkorn. 1995. Physical mapping of bacterial genomes. *J. Bacteriol.* 177:3361–3369.

Schwartz, D. C., and C. R. Cantor. 1984. Separation of yeast-chromosome sized DNAs by pulsed field gradient gel electrophoresis. *Cell* 37:67–75.

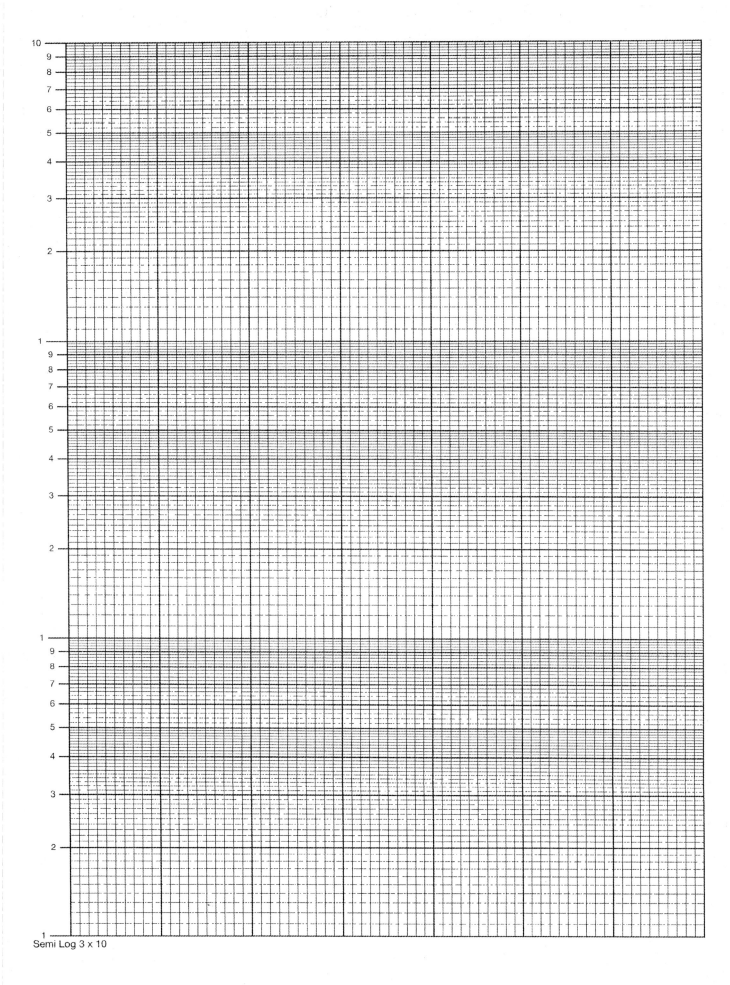

Semi Log 3 x 10

EXERCISE 28
Independent Projects in Molecular Biology

Up to this point in this course you have completed a series of exercises that has provided you with experience in most of the fundamental techniques used in modern molecular biology. You also should have gained some sense of integrating a series of protocols to reach a defined objective. Although you have accomplished a great deal in cloning and analyzing the *lux* operon from *Vibrio fischeri,* much more can be done to further our knowledge of this unique biological system. In fact, one might say we have just begun to study this system at a molecular level.

One limitation of a course such as this is that the student largely follows prewritten protocols and most of the media, reagents, supplies, and equipment are provided prior to conducting each experiment. Although this approach has allowed you to gain experience in a wide variety of molecular techniques in a relatively short period of time, it has provided you with little experience in designing experiments or preparing required materials. These skills are a key part of any research in molecular biology.

Research molecular biologists, like other scientists, generally do not just follow standard protocols, but usually design their own experiments to meet their research objectives. Designing experiments requires creativity in formulating relevant questions and testable hypotheses. It also requires proficiency in the techniques needed to answer the questions posed. At this point in the course, you should have excellent technical expertise in the basic procedures used in molecular biology and perhaps have gained some appreciation and curiosity about the nature of bioluminescence. In this exercise, you will have the opportunity to design your own experiment(s) to address questions and test hypotheses you pose concerning the molecular biology of bioluminescence.

▶Procedure

1. Based on your experience in this course and interests, choose a project that can be reasonably completed in the remaining course time. Be careful not to be too ambitious—many projects, although very worthwhile, may be beyond the time, materials, and facilities available for your project. Prior to beginning your project, you should write a paragraph clearly stating the objective and rationale of the biological question you wish to answer. Submit your proposed project to your instructor for approval.

2. After your instructor has approved your project, write out your experimental design. This should include a detailed protocol describing how you will accomplish your experimental objective. Include a flowchart. Submit this protocol to your instructor for approval.

3. Compile a list of all media, reagents, enzymes, DNAs, supplies, and equipment you will need to complete your project. In addition, be sure you have protocols for preparing any media or reagents you may need (refer to Appendix XV or other references). After compiling your list, have it approved by your instructor before proceeding.

4. After your instructor has approved your experimental design and list of materials, begin to compile the materials you need to conduct your project. Your instructor will assist you in finding necessary reagents and other materials and will inform you of any excess reagents that are available from previous exercises.

5. Prepare all media and reagents that are not already available for your use.

6. Begin your project!

7. At the completion of your project, write up your results in the format provided by your instructor.

Results and Discussion

1. This exercise has provided you with the opportunity to gain real-world experience in conducting research in molecular biology. What have you learned about the relative amount of time spent **preparing** to conduct research as opposed to actually **doing** the research? Do you have a newfound appreciation for the time and effort of your instructor and assistants in preparing the materials used in previous exercises? If so, feel free to express your appreciation!

2. What were the results of your project? Did accomplishing one objective lead to other ideas that you could use to design and conduct additional experiments if you had time? If so, what are they?

3. Based on what you have accomplished, design at least one other experiment that will yield new information related to the molecular biology of bioluminescence.

4. Although much research involves the use of existing scientific procedures, equipment, and so on, the lack of technologies often limits what scientists may accom-

plish. Hence, a significant amount of scientific research involves development of new and better ways to conduct scientific investigations. For example, one of the great advantages of the development of recombinant DNA techniques was that it provided a powerful vehicle to build **better** tools (e.g., new vectors with useful features such as α-complementation, and so on).

Based on your experience in conducting a project of your own design, can you envision any new procedures, vectors, or equipment that would enhance our ability to conduct research in molecular biology? If so, write a brief description of your idea(s) and the potential advantage(s).

APPENDIX I
The Metric System and Units of Measure

The metric system is used in all the sciences, and any scientist or student of science should be well versed in it. The units are defined by a universal convention, the "Systeme International d'Unites," and are commonly called SI units. It is important to have a good understanding of the metric system and the interconversion of metric units of measure in this course. Use this appendix to familiarize yourself with the abbreviations of units of measure (see Table A1.1) and the prefixes used with the abbreviations (see Table A1.2). Metric units are commonly used with prefixes to avoid using large exponentials (e.g., 10^{-6} g = 1 µg) or to convert concentrations into more revelant units (e.g., 2 g/l = 2 mg/ml = 2 µg/µl).

TABLE A1.1 Common SI Units of Measure and Accepted Abbreviations

Measurement	Units	Abbreviation	Definition
distance	meter	m	39.37 in.
weight	gram	g	0.035 oz
volume	liter	l	33.8 oz
time	second	s	
	minute	min	
	hour	h	
	year	yr	
chemical amount	mole	mol	6.02×10^{23} molecules
chemical concentration	molar	M	moles/liter
	normal	N	equivalents/liter
temperature	Celsius	°C	0°C = 273.15 K
	kelvin	K	
electrical current	ampere	A	
electrical potential	volt	V	
radioactivity	curie	Ci	2.2×10^{12} dpm
energy	calorie	cal	
	joule	J	1 J = 4.18 cal
electrical conductance	siemens	S	$ohms^{-1}$

276

TABLE A1.2 Prefixes and Abbreviations for SI Units

Prefix	Abbreviation	Definition (value unit multiplied by)
atto	a	10^{-18}
femto	f	10^{-15}
pico	p	10^{-12}
nano	n	10^{-9}
micro	μ	10^{-6}
milli	m	10^{-3}
centi*	c	10^{-2}
deci*	d	10^{-1}
deka*	da	10
hecto*	h	10^{2}
kilo	k	10^{3}
mega	M	10^{6}
giga	G	10^{9}
tera	T	10^{12}
peta	P	10^{15}
exa	E	10^{18}

*SI recommends prefixes denoting multiples of 10^{-3} and 10^{3} only. These prefixes are acceptable in certain conditions (for example, cm is a commonly used unit)

A P P E N D I X   I I
Centrifugation

Centrifugation is a commonly used tool in molecular biology. A wide variety of centrifuges are available with varying functions and applications. You will use several different types of centrifuges in this course, thus you should be familiar with the principles of centrifugation and proper operation procedures.

The most common purpose of centrifugation is to rapidly sediment particles in solution by applying a centrifugal force far in excess of gravitational force. This results in the separation of the sedimented particles from soluble portions of the solution or from particles that do not sediment out during the centrifugation. The rate that particles sediment in suspension can also be used to characterize biological molecules. The rate is dependent (in part) on the centrifugal force applied. The relative centrifugal force (RCF) is calculated from the following equation:

$$RCF = \frac{\omega^2 r}{g}$$

where ω is the angular velocity in radians per second, r is the distance of the particle from the axis of rotation, and g is the gravitational force. Because it is inconvenient to measure the angular velocity, RCF is generally expressed in terms of revolutions per minute (rpm) by the following equation:

$$RCF = 11.2 \times r \, (rpm/1000)^2$$

where r is given in cm. The relative centrifugal force is usually given in units of gravitational force and written as "$\times g$." For example, an RCF of 10,000 is written as $10,000 \times g$.

Note that the centrifugal force on the particle is related to the **square** of the speed. Therefore, doubling the centrifugation speed increases the RCF by a factor of four. The centrifugal force on a particle also increases with its distance from the axis of rotation (r). Thus, the relative centrifugal force varies at different positions in a centrifuge tube, and particles in a homogeneous medium will accelerate as the radius increases. This may be compensated (e.g., in a sucrose gradient) by increasing the density (and viscosity) of the medium as the radius increases. Most manufacturers specify the g force at the maximum and minimum radius (r_{max} and r_{min}) for a given rotor. Because the rpm rather than the g force is adjusted when using a centrifuge, you must look up the appropriate rpm for each rotor that corresponds to a given g force.

Another parameter that affects the pelleting efficiency is called the k-factor, a characteristic of a specific rotor, and is calculated from the following equation:

$$k = \frac{2.53 \times 10^{11} \, [\ln(r_{max}/r_{min})]}{(rpm^2)}$$

The smaller the k-factor, the greater the pelleting efficiency of the rotor. The k-factors supplied by manufacturers are calculated from the maximum speed of the rotor and increase with decreasing rpm. The k-factor calculated by a manufacturer also assumes that the tubes are full and that the centrifugation medium has the density and viscosity of water. Using partially filled tubes will decrease the k-factor, and use of a medium of greater viscosity or density than water will increase the k-factor. If the sedimentation coefficient (s; see below) is known, then the k-factor can be used to calculate the time (t) in hours required to pellet a given particle:

$$t = \frac{k}{s}$$

In addition to centrifugal force, the nature of particles (such as their size, buoyancy, symmetry, and stability) and the nature of the medium in which particles are suspended (its density and viscosity) will affect the rate of sedimentation. By taking all of these factors into account, you can determine the rate of sedimentation (dr/dt). A detailed discussion of these factors and how sedimentation rates are calculated is given by Rickwood (1984). From the sedimentation rate, a given particle can be characterized by a sediment coefficient (s). This has been defined by Svedberg and Pederson (1940) by the following equation:

$$s = \frac{1}{\omega^2 r} \times \frac{dr}{dt}$$

For most biological molecules, s is about 10^{-13}. Hence, the sedimentation coefficients are expressed in Svedberg units (S) where

1 Svedberg unit = 10^{-13} seconds

A Svedberg unit is dependent not only on the characteristics of the molecule but also the experimental conditions and the properties of the centrifugation medium. Thus, Svedberg and Peterson defined a standard sedimentation that would be obtained for a molecule in water at 20°C ($S_{20,w}$). In practice, sedimentation coefficients are not determined in water but in sucrose density gradients and then standardized to an $S_{20,w}$ value.

There are three fundamental types of centrifugal separations: (i) differential pelleting; (ii) rate-zonal centrifugation; and (iii) isopycnic centrifugation. **Differential pelleting** separates particles of different sizes and densities. This method is used, for example, to separate bacterial cells from culture media or cell debris from soluble portions of a cell lysate. This method works well if the difference between the size and density of the particles you wish to separate is great. However, if the size and/or density of the particles to be separated is relatively similar, many of the smaller particles will also be sedimented with the larger particles. Thus, differential centrifugation is generally used for initial processing of heterogeneous mixtures to enrich the fractions of interest prior to further purification.

Rate-zonal centrifugation prevents the problem of co-sedimentation of particles of different size by layering the sample in a narrow zone on top of a density gradient (a density gradient is simply a solution that increases in density with increasing distance down the tube). The density gradient minimizes mixing of the sample with the centrifugation medium and enhances the resolution of the bands of particles once centrifugation starts. During centrifugation, the sample particles separate as a series of bands depending on their shape, density, and size. The centrifugation is halted before the particles are pelleted, and the bands can be collected by fractionating the gradient. Restriction digests of chromosomal DNA are sometimes size-fractionated by density gradient centrifugation in sucrose gradients.

Isopycnic centrifugation separates particles solely on the basis of their densities. Isopycnic separations involve the centrifugation of particles through a density gradient until they reach an equilibrium density that is identical to their particulate density (referred to as their isopycnic position). Because the separation is based on reaching an equilibrium, prolonged centrifugation will not affect the separation as long as the gradients remain stable. For some isopycnic separations, the sample is loaded onto a preformed density gradient or mixed within a preformed gradient. However, certain types of gradient media (such as cesium chloride) form gradients when centrifuged. For such self-forming gradients, the sample can be mixed directly with the gradient medium, and the separation will occur when the gradient forms during centrifugation. In isopycnic centrifugation, the maximum density of the final gradient must exceed that of the densest particles. This is in contrast to rate-zonal centrifugations, where the maximum density of the gradient must not exceed that of the sample particles. Isopycnic centrifugation with cesium chloride has been widely used in molecular biology in such applications as Messelson and Stahl's demonstration of semiconservative DNA replication and purification of high-quality plasmid DNA (see Exercise 6).

The most common method of classifying centrifuges is based on their maximum speed. **Low-speed centrifuges** usually have maximum speeds of 2000–6000 rpm, and are commonly used for initial pelleting of cells or large organelles. **High-speed centrifuges** have maximum speeds of 18,000–25,000 rpm and are generally refrigerated. They are used to pellet cells and for subcellular fractionations requiring high g forces. A variation on the high-speed centrifuge is the **microcentrifuge**, which can rapidly centrifuge small volumes. Microcentrifuges are designed for tubes ranging from 0.2 to 2.0 ml (but commonly use 1.5-ml tubes). Microcentrifuges are widely used for pelleting cells from small volumes, small-scale DNA and protein isolations, precipitating DNA, and pooling small volumes of reagents. Most microcentrifuges have maximum speeds of 12–14,000 rpm and RCFs of up to 16,000 × g. Several models of low-speed microcentrifuges are also available.

Ultracentrifuges have maximum speeds of 40,000–100,000 rpm and reach RCFs of >600,000 × g. They are subdivided into two types: preparative and analytical ultracentrifuges. **Preparative ultracentrifuges** are used primarily for isolation of macromolecular fractions by density gradient centrifugation, whereas **analytical ultracentrifuges** are designed to provide very accurate data on the sedimentation properties of particles in density gradients. Both types are refrigerated and operate in a vacuum to minimize friction on the rotor during their high operational speeds. As might be expected, cost increases dramatically as maximum speed increases. Low-speed centrifuges are available for as little as $400 to $500, whereas ultracentrifuges generally cost in excess of $50,000, with rotors exceeding $10,000 each.

Three different types of rotors are commonly used in centrifuges: (i) swinging bucket rotors; (ii) fixed angle rotors; and (iii) vertical rotors. In **swinging bucket rotors**, each centrifuge tube is placed in individual buckets that are attached to the rotor such that they can swing out perpendicular to the axis of rotation. Once centrifugation starts, the buckets swing out to a horizontal position so that the centrifugal force is exerted along the axis of each centrifuge tube. Swinging bucket rotors provide a long path length and minimum wall effects and are thus commonly used for rate-zonal separations.

In **fixed angle rotors**, the tubes are placed in a solid rotor at a fixed angle. Angles range from about 14 degrees to 40 degrees; rotors with shallow angles are more efficient at pelleting because the sedimentation path length is shorter. Fixed angle rotors can be designed to withstand very high centrifugal forces ($>600,000 \times g$) and have relatively low k-factors. These rotors are useful for pelleting because the particles hit the side of the tube and slide down to form a pellet at the bottom. Fixed angle rotors are also commonly used for isopycnic separations.

In recent years, **vertical rotors** have become increasingly popular in high-speed and ultracentrifuges. Because the tubes are held perfectly vertical, the sedimentation path is extremely short (the diameter of the centrifuge tube), which results in low k-factors. In addition, the minimum radius is greater in a vertical rotor than in a fixed angle rotor, which means that the centrifugal force is greater. These factors combine to allow much more rapid separations in a vertical rotor than when using a fixed angle rotor operated at the same speed. For example, a plasmid isolation in a cesium chloride gradient that takes about 36 hours in a fixed angle rotor can be done in 4 to 5 hours in a vertical rotor.

Guidelines and Precautions in Using Centrifuges

Specific instructions for use of various centrifuges are provided by the manufacturer, and you should be familiar with proper operation before using a given centrifuge. The following are general rules that should always be adhered to when using any centrifuge.

1. Be sure that rotors are properly seated on the drive shaft.

2. Always check to be sure that the tubes opposite each other are balanced and are placed in a balanced orientation in the rotor. Place tubes in microcentrifuges with the hinges on the lids pointing out to facilitate finding pellets.

3. Be sure the lid is on the rotor (if one is required) prior to starting the centrifuge.

4. Keep the centrifuge clean, and do not allow frost or condensation to accumulate in refrigerated centrifuges.

5. Keep the rotors clean and dry. Store rotors for high-speed and ultracentrifuges inverted in a storage cabinet; do not leave them in the centrifuge.

6. When using high-speed and ultracentrifuges, always enter each centrifuge run in the log book.

Literature Cited

Rickwood, D., ed. 1984. *Centrifugation.* 2d ed. Washington D.C.: IRL Press.
Svedberg, T., and K. O. Pederson. 1940. *The ultracentrifuge.* New York: Claredon Press.

APPENDIX III
Spectrophotometry

An absorbance spectrophotometer measures the fraction of the light transmitted through a solution. Because it is a commonly used instrument in the laboratory, scientists should be thoroughly versed in its principles and operation. Spectrophotometers are used to determine concentrations of a wide number of organic and inorganic compounds in solution, to estimate cell numbers, and to quantify macromolecules such as nucleic acids and proteins in solution.

A reference blank is always required when using a spectrophotometer. The reference blank is a cuvette that holds an identical solution to that containing the compound to be assayed, but lacking the compound. Any absorption of light by the sample containing the compound decreases the amount of transmittance compared to the reference blank. The amount of light that is transmitted through a solution is referred to as the **transmittance (T)** and is defined by the following equation:

$$T = \frac{P}{P_0}$$

and,

$$\%T = \frac{P}{P_0} \times 100$$

where P is the amount of light passing through the sample and P_0 is the amount of light passing through the reference blank. By definition, the transmittance of the reference blank is 1.0 (100%). Note that T and $\%T$ are not equal!

The **absorbance** (abbreviated **A** or **Abs**) is the amount of light that is absorbed by a solution and is defined by the following equation:

$$A = -\log T$$

Because the reference is defined as 100% T (or $T = 1.0$), the absorbance of the reference blank is defined as zero. Note that the relation between A and T is logarithmic.

Quantitative spectrophotometry is based on Beer's Law, which states that the absorbance of a light-absorbing material is proportional to its concentration in solution:

$$A = \epsilon bc$$

where ϵ is the **extinction coefficient**, or absorbitivity (an intrinsic characteristic of the absorbing substance), b is the length of the light path passing through the sample (the interior thickness of the cuvette; generally in cm), and c is the concentration (usually in moles per liter). The extinction coefficient (ϵ) is frequently subscripted to reflect both the units of concentration and the wavelength of light used in the measurement. For instance, if a 1 M solution of compound Z assayed at 550 nm in a standard cuvette (1 cm wide) has an absorbance of 12, then the extinction coefficient

$\epsilon_{M,550}$ = 12 liters $\times$ moles^{-1} $\times$ centimeters^{-1}. The "strange" units of ϵ reflect the fact that the absorbance is a "unitless" measurement. Therefore, the units of ϵ are defined to cancel both concentration and length and thereby give a unitless absorbance.

If the absorbance of a compound obeys Beer's Law and the absorbance is plotted graphically (absorbance versus concentration), a straight line will result. In practice, the relationship between absorbance and concentration must be determined for each substance by absorbance spectrophotometry. This is referred to as a **standard curve**. If a portion of the plot is linear, you can determine the equation for the line by linear regression:

$$Y = mX + b$$

where Y = absorbance, m = the slope, X = the concentration, and b = the y-intercept of the line. The linear regression equation may be calculated using a computer software package, most spreadsheet programs, and many programmable calculators. The concentration of a substance in solution can now be determined by measuring its absorbance and using the regression equation. The concentration can also be calculated from the extinction coefficient (ϵ) and Beer's Law if ϵ is known.

There are frequent deviations from Beer's Law that result in a nonlinear curve. For example, linearity often ceases at high concentrations of the absorbing substance. It is important that you always plot and observe your data to determine if it is linear. It is inappropriate to calculate a linear regression equation from nonlinear data!

Spectrophotometric measurements are done with specific wavelengths of incident light to achieve the maximum absorbance of the compound of interest without interference from other components in solution. All substances transmit some portions of the light spectrum and absorb others. For example, a red color results because the solution absorbs short wavelengths of light (green and blue light) and transmits longer wavelengths (red light). An absorbance spectrum (a plot of absorbance as a function of the wavelength of the incident light) is normally measured to determine the optimal wavelength for measuring the absorbance of a given compound in solution. This can readily be done with the use of a scanning spectrophotometer that will automatically measure and plot the absorbance over a given range of wavelengths. Nucleic acids have an absorbance maximum in the ultraviolet (UV) range at 260 nm, whereas proteins absorb maximally at 280 nm. This characteristic allows UV spectroscopy to be used to quantify DNA and RNA in solution. However, because protein absorbs maximally at a nearby wavelength, nucleic acid quantification with a UV spectrophotometer requires highly purified samples.

The light from a spectrophotometer light source does not consist of a single wavelength but a continuous portion of the electromagnetic spectrum. The incident light is separated into specific portions of the spectrum by filters, prisms, or a diffraction grating. A small portion of the separated spectrum is allowed to pass through a slit, creating incident light of a narrow range of wavelengths. When you adjust the wavelength on a spectrophotometer, you are changing the position of the prism or diffraction grating, which directs different wavelengths of light toward the slit. The range of wavelengths in the incident light depends on the actual width of the slit and the quality of the diffraction grating or prism. The smaller the slit width, the better the ability of the instrument to resolve various compounds. Good instruments are able to resolve the incident light into a very small part of the spectrum. For example, a Spectronic 20 has a slit width of only 20 nm, whereas a Beckman DU-6 (with a ten-fold higher cost) has a 2 nm slit width. Expensive spectrophotometers frequently have adjustable slit widths.

Different light sources emit different portions of the electromagnetic spectrum. A deuterium lamp is used for measurements in the UV portion of the spectrum (200–340 nm) and a tungsten halogen lamp is generally used for measurements in the visible portion of the spectrum and beyond (340–800 nm). After passing through the solution, the incident light strikes a photomultiplier tube (PMT), which transforms the transmitted light into a current that is read on a detector. Photomultiplier tubes are not as sensitive in the far-red portion of the spectrum, and measurements above 800 nm are only possible on very high-quality instruments.

Special cuvettes must be used for different measurements. Relatively inexpensive silica glass cuvettes are used for measurements above 320 nm, and disposable plastic cuvettes can also be used in this range. Disposable cuvettes are useful when measuring the absorbance of bacterial cultures as they can be disposed and autoclaved rather than being disinfected prior to future use. Glass and plastic cuvettes absorb shorter wavelengths of light, and quartz or other special commercial glass must be used for measurements below 320 nm. Cuvettes, even glass ones, are fragile and expensive and should be handled carefully. Quartz cuvettes are **very** expensive (several hundred dollars per set) and should be treated with **extreme care**. High-quality cuvettes are sold in matched sets that have an identical absorbance over a defined wavelength range. This allows you to blank the instrument with one cuvette and use the other for measurements.

Guidelines and Precautions When Using Spectrophotometers

Specific instructions for use of various spectrophotometers are provided by the manufacturer, and you should be familiar with proper operation before using the instrument. The following are general rules that should always be adhered to when using spectrophotometers and cuvettes:

1. Turn on spectrophotometers 20 to 30 minutes prior to use. This allows the electronics to warm up and stabilize. When using a UV-visible spectrophotometer, only turn on the light source being used. Lamps are expensive (especially UV lamps) and have finite lifetimes.

2. Only handle cuvettes by the nonoptical surfaces (these are generally frosted glass or plastic). Fingerprints on the optical surfaces will absorb light and may scratch the glass. Wipe the optical surfaces with a Kimwipe™ immediately before reading an absorbance.

3. Always place glass and quartz cuvettes in a cuvette rack when not in use. Expensive and fragile cuvettes standing on a bench are easily knocked over and broken.

4. Always clean cuvettes with distilled water immediately after use. Dry the exterior of the cuvettes with a Kimwipe. Store inverted over a Kimwipe in a cuvette rack until dry. Return them to their storage box when dry.

APPENDIX IV
Agarose and Polyacrylamide Gel Electrophoresis

Electrophoresis is a widely used procedure to separate, identify, or purify biological molecules. In molecular biology, electrophoresis applications range from separating DNA fragments several dozen base pairs long in DNA sequencing to separating entire chromosomes by pulsed field electrophoresis. Although there are many types of electrophoresis, agarose gel electrophoresis and polyacrylamide gel electrophoresis are the most commonly used in molecular biology.

Basic Principles

Electrophoresis is the movement of a charged particle (such as dyes, small ions, macromolecules, and so on) in response to an electrical field. The rate of migration, or velocity, of a particle is a function of the electrical field strength, the net charge of the particle, and the viscosity of the electrophoretic medium. Thus, an increase in the voltage or charge on the particle will increase the electrophoretic mobility, whereas an increase in the frictional drag (caused by a denser medium) will decrease the mobility.

Many biologically important macromolecules are charged and therefore can be separated by electrophoresis. For example, DNA and RNA have a net negative charge at neutral pH due to the phosphate groups in the backbone and are readily separated by electrophoresis. Electrophoresis of nucleic acids is relatively simple because the charge in a nucleic acid is directly proportional to the size of the molecule. This is not true with proteins, however, and different proteins may have either positive, negative, or no charge at a given pH.

Frictional drag is created when performing electrophoresis of biological macromolecules in a gel medium. Gels create a matrix ("molecular sieve") that impedes the migration of macromolecules to varying degrees based on size. The gel matrix is relatively inert so it does not interact with the molecules being separated. The most commonly used gel materials for the separation of proteins and nucleic acids are polyacrylamide and agarose. Due to increased frictional drag, larger molecules migrate through the gel matrix slower than small molecules, resulting in their separation. The gel concentration can be varied to provide the desired degree of separation for the molecule(s) of interest. Increasing the gel concentration reduces the mobility of molecules, which is useful for separating small molecules, whereas lowering the concentration more effectively separates larger molecules. Gels also minimize diffusion of the molecules during and after separation so that they can be visualized or further manipulated. Gel electrophoresis is usually done analytically to visualize the individual molecules separated but may also be done preparatively to purify a given macromolecule for further study.

Several additional factors affect the mobility of macromolecules during electrophoresis, such as the conformation of the macromolecule being separated and the presence of molecules that interact with the macromolecules of interest. This is particularly true of plasmid DNA. Plasmids in a cell exist in a supercoiled form, and carefully isolated plasmids are largely supercoiled. If the plasmid has a single-stranded nick, the supercoiling is relaxed and the plasmid forms an open circle. If a double-strand break occurs, the plasmid will exist in a linear form. Although each of these molecules contain the same number of base pairs, their electrophoretic mobilities are significantly different (see Exercise 4). This differential mobility is often useful, however, as it allows you to visualize if a digestion of a supercoiled plasmid is complete when cloning in plasmid vectors (see Exercises 8 and 17). Molecules that interact with the macromolecules being separated also affect mobility. Ethidium bromide (a DNA intercalating compound used to stain DNA in agarose gels) is sometimes added to a gel and the buffer prior to electrophoresis. This allows visualization of the DNA migration during electrophoresis and eliminates the need for staining. However, the intercalation of ethidium bromide in the DNA decreases the mobility by approximately 15%.

The apparatuses that are used for gel electrophoresis are simple electric circuits defined by Ohm's law:

$$V = I \times R$$

where V is the electric potential (expressed in volts), I is the current (the movement of electrons, expressed in milliamps), and R is the resistance (anything present in any circuit that impedes current, expressed in ohms). Familiarity with Ohm's law will help you understand the practical considerations when performing electrophoretic separations. The resistance in an electrophoretic apparatus is created by the ionic strength of the buffer and the length and cross sectional area of the gel. Thus, Ohm's law states that if the resistance is constant, the voltage will be directly proportional to the current. However, if the resistance of an electrophoresis apparatus is increased (for example by decreasing the gel thickness or decreasing the ionic strength of the buffer) in a gel run at constant current, the voltage will increase and thus cause an increase in electrophoretic mobility.

Although an increase in voltage increases the mobility of the macromolecules being separated, the upper voltage limit is determined by the ability of the electrophoresis apparatus to dissipate heat. The following equation defines the power produced by the system (the ability to do work):

$$P = V \times I$$

where P is the power (expressed in watts).

The power produced during electrophoresis is used to move the molecules through the gel matrix, which creates heat. A given gel electrophoresis system is able to dissipate only a certain amount of heat and the voltage applied to a gel cannot exceed the capacity for heat dissipation. Heat is usually undesirable because it increases diffusion of bands in the gel and can even melt agarose gels. Uneven heat distribution in the gel leads to non-uniform electrophoretic migration. For example, heat is dissipated more effectively from the edges of a gel than in the center. Because heat causes increased mobility of molecules in the center of the gel, molecules of the same molecular weight run in a U-shaped pattern referred to as "smiling." Smiling is common in DNA sequencing gels if thermal plates (aluminum or ceramic) are not used to dissipate heat, and can make the analysis of samples separated on a gel difficult. However, heat generation can sometimes be beneficial. For example, when running sequencing gels, heat prevents intrastrand base pairing of the DNA fragments, and sequencing gels are commonly run using constant power to maintain temperatures of 45–55°C.

Electrophoresis Buffers

Electrophoresis is performed in buffers with defined pH and ionic strength. The pH of electrophoresis buffers affects the charge of the molecules being separated by protonation and deprotonation of ionizable groups on the molecule. Because the net charge is one factor affecting electrophoretic mobility, it is important that the pH (and thus the charge on the molecules being separated) remains constant during electrophoresis.

Electrophoresis buffers have low ionic strength, which creates resistance to the current flow. Recall that increased resistance results in greater electric field strength (Ohm's law), which in turn increases mobility. In addition, increased resistance decreases the current, which results in less heat generated.

The most commonly used buffers in electrophoresis of DNA molecules are Tris, acetate, EDTA (TAE) and Tris, borate, EDTA (TBE). TAE buffers have historically been used most commonly for agarose gel electrophoresis, but they have a low buffering capacity that can be exhausted after long periods of electrophoresis. The buffer near the negative electrode (anode) becomes alkaline, and the buffer near the positive electrode (cathode) becomes acidic. Most agarose gels are run for a relatively short period of time to minimize buffer breakdown, but the buffer should be circulated during long runs. TAE buffer can be reused for 4 to 5 electrophoretic separations, but the buffer in the electrophoresis apparatus should be recirculated between runs by rocking the electrophoresis chamber back and forth.

TBE buffers provide more buffering capacity than TAE but decrease electrophoretic mobility in agarose gels by about 10%. For electrophoretic separations such as DNA sequencing, TBE buffers are used to provide more buffering capacity and hence better stability during long runs.

Agarose Gel Electrophoresis

Agarose is the most common matrix used to separate DNA molecules electrophoretically. Agarose is a highly purified form of agar, a polysaccharide extracted from the marine red algae *Gelidium* or *Gracilaria*. It is a complex polymer composed of galactopyranose sugars complexed with sulfate and pyruvate esters. It forms a gel with relatively large pore sizes that makes it useful for separation of large macromolecules such as DNA. Because of the large pore size, agarose is less useful for analyzing proteins (which are considerably smaller than most nucleic acids) or very small fragments of nucleic acids (<300 bp). Agarose is relatively inexpensive, nontoxic, has a high gel strength at low concentrations, and is easy to use, making it ideal for routine electrophoretic separations of DNA.

Electroendosmosis (EEO) is an important property of agaroses. EEO is a measure of the flow of water through the gel, to the cathode during electrophoresis. EEO is largely related to the presence of sulfate and pyruvate esters on the polymer and varies among commercially available agaroses. A high EEO is useful for the separation of highly charged proteins from relatively uncharged proteins, but DNA separations are done with low EEO agarose to minimize run times. A detailed discussion of this property of agarose is given in *The Agarose Monograph,* published by FMC BioProducts (FMC 1988).

Agarose gels are run in horizontal gel chambers submerged in buffer and are typically run at field strengths of 1–5 V/cm (the field strength is the applied voltage divided by the distance between the cathode and anode in cm). During electrophoresis, the DNA molecules orient parallel to the field and migrate through the gel in an end-on, snake-like fashion called reptation. Because of this orientation, the friction encountered by large molecules is not as great per unit length as it is for small molecules, although the net charge is proportional to length. This results in poor resolution of large fragments (>20–40 kb), and large molecules are typically compressed in the upper region of a gel (note gel photographs in Exercises 4, 8, and 14). The decrease in resolution is more pronounced with increasing field strengths, and low field strengths (0.5–1 V/cm) should be used when high resolution of large fragments is required. A recently developed technique, pulsed field gel electrophoresis (see Exercise 27), eliminates the problem of lack of resolution of large fragments and is capable of resolving fragments exceeding 1000 kb and even whole chromosomes.

You may encounter a number of problems when running agarose gels that can cause poor results. The more common of these are summarized below:

- If the agarose is not completely molten when the gel is poured, agarose granules will act like boulders in a current and the bands of DNA will migrate in an inverted V around the agarose boulder rather than as clear linear bands.

- The gel must be completely solidified when the comb is removed, otherwise the wells will be improperly formed and bands will be distorted.

- If the leads from the power source are reversed or if you place the gel in the chamber backwards, you will perform that time-honored procedure called retrophoresis, which has been the scourge of students and molecular biologists everywhere. In a few short minutes the DNA will migrate to the top of the gel and be lost forever in the buffer!

- If lanes are overloaded with DNA, the bands will not be clearly resolved and result in "flaming" bands with streaks of DNA smearing toward the wells.

- A final mistake made by most molecular biologists at least once is to prepare an agarose gel in water or a different buffer than that used in the electrophoresis chamber. This causes the DNA to migrate in a very irregular pattern, resulting in smeared bands.

Despite these potential problems, agarose gel electrophoresis is a simple, yet powerful technique that is done on a daily basis in all molecular biology labs.

Polyacrylamide Gel Electrophoresis

Polyacrylamide gels are made by polymerizing acrylamide monomers and a bifunctional crosslinker N,N'-methylenebisacrylamide (bisacrylamide or "bis"). This reaction results in long strands of polyacrylamide crosslinked periodically by the bisacrylamide molecules. Like agarose, the molecular sieving characteristics of polyacrylamide gels can be enhanced by increasing the acrylamide concentration (polyacrylamide gels are typically made at concentrations of 4–20%). In addition, further increase in molecular sieving results from varying the concentration of the crosslinker, which is typically 2–4% of the total acrylamide concentration. The gels are polymerized in the presence of N,N,N',N'-tetramethylenediamine (TEMED) as a catalyst and ammonium persulfate (APS) as the initiator. The TEMED catalyzes the formation of free radicals from the APS, which in turn initiates polymerization.

The pore size of acrylamide gels is much smaller than that in agarose gels, which allows more efficient separation of smaller macromolecules. Therefore, polyacrylamide gels are commonly used to separate proteins, which are generally smaller than nucleic acids. Because the charge on proteins varies widely (and may be either positive or negative), proteins are often denatured in a detergent (SDS) and reducing agent (β-mercaptoethanol) prior to being separated electrophoretically. The SDS binds to the denatured protein, giving it a net negative charge that is proportional to the protein's size and allows efficient separation by polyacrylamide gel electrophoresis. Polyacrylamide gels are also commonly used to resolve small DNA fragments resulting from restriction digestions but are most widely used in molecular biology to separate the short nucleic acid fragments generated in DNA sequencing reactions.

The unpolymerized acrylamide and bisacrylamide monomers are potent neurotoxins; and extreme care must be taken when preparing acrylamide solutions or polyacrylamide gels. Gloves, a lab coat, and safety glasses should always be worn when working with acrylamide. If powdered acrylamide is used, a dust mask must also be worn. Fortunately, most suppliers now market premixed acrylamide and bisacrylamide solutions, which minimizes handling of the toxic powder.

Visualization of DNA in Agarose and Polyacrylamide Gels

The most common way to visualize nucleic acids is by staining with ethidium bromide. Ethidium bromide intercalates between the bases of double-stranded DNA molecules in the gel and can be visualized on a UV light transilluminator. Because the fluorescence of ethidium bromide is significantly greater when it is intercalated than when it is free in solution, the location of the DNA is readily observed by its orange fluorescence. Ethidium bromide staining is extremely sensitive and bands with as little as 2 ng of DNA can be detected in agarose gels. The sensitivity in polyacrylamide gels is less due to quenching of the fluorescence by the acrylamide. DNA fragments in DNA sequencing gels are detected by autoradiography in radioactive sequencing techniques or by laser detectors in automated fluorescent sequencing. Details of these procedures are given in Exercise 25.

Literature Cited

Bier, M., O. A. Palusinski, R. A. Mosher, and D. A. Saville. 1983. Electrophoresis: Mathematical modeling and computer simulation. *Science* 219:1281–1287.

FMC BioProducts. 1988. *The agarose monograph,* 4th ed. Rockland, ME.

Hames, B. D. 1986. An introduction to polyacrylamide gel electrophoresis. In *Gel electrophoresis of proteins—a practical approach,* ed. Hames, B. D., and D. Rickwood, 1–86. Washington D.C.: IRL Press.

Smith, S. S., T. E. Gilroy, and F. A. Ferrari. 1983. The influence of agarose-DNA affinity on the electrophoretic separation of DNA fragments in agarose gels. *Anal. Biochem.* 128:138–151.

Stelwagen, N. 1987. Electrophoresis of DNA in agarose and polyacrylamide gels. In *Advances in electrophoresis,* ed. Chrambach, A., M. J. Dunn, and B. J. Radola, 177–224. vol. 1. New York: VCH Publishers.

APPENDIX V
Methylene Blue Staining of Agarose Gels

DNA fragments separated on agarose gels are generally stained with ethidium bromide and then visualized on a UV transilluminator. Although very sensitive and effective, this procedure involves potential exposure to a mutagen (ethidium bromide) and use of hazardous UV light. An alternative technique is to stain agarose gels with methylene blue. This procedure is nonhazardous and you can visualize the fragments on a simple light box that is considerably less expensive than a UV transilluminator. These features make this technique useful for teaching applications where expense and minimizing exposure to hazardous materials are of concern. Limitations of the methylene blue technique, however, are that it is less sensitive (by about one-fifth) and requires longer staining and destaining times in order to visualize the DNA fragments.

Reagents

❑ DNA fragments separated on an agarose gel   ❑ distilled water
❑ methylene blue stain (0.025% aqueous solu-
   tion)

Supplies and Equipment

❑ gel staining tray   ❑ Polaroid camera with methylene blue filter
❑ white light box or overhead projector   ❑ type 667 film
❑ clear plastic wrap

➡ Procedure

1. Load DNA samples into an agarose gel and electrophorese until the desired separation is obtained.

2. After electrophoresis, remove the gel and transfer it to a staining tray. Flood the gel with a solution of 0.025% methylene blue. Wear rubber gloves during this step to avoid staining your hands.

3. Stain for 20 to 30 minutes. Gentle shaking will enhance staining.

4. Using a funnel, pour the methylene blue stain into the stain container. The stain may be reused numerous times.

5. Rinse the gel several times in distilled or tap water and allow it to destain while flooded with water. Gels may be observed in a few minutes, although better results are obtained after several hours or overnight destaining.

6. Cover the glass surface of the light box or overhead projector with plastic wrap (the methylene blue can stain the glass). Place the gel on the plastic wrap and turn on the light. DNA bands will stain dark blue against a lighter blue background.

7. Gels may be photographed using a handheld Polaroid camera with a hood (similar to those used to photograph ethidium bromide stained gels). Use the type 667 black-and-white Polaroid film and a methylene blue filter (available from FOTO-DYNE, Inc.). You will have to experiment with the exposure depending on the intensity of your light source, but a 1/30 second exposure with an aperture of f/32 is a good starting point. The area around the gel can be blocked out with dark paper to avoid a bright background around the photograph of the gel.

8. Following observation, gels may be discarded in the trash. Clean up any spilled methylene blue with a wet paper towel.

APPENDIX VI
Nucleic Acid Hybridization

Nucleic acid hybridizations are powerful tools used in many aspects of molecular biology (see Exercise 16). They are based on the fundamental characteristics of nucleic acid structure and chemistry. When double-stranded DNA molecules are heated to 90–100°C or subjected to very alkaline conditions (pH >12), the complementary base pairs that hold the two strands together are disrupted and the DNA **denatures**. If the solution is neutralized or is allowed to cool slowly, the complementary single strands of DNA will reform double helices in a process called renaturation, or **hybridization**. Hybridization reactions can be performed with any complementary single-stranded nucleic acid molecules (DNA:DNA, RNA:DNA, RNA:RNA) and have proven to be a very sensitive method for detecting specific nucleic acid sequences. Most hybridization procedures involve the use of a purified, single-stranded DNA or RNA fragment called a **probe**. The probe is complementary to the nucleic acid sequence (RNA or DNA) that you wish to detect (the **target** sequence). Due to the high binding specificity of complementary nucleic acid strands, probes are able to find a "needle in a haystack" of millions to billions of base pairs of DNA. Probes can be produced from naturally derived DNA up to several kilobases in length or can be chemically synthesized oligonucleotides (commonly 20 to 40 bases long). Because it is not possible to "see" a hybridization, probes must be labeled with a reporter group that allows visualization of hybridization events. The reporter group may be a radioactive compound (such as ^{32}P), an enzyme that produces a colored product, or a chemical group that fluoresces or produces chemiluminescence.

The hybridization exercises in this manual involve the use of one specific probe and predetermined hybridization conditions. However, there are an indefinite number of target sequences and probes that are used in hybridization techniques and each procedure must be individually optimized. The hybridization conditions are established based on a knowledge of the structure, chemistry, and interactions of nucleic acids as well as an understanding of the parameters affecting the specificity of hybridization. By controlling a variety of conditions, specific hybridization of the probe to the target of interest can be maximized while preventing nonspecific hybridization to nontarget sequences. Unfortunately, hybridization techniques involve more mythology and speculation than many other areas of molecular biology. This appendix attempts to clarify the underlying principles and to provide a rational approach to this important technique. In addition, the principles of nucleic acid hybridization relate to any process in which single-stranded nucleic acids combine with complementary base sequences. These include such common techniques as the polymerase chain reaction (PCR) and Sanger (dideoxy) DNA sequencing.

DNA Structure

In most biological systems, DNA exists in a double-stranded helix. Double-stranded DNA is stable under physiological conditions, but the strands are denatured in local regions during replication, transcription, and recombination. This separation is possible because the DNA strands are held together and stabilized by hydrogen bonds formed between the complementary bases on opposite strands and by interactions between the planar rings of adjacent bases located on the same strand (base-stacking).

Although the individual bonds and forces that hold complementary strands together are weak, the large number of interactions result in a stable molecule under physiological conditions. Complementary DNA strands are analogous to Velcro strips, where a single hook (representing a hydrogen bond) holds the strips together poorly, but a large number of hooks results in a strong binding of the two strips. Nucleic acid hybridization is based on manipulation of the weak forces that hold double-stranded nucleic acids together and also on the specificity of single-stranded nucleic acids for their complementary sequence. Because adenine nucleotides only bond with thymine, and guanine nucleotides only bond to cytosine, a single-stranded nucleic acid (probe) can bond perfectly only to a polynucleotide strand that has a complementary sequence (target sequence).

Because nucleic acids have a net negative charge due to the phosphates in the backbone of the molecule, complementary strands have a tendency to repel each other. Therefore, in order for complementary strands to hydrogen bond to form a double helix, these repelling charges must be shielded. This shielding is done by cations in solution that associate with the phosphate groups, effectively minimizing the charge repulsion.

Another factor preventing complementary strands of DNA from forming a double helix is heat. A phenomenon used to characterize DNA from different sources is the thermal melting point (T_m) of the molecule. When double-stranded DNA is heated, the strands will separate over a relatively narrow temperature range. The T_m is defined as the temperature at which half of the DNA molecules are in the single-stranded state and half are in the double-stranded state. Because guanine-cytosine (G-C) pairs are held together with three hydrogen bonds, they are stronger than adenine-thymine (A-T) pairs, which are held together with two hydrogen bonds. Therefore, the more G-C pairs in a DNA molecule, the higher the T_m will be. This relationship is so quantitative that the T_m is used to determine the G + C content of organisms. For a hybridization with a given probe and its target sequence, the T_m is always the same and can be used, in part, to determine the hybridization conditions (Sambrook, et al. 1990).

Hybridization Stringency

When performing a hybridization reaction, it is important to control the specificity of the probe for the target sequence. The level of specificity in a hybridization reaction is referred to as the **stringency**. If the stringency is high, the probe will hybridize to only perfectly matched target sequences, whereas under conditions of low stringency, some number of mismatched bases between the probe and target are tolerated. If a probe is exactly complementary to a target sequence, then high stringency conditions are used which minimizes background and detection of mismatched sequences. However, low stringency hybridizations are often used when it is not known if the probe is an exact match to the target. For example, if you prepared a probe complementary to the *lux*I gene of *Vibrio fischeri* and wished to probe other species of bacteria for a *lux*I gene, you would perform the hybridization at low stringency. This would allow the probe to hybridize to *lux*I genes in other organisms whose base sequence was similar but not identical to that of *V. fischeri.*

The stringency of a hybridization is adjusted by varying chemical and physical parameters that alter the hydrogen bonding allowing complementary nucleic acids to hybridize. Anything that reduces the ability of the bases to hydrogen bond will increase the stringency, whereas conditions that promote hydrogen bonding between bases will decrease stringency. From a practical standpoint, the stringency of hybridizations are increased by increasing the temperature, decreasing the ionic strength, or by adding denaturants that decrease the ability of the probe and target nucleic acid to hybridize.

As the temperature of a hybridization increases toward the T_m of a probe and its complementary target sequence, there will be less tendency of the probe to hybridize with sequences not exactly complementary to the probe. Thus, at temperatures near the T_m (high stringency), the probe should only be able to hybridize to the exact complementary sequence. The hydrogen bonding between nucleic acids can also be disrupted by decreasing the concentration of cations in solution. Decreasing the cations reduces the shielding of the negative charges that can repel complementary DNA strands and thus increases stringency. The cation concentrations are carefully controlled in hybridization reactions to adjust the stringency.

Compounds that form hydrogen bonds with the bases in single-stranded nucleic acids (denaturants) can also reduce the tendency of the bases to bond to each other in a complementary fashion, and thus increase stringency. Formamide is such a denaturant, and is often added to hybridization reactions to increase the stringency of the reaction because formamide competes for hydrogen binding sites with the complementary bases of the probe and target.

Hybridization Kinetics

Hybridization between a probe and target sequence occurs in two steps (see Figure A6.1). The first (and rate limiting) step is **nucleation**, where a short complementary region(s) (20–30 bases) of the probe and target collide. After the initial nucleation, a second step, called "**zippering**," causes rapid base pairing of the regions flanking the nucleation site in a fashion analogous to closing a zipper. The process of hybridization in solution is essentially a reaction in which the probe and target DNA are reactants. Therefore, the rate of hybridization is a function of the concentration of the probe and target DNA. Because the probe is often expensive and available in limited quantities, hybridizations are generally done in the smallest possible volumes to increase the concentration of probe and target. The effective concentration of the probe can also be increased by adding excluding agents such as dextran sulfate or polyethylene glycol. These reagents exclude water, which decreases the aqueous volume in the hybridization and can increase hybridization rates up to tenfold. However, they also increase the viscosity of the buffer, which slows the diffusion (and hence hybridization rate).

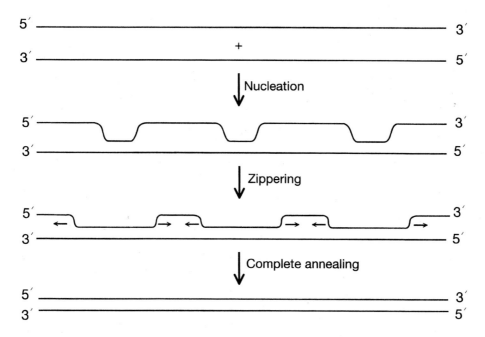

Figure A6.1
Steps in nucleic acid hybridization.

Several factors in addition to concentration affect the hybridization rate. The reaction rate is a function of temperature because increased temperature will speed up the movement of the reactants in solution and thereby increase the number of collisions between the probe and target sequences. Viscosity of the hybridization solution affects the rate because increased viscosity slows the movement of the DNA molecules and hence decreases the number of collisions of probe and target. The hybridization rate is also a function of the length of the DNA strands; as the probe length increases, the hybridization rate decreases. Short oligonucleotide probes (20–40 bp) hybridize very rapidly, providing a significant advantage compared to longer probes.

Hybridization Procedure

Preparation of target nucleic acids. The target in a hybridization may be genomic or plasmid DNA, DNA restriction fragments separated on an agarose gel, RNA (messenger or ribosomal), or even bulk nucleic acids isolated from natural habitats. In order to perform a hybridization, the target must be denatured (usually with alkali) to a single-stranded form. Most hybridization reactions in molecular biology are performed with the single-stranded target DNA or RNA immobilized on solid substrates such as a nitrocellulose or nylon membrane. Nylon has become popular in recent years because of its increased durability, better binding characteristics, and resistance to alkaline denaturants. The nucleic acid may be transferred to the membranes from an agarose gel by Southern blotting (for DNA), northern blotting (for RNA), or by directly loading the sample onto the membrane in spots using a "slot" or "dot" blotter.

Preparation of probes. As previously mentioned, probes may be either long (produced from biologically derived nucleic acids; usually >100 bp) or short (20–40 bp) synthetic oligonucleotides. The use of synthetic DNA probes has become more widespread with improvements in DNA synthesis technology and decreased cost. Oligonucleotide probes have several advantages over large, naturally derived sequences. The hybridization is faster due to their short length, increased diffusion, and because they can be used in higher concentrations. They are synthesized as single-stranded molecules, eliminating the need for denaturization prior to use. Finally, the stringency can be more tightly controlled with short oligonucleotide probes because they contain only the sequence of interest. A disadvantage of oligonucleotide probes is that the sequence of interest must be known in order to have the probe synthesized.

Long probes are usually associated with their complementary strand and must be denatured (with alkali or heat) prior to use. Because it is inconvenient to remove the strand complementary to the probe, this strand competes with the target DNA for binding of the probe during the hybridization reaction. This problem can be eliminated by synthesizing single-stranded RNA probes from DNA cloned in vectors such as the pGEM™ series, which have promoters allowing transcription of the cloned DNA. Controlling the stringency of a hybridization is difficult when using long probes because they essentially behave like a series of short probe sequences, each with its own optimum hybridization conditions.

All probes must be labeled with a reporter to allow detection. There are a variety of ways to label nucleic acid probes with both radioactive and nonradioactive reporters. Labeling methods for long probes often use polymerases to incorporate nucleotides that have been modified to contain a radioactive or nonradioactive label. Short oligonucleotide probes are commonly end-labeled with ^{32}P by kinase reactions. Nonradioactive reporters are added to short probes by inserting a modified base in the oligonucleotide during synthesis. The reporter (enzyme or chemiluminescent chemical) is then conjugated to the modified base.

Prehybridization. Prior to addition of the probe, membranes containing immobilized single-stranded target sequences are treated in a prehybridization step to block nonspecific probe binding sites. Prehybridization buffers contain one or more materials that bind tightly to the membrane to block regions on the membrane where target DNA has not bound. Blocking agents include nucleic acids (salmon sperm DNA, tRNA, and so on) or proteins (bovine serum albumin, skim milk, or even Bailey's Irish Cream!). This step is essential prior to performing a hybridization because the probe, being a nucleic acid, will also bind to the membrane, resulting in a hybridization signal from the entire membrane rather than the discrete sites where the target DNA is located. Prehybridization solutions also frequently contain detergents such as sodium dodecyl sulfate (SDS) that aid in wetting the membrane and preventing nonspecific binding of the probe.

Prehybridization lowers the background signal during detection and thus increases the sensitivity of the detection of target sequences that hybridize with the probe. Prehybridization is usually conducted for 30 minutes to several hours using the same buffer (minus the probe) and the same temperature as the subsequent hybridization step.

Hybridization. In this step, the probe binds to the target sequences immobilized on the membrane. In most instances, the probe is simply added to the prehybridizaton buffer. The volume of hybridization buffer is kept as small as possible (usually 100 µl/cm² of membrane area) to maximize the probe concentration, which will increase the hybridization rate. Most hybridization buffers have a high concentration of cations (to prevent intrastrand charge repulsion) and may contain formamide (to moderately increase the stringency of the reaction). Hybridizations are best done at elevated temperatures (20-25°C below the calculated T_m for long probes and 5-10°C below the T_m for short probes) to increase the hybridization rate and to further increase stringency. In general, hybridizations should be conducted in the shortest possible time using the minimum amount of probe required to limit the background resulting from nonspecific probe binding. Traditionally, many molecular biologists have used long hybridizations (overnight at room temperature), which may actually limit sensitivity and increase background.

Washes. Although the hybridization step is done at moderately high stringency, the wash steps provide the stringency necessary for detection of only the target sequence. After hybridization, the hybridization buffer is removed from the membrane (this may be saved and reused for subsequent hybridizations). The washes are then done in a series of steps at increasing stringency to remove the probe from all but the target sequence(s). In general, this is accomplished by increasing the temperature and lowering the concentration of cations in the wash solutions. The concentration of detergents is often increased to aid removal of nonspecifically bound probe from the membranes. The wash steps are generally performed at 3-20°C below the T_m of the probe, but this temperature should be determined empirically for each probe.

Detection of Hybridization Signal. The method of detecting probe bound to the target sequence is dependent upon the method used to label the probe sequence. If the probe is labeled with a radioisotope, autoradiography is used to produce an image of the membrane on X-ray film. Radioactive probes also allow detection of very weak hybridization signals because the exposure to X-ray film can be extended until the desired signal is obtained. Use of radioactive probes also allows the membrane to be exposed to film between wash steps of increasing stringency to determine which wash conditions (stringency) give the best signal to background ratios. If nonradioactive labeled probes are used, the conditions used in the wash steps must be optimized in separate experiments because the detection step often prohibits subsequent detection after additional wash steps. Nonradioactive probes are commonly linked to enzymes, and the detection is based on an enzymatic reaction that produces a colored precipitate that remains bound to the membrane (such as in the alkaline phosphatase conjugated probe used in this manual). Alternately, other reporters produce chemiluminescence, which can be detected by autoradiography.

Literature Cited

De Ley, J., H. Cattoir, and A. Reynaerts. 1970. The quantitative measurement of DNA hybridization from renaturation rates. *European J. Biochem.* 12:133.

Maas, R. 1983. An improved colony hybridization method with significantly increased sensitivity for detection of single genes. *Plasmid* 10:296.

Mienkoth, J., and G. Wahl. 1984. Hybridization of nucleic acids immobilized on solid supports. *Anal. Biochem.* 138:267.

Southern, E. 1975. Detection of specific sequences among DNA fragments separated by gel electrophoresis. *J. Mol. Biol.* 98:503.

Wetmur, J. G., 1975. Acceleration of DNA renaturation rates. *Biopolymers* 14:2517.

Wetmur, J. G. and N. Davidson. 1968. Kinetics of renaturation of DNA. *J. Mol. Biol.* 31:349.

APPENDIX VII
Alcohol Precipitation of Nucleic Acids

In many of the exercises used in this course, nucleic acids are precipitated with ethanol or isopropanol. These procedures are widely used in molecular biology and work equally well for DNA or RNA. Alcohol precipitations are valuable because they allow the nucleic acids to be concentrated by removing them from solution as an insoluble pellet. The pellet can then be resuspended in a buffer of choice. These precipitations also provide a convenient method to remove salt, nucleotide and protein contaminants, and allow one to change the composition of the buffer in which the nucleic acids are suspended.

Precipitation of nucleic acids occurs in two steps: (i) formation of the precipitate; and (ii) collection (or pelleting in a centrifuge). There are a number of factors that affect the success of a precipitation. First, there must be at least a 0.2 M concentration of a monovalent cation present to shield the high negative charge of the phosphate present in nucleic acids. Without this shielding, the intrastrand repulsion is too great for the nucleic acids to aggregate and form a precipitate. Sodium acetate, sodium chloride, lithium chloride, or ammonium acetate are commonly used for this purpose in alcohol precipitations. The salts are commonly made in 3–10 M stock solutions that are diluted in the nucleic acid solution to give the desired final concentration (see Table A7.1). Sodium acetate is commonly used for routine DNA precipitations, but other salts have advantages under certain conditions. Ammonium acetate is used to reduce the co-precipitation of free nucleotides (dNTPs). This salt should not be used, however, if the DNA is to be phosphorylated by polynucleotide kinase, as this enzyme is inhibited by ammonium ions. Lithium is very soluble in ethanol solutions, and lithium chloride is used when precipitating RNA where increased concentrations of ethanol are used. Sodium chloride can be used if the sample contains SDS, as NaCl allows the detergent to remain soluble in ethanol solutions.

TABLE A7.1 Salt Solutions Commonly Used to Precipitate Nucleic Acids

Salt	Concentration of stock (M)	Final concentration (M)
Ammonium acetate	10.0	2.5
Lithium chloride	8.0	0.8
Sodium chloride	5.0	0.2
Sodium acetate (pH 5.2)	3.0	0.3

Modified from Sambrook, J, E F. Fritch, and T Maniatis. 1989 *Molecular cloning: A laboratory manual,* 2d. ed, p E 11. New York: Cold Spring Harbor Laboratory Press

After the addition of the salt, 2 volumes of cold ethanol are added to precipitate the nucleic acid (2.5 volumes are used for RNA precipitations). Isopropanol may also be used to precipitate nucleic acids in place of ethanol. When isopropanol is used, the volume is decreased to between 0.6 and an equal volume of alcohol. This greatly reduces the volume of the precipitation reaction, which is advantageous if the capacity of the tubes being used is limited. Isopropanol, however, results in more protein contamination of the nucleic acid solution and is, therefore, generally used with fairly purified nucleic acid preparations.

A second factor affecting nucleic acid precipitations is the temperature during the precipitation step. In the past, it was commonly thought that precipitations occurred best at low temperatures (-20 or $-70°C$), although the recovery at these temperatures is actually much less than at $0°C$ or room temperature (see Table A7.2). It is now recommended that precipitations be done in a range from $0°C$ (on ice) to room temperature for optimal recovery.

The third factor affecting alcohol precipitations is the concentration of the nucleic acids. Precipitation occurs effectively within 10 minutes or less in solutions with concentrations >0.5 µg/ml (see Table A7.2). But as the concentration of nucleic acid in the sample decreases, the incubation time should be increased up to 30 minutes or even overnight for very low concentrations (<0.05 µg/ml). Precipitations of very low concentrations of nucleic acids can also be increased by the addition of a carrier such as RNA or glycogen.

TABLE A7.2 Effect of Time and Temperature on Ethanol Precipitation with Ammonium Acetate

| | Percent DNA recovered | | | | | | | | | | | | | | | |
| | -70°C | | | | -20°C | | | | 0°C | | | | 22°C | | | |
DNA concentration	0 min	10 min	30 min	Over-night	0 min	10 min	30 min	Over-night	0 min	10 min	30 min	Over-night	0 min	10 min	30 min	Over-night
5 µg/ml	85	80	89	91	87	78	91	96	88	94	94	96	88	97	93	100
0.5 µg/ml	62	46	52	50	57	52	65	83	60	58	63	98	62	64	65	92
0.05 µg/ml	28	29	30	32	35	33	49	69	36	33	38	92	47	40	36	87
0.005 µg/ml	25	27	38	33	41	38	49	72	37	33	39	86	40	35	38	85

From Crouse, J, and D Amorese 1987 Ethanol precipitation: Ammonium acetate as an alternative to sodium acetate *BRL Focus* 9(2):3. Reprinted by permission

A final factor affecting recovery in nucleic acid precipitations is the time and speed of centrifugation. For centrifugations performed in a microcentrifuge, nucleic acids are routinely collected at the maximum speed of the centrifuge (12,000–14,000 $\times$ g) at a temperature between $4°C$ and ambient. Ten-minute centrifugations give good recoveries of nucleic acid if the concentration of DNA is high (>0.5 µg/ml). If the concentration of nucleic acid is low or the size is small (<100 bp), longer centrifugations (30 to 60 minutes) will enhance recovery. Alternatively, centrifugations may be carried out in an ultracentrifuge to generate higher g forces. Many newer ultracentrifuges have rotors or adapters that will accommodate microcentrifuge tubes. For the precipitations used in this course, the nucleic concentrations are generally high, which allows precipitated nucleic acids to be collected by 10-minute centrifugations in a standard microcentrifuge.

Following the precipitation of nucleic acids, the pellet is commonly washed with a solution of 70–80% ethanol followed by another centrifugation. This removes traces of salts from the nucleic acid, but care must be taken to prevent loss of the pellet (which is often invisible). The following is a standard procedure for precipitating relatively high concentrations of DNA from an aqueous solution:

1. Measure the volume of DNA solution to be precipitated. If it is greater than 400 μl, divide the solution between as many microcentrifuge tubes as needed.

2. Add 0.1 volumes of 3 M sodium acetate (pH 5.2) and mix thoroughly by vortexing. Other salts may be used if desired.

3. Add 2.0 volumes of ice-cold ethanol and mix thoroughly. Store on ice or at room temperature for 10 minutes to precipitate the DNA.

4. Centrifuge for 10 minutes in a microcentrifuge at full speed (12,000–14,000 × g). Be sure the hinge on the microcentrifuge tube is pointing out, as you may not be able to see the pellet.

5. Carefully decant the supernatant and remove the last traces with a micropipetter. Be careful not to disrupt the pellet with the pipet tip (the pellet will be along the side of the tube below the hinge and may be invisible).

6. Add 0.5–1.0 ml of 70% ethanol and **gently** rock the tube once to rinse the pellet. Centrifuge again for 5 minutes in a microcentrifuge at full speed.

7. Remove the supernatant as described in step 5. Set the tube open on your bench for 10 to 15 minutes until the last traces of ethanol are removed (as indicated by smell).

8. Dissolve the pellet (which is often invisible) in the desired volume of TE buffer. Run the buffer along the side of the tube where the pellet is to ensure that the DNA is dissolved.

Literature Cited

Crouse, J., and D. Amorese. 1987. Ethanol precipitation: Ammonium acetate as an alternative to sodium acetate. *BRL Focus* 9(2):3–5.

Sambrook, J., E. F. Fritch, and T. Maniatis. 1989. *Molecular cloning: A laboratory manual.* 2d ed. New York: Cold Spring Harbor Laboratory Press.

Zeugin, J. A., and J. L. Hartley. 1985. Ethanol precipitation of DNA. *BRL Focus* 7(4):1–2.

APPENDIX VIII
Care and Handling of Enzymes

Restriction endonucleases (restriction enzymes) are essential tools in molecular biology because of their ability to recognize and cut specific DNA sequences. Unlike the early days of the recombinant DNA era (1970s and 1980s), restriction enzymes are now commercially available in a highly purified and stable form. Restriction enzymes (and several other enzymes used in molecular biology) are concentrated to enhance stability during storage, and small volumes (usually 1 µl) are used in individual digestions. This results in one tube of enzyme potentially being entered as much as several hundred times, providing ample opportunity for contamination (with nucleases) or inactivation (by warming) of an enzyme preparation. Because a large number of people generally use a centralized enzyme stock, it is important that all users handle the enzymes properly to prevent contamination and inactivation. A number of guidelines concerning the usage of enzymes are listed below.

1. Always wear protective gloves when handling the **stock** tubes of enzymes (gloves are not necessary to handle tubes containing small amounts aliquoted for use in setting up digestions). The gloves are to protect the enzyme from you and not vice versa. Your skin is covered with nucleases that, if introduced into an enzyme preparation, can cause nonspecific degradation of the DNA you are attempting to cut with the restriction enzyme.

2. Be sure the tube of enzyme is briefly centrifuged (for 2 to 3 seconds) before opening. The enzyme may stick to the cap and be lost or contaminated. We recommend centrifuging tubes when they are received from the supplier and storing them upright in the freezer to eliminate the need for repeated centrifugation. Use a refrigerated microcentrifuge if possible (or one in a cold room).

3. Exercise extreme caution to avoid contact of the inside of the cap and tube with anything (fingers, desktop, floor, and so on).

4. Temperature variation is the worst enemy of an otherwise properly handled enzyme preparation, and the range and duration of temperature shifts should be minimized. When removing enzymes from the freezer, do not hold the vial at or below the level of the enzyme—the heat from your fingers will rapidly warm the enzyme preparation. When removing enzyme stocks from the freezer for use, place them directly on ice or preferably in a microcooler that can maintain $-20°C$ (microcoolers are stored in a $-20°C$ freezer and are able to maintain $-20°C$ for several hours at room temperature). Return enzymes to the freezer **immediately** after use.

   DNA metabolizing enzymes, such as T4 DNA ligase, DNA polymerases, T4 polynucleotide kinase, and reverse transcriptase, are particularly temperature sensitive, and extra care should be taken to minimize warming of these enzymes.

5. Use a fresh, sterile pipet tip every time you dispense enzyme from the stock tube. Even if the pipet was used only to transfer enzyme to a sterile tube, discard the tip; pipet tips are cheaper than enzymes and the labor of your coworkers.

6. If you need multiple aliquots of the same enzyme, remove the total amount needed (plus a small excess) at one time, and immediately return the enzyme stock to the freezer.

7. When possible, use digestion conditions recommended by the manufacturer (type of buffer, incubation temperature, and so on). Enzyme manufacturers supply 10X restriction buffer with each enzyme that provides optimal reaction conditions when diluted to a 1X concentration. Even though enzymes can function under a range of conditions, their efficiency is often markedly reduced at altered salt concentrations or pH. The catalogs from enzyme suppliers contain tables indicating the activity of enzymes in different buffers. Information from these tables will enable you to determine whether multiple enzyme digests can be done in a single buffer with sufficient activity and specificity.

8. Most enzyme preparations are supplied as a 50% glycerol solution. This allows them to be stored at −20°C without forming ice crystals. Avoid storing any enzyme below this temperature unless the supplier recommends otherwise. Temperatures below −20°C result in formation of ice crystals, which may inactivate some enzymes.

   Note that most freezer compartments in refrigerators do not maintain −20°C. The actual temperature varies widely (especially in frost-free freezers that go through daily heating cycles to maintain the frost-free condition). This can have a detrimental effect on the activity of the enzyme over time. Use of a non–frost-free freezer is highly recommended.

9. Glycerol affects the specificity and activity of most enzymes. Therefore, enzymes should be diluted to give a glycerol concentration of 5% or less (i.e., the volume of enzyme added should not exceed 10% of a total reaction volume).

10. Certain restriction enzymes may exhibit "star" activity under certain conditions. Star activity refers to an enzyme recognizing and cutting base sequences other than its normal recognition sequence. For example, *Eco*R I has a recognition sequence of 5' GAATTC 3' but, at high enzyme concentrations, it will also recognize and cut at the sequence 5' AATT 3'. Star activity can be induced by low ionic strength (i.e., from an improper buffer), high concentration of glycerol or enzyme, or elevated pH. Catalogs from enzyme suppliers provide tables indicating which enzymes are prone to star activity and under what conditions.

11. Enzyme preparations in 50% glycerol are very viscous, and a micropipetter can be inaccurate in measuring enzymes, especially when operated quickly. Only the very end of the pipet tip should touch the surface of the enzyme preparation, and the plunger on the pipetter should be released slowly. Submersing the tip in the enzyme preparation will result in the transfer of more enzyme than intended, which wastes enzyme and may inhibit activity or induce star activity.

12. When preparing a digest, **always add the enzyme last** so that it is placed directly into the correct concentration of buffer.

13. The most reliable method to inactivate restriction enzymes following a digest is by extraction with phenol and/or chloroform. Some enzymes can be inactivated by heat (65°C), which is less time-consuming and avoids the use of toxic solvents. Catalogs from enzyme suppliers provide tables indicating which enzymes are heat-inactivatable.

APPENDIX IX
Restriction Endonucleases

The following table includes a sampling of restriction endonucleases, their source, and recognition and cleavage sites. The list includes all enzymes used or referenced in this manual but is a small fraction of the more than 2500 restriction endonucleases identified to date. More complete lists are available in the catalogs from any supplier of enzymes (see Appendix XVII). These catalogs also contain a wealth of information on the conditions for digestions with the enzymes, conditions for doing simultaneous digests with multiple enzymes, factors that interfere with enzyme activity, and so on.

Note the following in interpreting the table:

- Two restriction enzymes that recognize the same sequence are referred to as isoschizomers. For enzymes in the table that have isoschizomers, the isoschizomer is listed in parentheses.

- The recognition sequences for each enzyme are written from 5' → 3' with only one strand given. The point of cleavage is indicated by an arrow (↓). For example, G↓TCGAC is an abbreviation for:

    5' G ↓ T C G A C 3'
    3' C A G C T ↑ G 5'

    Note that various enzymes have recognition sequences ranging from 4 to 8 bases long. Some enzymes cleave at the axis of symmetry (resulting in blunt ends), whereas others cleave asymmetrically, resulting in single-stranded complementary ends ("sticky ends").

- The enzyme *Fok* I is a type III restriction enzyme and cleaves away from its recognition sequence. The sites of cleavage are indicated in parentheses following the recognition sequence. For *Fok* I, GGATG (9/13) indicates cleavage as shown below, where N represents any nucleotide (A, C, G, or T):

    5' G G A T G N N N N N N N N N ↓ 3'
    3' C C T A C N N N N N N N N N N N N N ↑ 5'

- Bases appearing in parentheses signify that either base may occupy that position in the recognition sequence. For example, *Acc* I [GT↓(A/C)(T/G)AC] cleaves the sequences GT↓AGAC, GT↓ATAC, GT↓CGAC, and GT↓CTAC.

- Pu and Py indicate that either purine (A, G) or either pyrimidine (C, T), respectively, may occupy that position in the recognition sequence. For example, *Hinc* II (GTPy↓PuAC) cleaves the sequences GTC↓GAC, GTC↓AAC, GTT↓GAC, and GTT↓AAC.

Enzyme	Microorganism	Sequence	# cleavage sites	
			Lambda	pBR322
Acc I	Acinetobacter calcoaceticus	GT↓(A/C)(T/G)AC	9	2
Apa I	Acetobacter pasteurianus	GGGCC↓C	1	0
Ase I	Aquaspirillum serpens	AT↓TAAT	17	1
Ava I	Anabaena variabilis	C↓PyCGPuG	8	1
BamH I	Bacillus amyloliquefaciens H	G↓GATCC	5	1
Bcl I	Bacillus caldolyticus	T↓GATCA	8	0
Bgl II	Bacillus globigii	A↓GATCT	6	0
BssH II	Bacillus stearothermophilus H3	G↓CGCGC	6	0
BstE II	Bacillus stearothermophilus ET	G↓GTNACC	13	0
Dra I	Deinococcus radiophilus	TTT↓AAA	13	3
EcoR I	Escherichia coli RY 13	G↓AATTC	5	1
EcoR V	Escherichia coli J62 pLG74	GAT↓ATC	21	1
Fok I	Flavobacterium okeanokoites	GGATG (9/13)	149	12
Fse I	Frankia species	GGCCGG↓CC	0	0
Hha I	Haemophilus haemolyticus	GCG↓C	215	31
Hinc II	Haemophilus influenzae Rc	GTPy↓PuAC	35	2
Hind III	Haemophilus influenzae Rd	A↓AGCTT	7	1
Kpn I	Klebsiella pneumoniae OK8	GGTAC↓C	2	0
Mbo I (Sau3A I)	Moraxella bovis ATCC 10900	↓GATC	116	22
Nae I	Nocardia aerocolonigenes	GCC↓GGC	1	4
Nar I	Nocardia argentinensis	GG↓CGCC	1	4
Nde I	Neisseria denitrificans	CA↓TATG	7	1
Nhe I	Neisseria mucosa heidelbergensis	G↓CTAGC	1	1
Not I	Nocardia otitidis-caviarum	GC↓GGCCGC	0	0
Pvu I	Proteus vulgaris	CGAT↓CG	3	1
Pvu II	Proteus vulgaris	CAG↓CTG	15	1
Pst I	Providencia stuartii 164	CTGCA↓G	28	1
Sac I (Sst I)	Streptomyces achromogenes	GAGCT↓C	2	0
Sal I	Streptomyces albus G	G↓TCGAC	2	1
Sca I	Streptomyces caespitosus	AGT↓ACT	5	1
Sau3A I (Mbo I)	Staphylococcus aureus 3A	↓GATC	116	22
SexA I	Streptomyces exfoliatus	A↓CC(A/T)GGT	5	0
Sfi I	Streptomyces fimbriatus	GGCCNNNN↓NGGCC	0	0
Sma I	Serratia marcescens	CCC↓GGG	3	0
SnaB I	Sphaerotilus natans	TAC↓GTA	1	0
Sph I	Streptomyces phaeochromogenes	GCATG↓C	6	1
Sst I (Sac I)	Streptomyces stanford	GAGCT↓C	2	0
Stu I	Streptomyces tubercidicus	AGG↓CCT	6	0
Xba I	Xanthomonas badrii	T↓CTAGA	1	0
Xho I	Xanthomonas holcicola	C↓TCGAG	1	0

A P P E N D I X X
Enzymes Used in Molecular Biology

Type II Restriction Endonucleases (Restriction Enzymes)

These enzymes catalyze the symmetric cleavage of double-stranded DNA at specific palindromic base sequences. When cleavage occurs at the axis of symmetry, blunt ends are generated; when cleavage occurs on opposite sides of the axis of symmetry, fragments are created which have protruding single-stranded termini that are often called "sticky ends." Depending on the enzyme used to cleave the DNA, the sticky end can have a 3' or a 5' overhang. Most restriction enzymes have a recognition sequence of 4, 5, or 6 bases. The size of the enzyme's recognition sequence will influence the frequency at which DNA is cut. The sequence for a 4 base pair recognition site occurs more frequently than a sequence for a 6 base pair recognition site (1 in 4^4 versus 1 in 4^6). A 5'-P and a 3'-OH are produced at the cleavage site; these ends can serve as substrates for DNA ligase. A large number of highly purified restriction enzymes are commercially available. The discovery of isoschizomers (a restriction enzyme that recognizes and cleaves the same base sequence as another enzyme) has resulted in the replacement of certain problematic enzymes with isoschizomers with improved performance at a reduced cost.

DNA Methylases

Type II restriction/modification systems are comprised of a restriction endonuclease and a DNA methylase that recognizes and methylates the same base sequence. This modification of the recognition sequence prevents the host organism from cutting its own DNA. Methylation involves the transfer of a methyl group from S-adenosyl methionine to a specific base within the recognition and cleavage sequence. Prokaryotes often modify the N^6 of adenine and guanine, the N^4 or C^5 of cytosine, or the C^5 of thymine. In molecular cloning, methylases are used to protect specific restriction sites within a DNA that is to be cloned by the addition of synthetic linkers containing the specific recognition sequence. After unmethylated linkers are ligated to the methylated DNA, the construct can be treated with a restriction enzyme. The internal (methylated) restriction sites will not be cleaved, but the linkers will be, allowing specific sticky ends to be produced. There are a number of DNA methylases commercially available.

T4 DNA Ligase

This enzyme is derived from the *E. coli* bacteriophage T4. T4 DNA ligase catalyzes the formation of a phosphodiester bond between adjacent 3'-OH and 5'-P termini with the hydrolysis of ATP to AMP + PP_i. The substrate for this enzyme is double-stranded DNA, RNA, or DNA:RNA heteroduplexes. This enzyme can ligate nucleic acids that have either cohesive or blunt ends.

T4 Polynucleotide Kinase

Polynucleotide kinase catalyzes the transfer of the γ-phosphate of ATP to a 5'-OH of RNA or DNA. T4 polynucleotide kinase also has 3'-phosphatase activity, and it can catalyze an exchange reaction with a 5'-P of an oligo- or polynucleotide, ATP and ADP. This allows a phosphorylated DNA to be radiolabeled with ^{32}P without performing a separate reaction to dephosphorylate the nucleic acid prior to labeling. T4 polynucleotide kinase is commonly used for: (i) end-labeling of probes with ^{32}P; (ii) labeling 5' termini of DNA fragments for use in Maxam and Gilbert DNA sequencing; and (iii) phosphorylating synthetic linkers and other nucleic acids prior to ligation.

Alkaline Phosphatase (Bacterial [BAP]; Calf Intestinal [CIP])

Alkaline phosphatase catalyzes the removal of the 5'-P from DNA, RNA, NTPs and dNTPs. It is used for: (i) removal of 5'-P prior to end-labeling with polynucleotide kinase; and (ii) to remove 5'-P from DNA fragments or vectors to prevent self-ligation. It is also often conjugated to probes to allow detection (see Exercises 16, 21, 24).

DNA Polymerase I (Holoenzyme)

This enzyme, derived from *E. coli,* catalyzes the polymerization of nucleotides into duplex DNA in the 5'→ 3' direction on a primed, single-stranded DNA template. It also has 5'→ 3' and 3'→ 5' exonuclease activity, and is capable of nick-translation. It is an accurate but weakly processive enzyme that requires Mg^{2+} and deoxyribonucleoside triphosphates. It is used to radiolabel double-stranded DNA by nick-translation. This enzyme was originally used for second-strand synthesis in cDNA cloning, but it has been replaced by a number of other polymerases.

Klenow Fragment of DNA Polymerase I

This enzyme is produced by cleaving *E. coli* DNA polymerase I with the proteolytic enzyme subtilisin or through the genetic engineering of DNA polymerase I. As a result of these modifications, the 5'→ 3' exonuclease activity is removed but the 5'→ 3' polymerase and 3'→ 5' exonuclease (proofreading) activities remain. This enzyme is commonly used for: (i) creating blunt-ended DNA by filling in the 3'-recessed termini produced by restriction digests of DNA; (ii) *in vitro* labeling of the termini of DNA fragments possessing 5' tails by using [^{32}P]dNTPs; (iii) second-strand synthesis in cDNA cloning; (iv) removal of protruding 3' single-stranded termini created by some restriction enzymes; (v) labeling DNA by random priming; and (vi) historically, sequencing by the Sanger dideoxy method (other, more effective enzymes are currently used).

T4 DNA Polymerase

This enzyme is derived from the *E. coli* bacteriophage T4. Like the Klenow fragment of DNA polymerase I, this DNA-dependent DNA polymerase has $5' \rightarrow 3'$ polymerase activity and a $3' \rightarrow 5'$ exonuclease activity (proofreading) but lacks the $5' \rightarrow 3'$ exonuclease activity. However, the $3' \rightarrow 5'$ exonuclease activity is about 200 times greater than that of DNA polymerase I. This enzyme is used for: (i) filling in recessed 3' termini; (ii) labeling the termini of DNA fragments with protruding 5' ends by an exchange reaction, (iii) digesting the single-stranded 3' end of DNA to form a blunt end; (iv) extension of primers on a single-stranded template for site-directed mutagenesis; and (v) labeling long probes by replacement synthesis.

Avian Myeloblastosis Virus RNA-Dependent DNA Polymerase (AMV Reverse Transcriptase)

This retroviral enzyme, isolated from purified AMV, contains two polypeptides that carry a $5' \rightarrow 3'$ polymerase activity, a $5' \rightarrow 3'$ ribonuclease activity, and potent RNase H activity. In addition to being an RNA-dependent DNA polymerase, this enzyme is also a DNA-dependent DNA polymerase. The polymerase activities require a primer, single-stranded template, and Mg^{+2}. This enzyme is commonly used for: (i) first- and second-strand synthesis of cDNA for molecular cloning; (ii) Sanger dideoxy sequencing of RNA and DNA templates; and (iii) labeling the termini of DNA fragments by filling in 5' overhangs.

Moloney-Murine Leukemia Virus RNA-Dependent DNA Polymerase (MMLV Reverse Transcriptase)

This retroviral enzyme has been cloned and expressed in *E. coli*. It can synthesize a DNA strand from a primed, single-stranded RNA or DNA template. However, it has a much lower RNase H activity than AMV-RT and, as a result, is the enzyme of choice for cDNA synthesis because it can produce cDNAs up to 10 kb. Other uses include: (i) synthesis of cDNA for Reverse Transcriptase Polymerase Chain Reaction; (ii) Sanger dideoxy sequencing of DNA templates; and (iii) labeling the termini of DNA fragments by filling in 5' overhangs.

T7 DNA Polymerase

This enzyme, derived from *E. coli* bacteriophage T7, is a tight complex of the polymerase and thioredoxin. It is a fast, highly processive DNA polymerase ($5' \rightarrow 3'$) that also has a potent $3' \rightarrow 5'$ exonuclease but lacks a $5' \rightarrow 3'$ exonuclease activity. It uses primed, single-stranded DNA as a template. It is commonly used for: (i) oligonucleotide-directed *in vitro* mutagenesis and other uses that require copying long stretches of DNA template; (ii) end-labeling DNA by exchange or fill-in reactions; and (iii) labeling large DNA fragments by random priming.

The potent $3' \rightarrow 5'$ exonuclease activity of T7 DNA polymerase has been removed by oxidation with iron and genetic engineering in **Sequenase**™ versions 1.0 and 2.0, respectively. This enzyme, which is patented by US Biochemical, is the best enzyme available for medium-temperature Sanger dideoxy sequencing of DNA templates.

Taq DNA Polymerase

This thermostable DNA-dependent DNA polymerase was initially isolated and purified from the extreme thermophile *Thermus aquaticus* but has since been cloned. The enzyme possesses a $5' \to 3'$ polymerase activity, a $5' \to 3'$ exonuclease activity, but no $3' \to 5'$ exonuclease activity. It requires a primed, single-stranded DNA template and Mg^{+2}. Because this enzyme is stable at 95°C and has optimal polymerization activity at 80°C, its most frequent use is in the polymerase chain reaction. *Taq* DNA polymerase is also used for DNA sequencing by cycling the reaction through a PCR containing a label (radioactive or nonradioactive) and dideoxyribonucleotides. This enzyme is also used for DNA sequencing at a single, elevated temperature; this can be especially useful when sequencing templates that contain significant levels of secondary structure.

Various engineered preparations of *Taq* polymerase are available through Perkin Elmer. These enzymes have been designed for different purposes. The most significant change is the production of the Stoffel fragment, which contains the $5' \to 3'$ polymerase activity but lacks the $5' \to 3'$ exonuclease activity. This modification also results in a more thermostable enzyme that is useful for performing PCR with a high number of cycles.

Pfu DNA Polymerase

This thermostable DNA-dependent DNA polymerase is derived from the extreme thermophile *Pyrococcus furiosus*. This enzyme possesses a $5' \to 3'$ polymerase, a $3' \to 5'$ exonuclease, but no $5' \to 3'$ exonuclease activity. It has greater fidelity, thermal stability, and a slightly lower optimum polymerization temperature than *Taq* DNA polymerase; hence it is marketed as an alternative enzyme for use in PCR.

Vent™ DNA Polymerase

This thermostable DNA-dependent DNA polymerase is derived from the extreme thermophile *Thermococcus litoralis*. It possesses a $5' \to 3'$ polymerase, a $3' \to 5'$ exonuclease, but no $5' \to 3'$ exonuclease activity. It has greater thermostability and fidelity than *Taq* DNA polymerase but a lower fidelity than *Pfu* DNA polymerase. This enzyme is marketed as an alternative for use in PCR. A derivative of this enzyme that lacks the $3' \to 5'$ exonuclease can be used for high-temperature and PCR cycle-sequencing of DNA.

Deep Vent™ DNA Polymerase

This high-fidelity thermophilic DNA polymerase is purified from *E. coli* carrying the Deep Vent™ DNA polymerase gene from a *Pyrococcus* species. This organism was isolated from a submarine thermal vent 2010 meters below the ocean surface and can grow at 104°C. The half-life of this enzyme at 95°C is 23 h (in comparison to 1.6 h for *Taq* polymerase), and it is used for high-temperature DNA sequencing and thermal-cycle sequencing. A commercially available form of this enzyme has been genetically engineered to eliminate the 3'→ 5' exonuclease activity.

Terminal Deoxynucleotidyl Transferase (Terminal Transferase)

Terminal transferase is a DNA polymerase derived from calf thymus that catalyzes the repetitive, template-independent addition of dNTPs onto the 3'-OH of a DNA or RNA molecule. The enzyme works best on molecules with a 3' protrusion and does not appear to possess nuclease activity. Uses include: (i) the addition of complementary homopolymeric tails to the 3' end of DNA cloning vectors and cDNA; and (ii) radio-labeling the 3' termini of nucleic acids fragments with [α-^{32}P]dNTPs.

SP6 RNA Polymerase

This is a DNA-dependent RNA polymerase isolated from *Salmonella typhimurium* cells infected with bacteriophage SP6. Transcription is initiated from double-stranded DNA templates that contain SP6-specific promoter sequences. Cloning vectors have been developed that direct transcription from phage-specific (e.g. SP6) promoter sequences that flank the multiple cloning sites of these vectors. Uses of these vectors in conjunction with SP6 RNA polymerase include: (i) the generation of single-stranded RNA probes from sequences cloned into plasmids containing the SP6 promoter adjacent to the multiple cloning site; (ii) the production of biologically active mRNA from *in vitro* transcription reactions; (iii) intron and exon mapping of genomic DNA; (iv) expression of cloned genes in bacterial and yeast hosts through the addition of an SP6 RNA polymerase gene into a host that contains the sequence of interest cloned into a vector containing an SP6 RNA polymerase promoter sequence; and (v) *in situ* production of RNA for antisense studies.

T7 RNA Polymerase

This enzyme is a DNA-dependent RNA polymerase isolated from *E. coli* cells infected with bacteriophage T7. Transcription is initiated from double-stranded DNA templates that contain T7-specific promoter sequences. T7 promotors have also been used in cloning vectors that direct transcription from phage-specific promoters that flank the multiple cloning sites. Uses of these vectors in conjunction with T7 RNA polymerase include: (i) the generation of single-stranded RNA probes; (ii) the production of biologically active mRNA from *in vitro* transcription reactions; (iii) intron and exon mapping of genomic DNA; (iv) expression of cloned genes in bacterial and yeast hosts through the addition of a T7 RNA polymerase gene into a host that contains the sequence of interest cloned into a vector containing a T7 RNA polymerase promoter sequence; and (v) *in situ* production of RNA for antisense studies.

Poly A Polymerase

This enzyme from *E. coli* polymerizes AMP (derived from ATP) onto free 3'-OH termini of RNA. It is used to: (i) label the 3' termini of RNA with [α-^{32}P] labeled ATP to generate hybridization probes; and (ii) to prepare poly A RNA for cDNA cloning.

Ribonuclease A

RNase A is a pyrimidine-specific endoribonuclease derived from bovine pancreas. It attacks single-stranded RNA at the 3'-phosphate group and cleaves the 5'-phosphate linkage to the adjacent nucleotide resulting in a 5'-OH and a 3'-P. The most common usage is to remove RNA from DNA preparations by degrading RNA into oligoribonucleotides, which are too small to be alcohol precipitated with the DNA.

Ribonuclease T1

This enzyme is an endoribonuclease, isolated from *Aspergillus oryzae,* that cleaves at the 3'-P of guanine nucleotides. It is commonly used to remove RNA from DNA preparations by degrading RNA into oligoribonucleotides, which are too small to be alcohol precipitated with the DNA. It is especially effective when used in conjunction with a pyrimidine-specific nuclease such as ribonuclease A.

Nuclease S1

Nuclease S1 is an endonuclease isolated from *Aspergillus oryzae,* which degrades single-stranded DNA and RNA resulting in 5'-phosphoryl mono- or oligonucleotides. It is used to: (i) remove unhybridized regions of heteroduplex nucleic acids; (ii) remove single-stranded tails from DNA to produce blunt ends; and (iii) previously used to open hairpin loops generated during cDNA synthesis (not used much today because better cDNA synthesis strategies are available).

Nuclease BAL-31

This enzyme is a calcium-dependent nuclease isolated from *Alteromonas espejiana* BAL-31 that progressively removes mononucleotides from duplex DNA by acting as a 3'→ 5' exonuclease. It is also a highly specific endonuclease that cleaves double-stranded RNA, DNA, and RNA:DNA heteroduplexes at nicks, gaps, and single-stranded regions. The activity of this enzyme is directly proportional to its concentration; therefore, it is useful for progressive shortening of double-stranded DNA fragments from both termini. This allows the generation of a set of fragments of defined lengths that are useful for: (i) restriction mapping; (ii) DNA sequencing; and (iii) subcloning. The endonuclease activity can also be used for analysis of RNA:DNA duplex structure.

Mung Bean Nuclease

This is a single-strand specific exonuclease that degrades single-stranded DNA to yield 5'-phosphoryl mono- and oligonucleotides. Double-stranded RNA, DNA, and RNA:DNA heteroduplexes are resistant to degradation unless excessive amounts of enzyme are present. It will not degrade the DNA strand opposite of nicks and is therefore the enzyme of choice for generating blunt ends by nuclease digestion.

Exonuclease III

This enzyme from *E. coli* catalyzes the step-wise removal of mononucleotides from 3'-OH termini of double-stranded DNA. It also acts on nicks in double-stranded DNA, generating single-stranded gaps. Because the enzyme does not act on single-stranded DNA, 3'-protruding termini of 4 bases or greater are resistant to cleavage. The preferred substrate is DNA with a recessed 3'-OH or a blunt end. This enzyme also has apurinic and apyrimidinic endonuclease activity, RNase H activity, and 3'-phosphatase activity. It is used for: (i) preparation of strand-specific probes; (ii) preparation of single-stranded template for Sanger dideoxy sequencing; (iii) site-directed mutagenesis; and (iv) the generation of unidirectional, nested deletions (in conjunction with S1 nuclease) useful for DNA sequencing and genetic analysis of cloned genes.

Deoxyribonuclease I

Bovine pancreatic deoxyribonuclease I (DNase I) is an endonuclease that degrades double-stranded or single-stranded DNA to oligo- and mononucleotides. The enzyme attacks sites adjacent to pyrimidine nucleotides in each strand. DNase I is used to: (i) generate nicks in double-stranded DNA prior to labeling by nick-translation; (ii) remove DNA from RNA preparations (must use RNase-free, DNase for this application); (iii) DNase footprinting studies to analyze sites where proteins interact with DNA; and (iv) produce random DNA fragments for generation of M13 clones to be used for DNA sequencing.

Proteinase K

Proteinase K is a serine protease with broad cleavage specificity isolated from the fungus *Tritirachium album* Limber. It hydrolyzes peptide bonds adjacent to the carboxylic group of aliphatic and aromatic amino acids and is useful for general digestion of protein in biological samples. It is stabilized by Ca^{2+}, and use of the enzyme in the presence of EDTA reduces its activity by 80%. The enzyme is active in 1% SDS, 2 M urea, and at temperatures from 37–56°C (it denatures above 65°C). Proteinase K is used for: (i) removal of nucleases from RNA and DNA preparations; (ii) preparation of DNA and RNA for a wide variety of uses; and (iii) protein fingerprinting.

Maps of Cloning Vectors and Bacteriophage Lambda

pBR322

pBR322 (see Table A11.1) was one of the original plasmid cloning vectors and served as a workhorse in molecular biology for many years. Although now largely replaced with vectors such as the pUC plasmids and their derivatives, some knowledge of this vector is useful because much of the literature in molecular biology and genes cloned in the 1980s involves use of this cloning vector. It was constructed *in vitro* by Bolivar, et al. (1977) using the tetracycline resistance gene *(tet^r)* from pSC101, the origin of replication *(ori)* from the plasmid ColE1, and the ampicillin resistance gene *(amp^r)* from transposon *Tn3*.

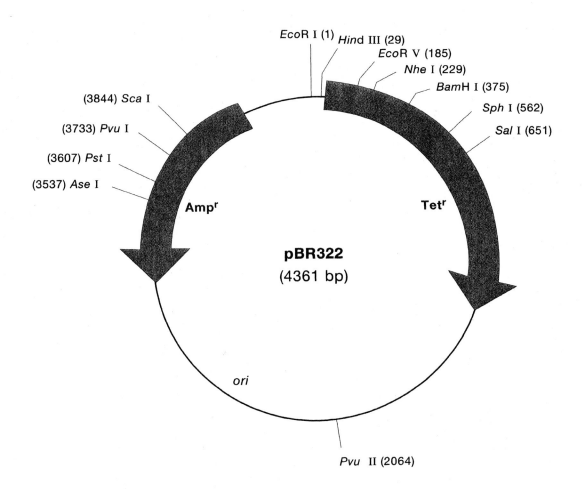

TABLE A11.1 Location of Selected Restriction Sites in pBR322

Enzyme	Number of sites	Locations								
Acc I	2	651	2246							
Apa I	0									
Ase I	1	3537								
Ava I	1	1425								
BamH I	1	375								
Bcl I	0									
Bgl II	0									
BssH II	0									
BstE II	0									
Cla I	1	23								
Dra I	3	3232	3251	3943						
EcoR I	1	4361								
EcoR V	1	185								
Fok I	12	112	133	987	1032	1681	1770	1848	2009	2150
		3348	3529	3816						
Fse I	0									
Hha I	31	101	233	261	414	435	495	549	701	776
		816	947	1206	1357	1419	1455	1645	1728	2076
		2179	2209	2350	2383	2653	2720	2820	2994	3103
		3496	3589	3926	4258					
Hinc II	2	651	3905							
Hind III	1	29								
Hpa I	0									
Kpn I	0									
Nae I	4	401	769	929	1283					
Nar I	4	413	434	548	1205					
Nde I	1	2297								
Nhe I	1	229								
Not I	0									
Pst I	1	3609								
Pvu I	1	3733								
Pvu II	1	2066								
Sal I	1	651								
Sau3A I (Mbo I)	22	349	376	467	826	1098	1129	1144	1461	1668
		3042	3117	3128	3136	3214	3226	3331	3672	3690
		3736	3994	4011	4047					
Sca I	1	3844								
Sfi I	0									
Sma I	0									
SnaB I	0									
Sph I	1	562								
Sst I (Sac I)	0									
Stu I	0									
Xba I	0									
Xho I	0									

pUC18 and pUC19

pUC18 and pUC19 (see Table A11.2) are small, high copy plasmid cloning vectors developed by Messing and coworkers (Vierra and Messing 1982). They include a 54-bp multiple cloning site that is flanked by sites complementary to primers used in DNA sequencing (pUC-F and pUC-R). The multiple cloning site is inserted into the gene encoding the α-fragment of β-galactosidase (*lacZα*), which allows screening clones by α-complementation. pUC18 and pUC19 are identical vectors with the exception that the multiple cloning site (but not the sequencing primer binding sites) is in opposite orientations. pUC vectors and their many derivatives have largely replaced pBR322 as a cloning vector.

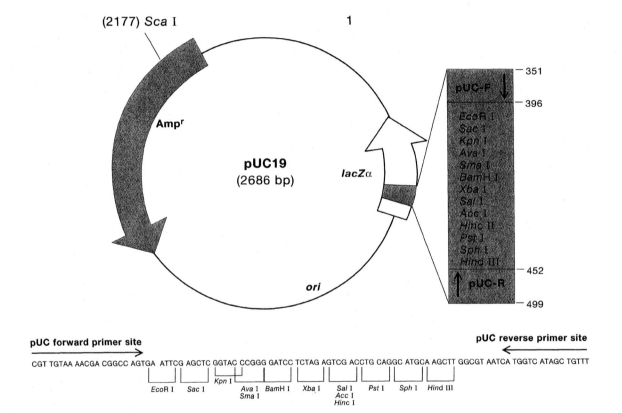

TABLE A11.2 Location of Selected Restriction Sites in pUC19

Enzyme	Number of sites	Locations								
Acc I	1	429								
Apa I	0									
Ase I	3	576	635	1870						
Ava I	1	412								
BamH I	1	417								
Bcl I	0									
Bgl II	0									
BssH II	0									
BstE II	0									
Cla I	0									
Dra I	3	1563	1582	2274						
EcoR I	1	396								
EcoR V	0									
Fok I	5	77	321	1679	1860	2147				
Fse I	0									
Hha I	17	3	106	236	257	588	653	681	714	984
		1051	1151	1325	1434	1827	1920	2257	2589	
Hinc II	1	429								
Hind III	1	447								
Hpa I	0									
Kpn I	1	408								
Nae I	0									
Nar I	1	235								
Nde I	1	183								
Nhe I	0									
Not I	0									
Pst I	1	435								
Pvu I	2	276	2066							
Pvu II	2	306	628							
Sal I	1	429								
Sau3A I (Mbo I)	15	277	418	1373	1448	1459	1467	1545	1557	1662
		2003	2021	2067	2325	2342	2378			
Sca I	1	2177								
Sfi I	0									
Sma I	1	412								
SnaB I	0									
Sph I	1	441								
Sst I (Sac I)	1	402								
Stu I	0									
Xba I	1	423								
Xho I	0									

pGEM™-3Zf(+)

pGEM-3Zf(+) (see Table A11.3) is one of a series of plasmid cloning vectors developed by Promega Corporation (1991). The pGEM vectors are derivatives of pUC19 that allow α-complementation and also have bacteriophage promoters (SP6 and T7) that flank the multiple cloning site. These promoters are flanked by the pUC forward and reverse primer sites as in pUC18 and pUC19. The bacteriophage promoters allow one to synthesize an RNA copy of the cloned DNA that may be used for transcriptional studies or as hybridization probes. pGEM-3Zf(+) also contains the origin of replication of the filamentous bacteriophage f1 that allows synthesis of a single-stranded copy of the cloned DNA to use as a template for DNA sequencing.

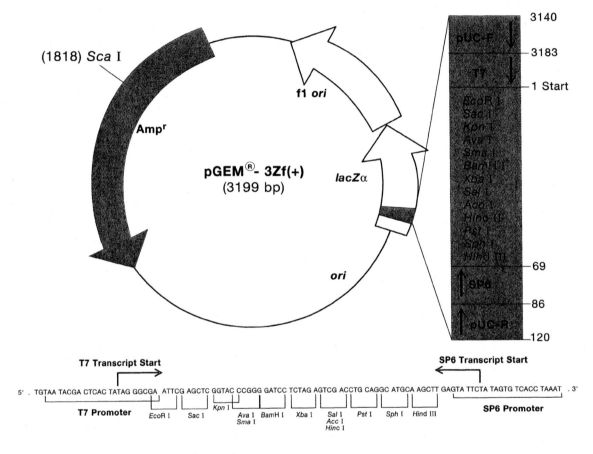

Modified from Promega Corporation with permission

TABLE A11.3 Location of Selected Restriction Sites in pGEM™-3Zf(+)

Enzyme	Number of sites	Locations								
Acc I	1	39								
Apa I	0									
Ava I	1	21								
BamH I	1	26								
Bcl I	0									
Bgl II	0									
BssH II	0									
BstE II	0									
Cla I	0									
Dra I	3	1204	1223	1915						
EcoR I	1	5								
EcoR V	0									
Fok I	5	1304	1485	1772	2415	3115				
Fse I										
Hha I	22	229	294	322	355	625	692	792	966	1075
		1468	1561	1848	2230	2330	2433	2938	2946	2972
		2994	3003	3016	3040					
Hinc II	1	40								
Hind III	1	56								
Hpa I	0									
Kpn I	1	21								
Nae I	1	2891								
Nar I	0									
Nde I	1	2509								
Nhe I	0									
Not I	0									
Pst I	1	48								
Pvu I	2	1708	3060							
Pvu II	2	269	3089							
Sal I	1	38								
Sau3A I (Mbo I)	15	26	1011	1086	1097	1105	1183	1195	1300	1641
		1659	1705	1963	1980	2016	3057			
Sca I	1	1818								
Sfi I	0									
Sma I	1	23								
SnaB I	0									
Sph I	1	54								
Sst I (Sac I)	1	15								
Stu I	0									
Xba I	1	32								
Xho I	0									

Bacteriophage Lambda (λ)

Lambda is a bacteriophage (virus that infects bacteria) that infects *Escherichia coli*. The lambda genome is a linear double-stranded DNA molecule that has 12 base single-stranded complementary ends. These complementary ends allow it to anneal to form a circular molecule after it infects an *E. coli* cell. The circular molecule is 48,502 bp in length. Lambda has been studied extensively as a model genetic system and has been modified to develop a variety of cloning vectors (Hendrix et. al., 1983). The figure below indicates the location of selected restriction sites, and Table A11.4 indicates the exact location of cleavage.

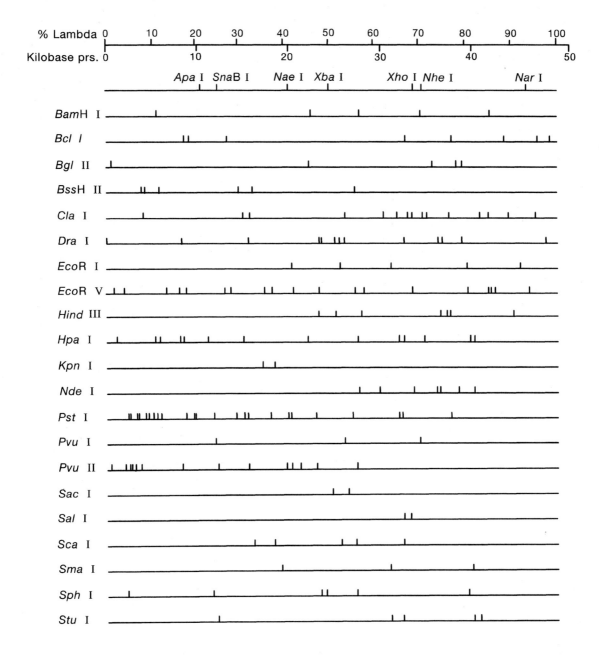

TABLE A11.4 Location of Selected Restriction Sites in Bacteriophage Lambda (Sites for Enzymes Cutting More than 15 Times Are Not Given)

Enzyme	Number of sites	Locations								
Acc I	9	2190	15260	18834	19473	31301	32745	33244	40201	42921
Apa I	1	10086								
Ava I	8	4720	19397	20999	27887	31617	33498	38214	39888	
BamH I	5	5505	22346	27972	34499	41732				
Bcl I	8	8844	9361	13820	32729	37352	43682	46366	47942	
Bgl II	6	415	22425	35711	38103	38754	38814			
BssH II	6	3522	4126	5627	14815	16649	28008			
BstE II	13	5687	7058	8322	9024	13348	13572	13689	16012	17941
		25183	30005	36374	40049					
Cla I	15	4198	15583	16120	26616	30289	31990	32963	33584	34696
		35050	36965	41363	42020	43824	46438			
Dra I	13	90	8460	16294	23110	23284	25436	26132	26665	32703
		36302	36530	38833	47429					
EcoR I	5	21226	26104	31747	39168	44972				
EcoR V	21									
Fok I	149									
Fse I	0									
Hba I	215									
Hinc II	35									
Hind III	7	23130	25157	27479	36895	37459	37584	44141		
Hpa I	14	732	5267	5708	7948	8199	11583	14991	21902	27316
		31807	32217	35259	39606	39834				
Kpn I	2	17053	18556							
Nae I	1	20040								
Nar I	1	45679								
Nde I	7	27630	29883	33679	36112	36668	38357	40131		
Nhe I	1	34679								
Not I	0									
Pst I	28									
Pvu I	3	11933	26254	35787						
Pvu II	15	209	1917	2385	2526	3058	3637	7831	12099	12162
		16078	19716	20059	20695	22991	27412			
Sal I	2	32745	33244							
Sau3A I (Mbo I)	116									
Sca I	5	16421	18684	25685	27263	32802				
Sfi I	0									
Sma I	3	19397	31617	39888						
SnaB I	1	12188								
Spb I	6	2212	12002	23942	24371	27374	39418			
Sst I (Sac I)	2	24772	25877							
Stu I	6	12434	31478	32997	39992	40596	40614			
Xba I	1	24508								
Xho I	1	33498								

Literature Cited

Bolivar, F., R. L. Rodriguez, P. J. Greene, M. C. Betlach, H. L. Heyneker, H. W. Boyer, J. H. Crosa, and S. Falkow. 1977. Construction and characterization of new cloning vehicles. II. A multipurpose cloning system. *Gene* 2:95–113.

Hendrix, R. W., J. W. Roberts, F. W. Stahl, and R. A. Weisberg, eds. 1983. *Lambda II.* New York: Cold Spring Harbor Laboratory Press.

Promega Corporation. 1991. Promega Protocols and Applications Guide, 2d ed. Madison, WI.

Vierra, J., and J. Messing. 1982. The pUC plasmids, an M13mp7-derived system for insertion mutagenesis and sequencing with synthetic universal primers. *Gene* 19:259.

APPENDIX XII
Proper Handling and Disposal of Hazardous Materials

Numerous reagents used in molecular biology pose potential health hazards. However, if appropriate precautions are taken, such materials can be used safely in teaching and research laboratories. The following materials should be handled with caution and properly disposed of in the laboratory. Proper procedures for use and disposal of radioisotopes are given in Appendix XIII.

Ethidium Bromide

Ethidium bromide is toxic, a strong mutagen, and a suspected carcinogen. Gloves, eye protection, and a lab coat should always be used whenever handling it, and all work must be done over plastic-backed absorbent paper to avoid contamination of working surfaces. The greatest risk is from inhaling the dry powder; wear a dust mask and take extreme care when preparing stocks from the powder. Risk can be minimized by purchasing ready-made solutions from a chemical supplier. Particular care should also be exercised when using concentrated solutions of ethidium bromide (10 mg/ml) to prepare the solution used for staining agarose gels (1 μg/ml). We recommend reusing the staining solution repeatedly to minimize handling and disposal. When the solution no longer stains effectively, a fresh aliquot of ethidium bromide stock solution can be added, which eliminates the need to decontaminate staining solutions.

All work with ethidium bromide should be limited to a contained area or sink. If solutions of ethidium bromide are spilled on absorbent paper, the paper should be disposed of as hazardous waste. Spills on equipment or bench tops should be wiped up with a wet paper towel followed by five rinses with additional wet paper towels. Decontaminate paper towels as described below or place in hazardous waste.

Ethidium bromide solutions and ethidium bromide stained gels should be decontaminated prior to disposal. Gels may also be dried (to reduce volume) and placed in a toxic waste disposal container. For many years, bleach (hypochlorite) was used to decontaminate solutions of ethidium bromide. This converts the ethidium bromide to another mutagenic compound, and this method is no longer recommended. The following method is easy to use and will effectively decontaminate solutions of ethidium bromide (mutagenicity is reduced about 3000-fold). Additional methods are given by Sambrook, et al. (1989).

1. Dilute the solution of ethidium bromide to less than 0.1 mg/ml with water.

2. Add one volume (equal to the volume of the ethidium bromide solution or the estimated volume of the gels) of 0.1 M $KMnO_4$ and mix carefully to avoid any spills.

CAUTION: $KMnO_4$ is an irritant and the powder is potentially explosive. Handle in a fume hood.

3. Add one volume of 0.5 M HCl and mix carefully.

4. Let the solution react for 2 to 4 hours (or overnight). Add one volume of 0.5 M NaOH and mix carefully. The solution may now be discarded down the sink and decontaminated gels may be placed in the regular trash.

Phenol and Chloroform

Phenol is toxic, corrosive, and can cause severe burns. Phenol is especially dangerous because it is also an anesthetic that may prevent you from realizing you are being burned. Gloves, eye protection, and a lab coat should always be worn when working with phenol or solutions that contain it. All manipulations with phenol should be carried out in a fume hood. If phenol is spilled on skin, the area should be rinsed with copious amounts of water and washed with soap and water.

Chloroform is highly volatile, toxic, an irritant, and a carcinogen in laboratory animals. Gloves, eye protection, and a lab coat should also be worn when working with chloroform, and phenol:chloroform, and all manipulations with these solutions should be done in a fume hood.

Phenol and chloroform waste should be stored in a sealed hazardous waste container in a fume hood and disposed of as hazardous waste. Federal air quality regulations forbid disposal of these chemicals by evaporation.

Biohazards

The microorganisms used in this course, although not pathogenic, should be treated as biohazards. All liquid and solid media, glassware, centrifuge tubes, and so on, containing bacterial cultures should be disinfected by autoclaving prior to washing or disposal. Standard bacteriological practices (see Laboratory Guidelines and Safety, pages xxvii–xxviii) should be followed in the laboratory at all times.

The exercises in this laboratory also involve the formation of bacteria containing recombinant DNA molecules. All work involving the formation of recombinant DNA molecules is regulated by the National Institutes of Health (NIH) and the Recombinant DNA Advisory Committee (RAC). Since 1982, the formation of recombinant DNA molecules involving solely bacterial DNAs and hosts has been exempt from regulation. Therefore, the experiments in this manual pose no harm and are not in violation of federal guidelines. All recombinant strains, however, should be treated as biohazards and disinfected by autoclaving prior to disposal.

Literature Cited

Sambrook, J., E. F. Fritsch, and T. Maniatis. 1989. *Molecular cloning: A laboratory manual.* 2d ed. New York: Cold Spring Harbor Laboratory Press.

APPENDIX XIII
Procedures and Precautions for the Use of Radioisotopes

Radioisotopes provide powerful tools to biological scientists and are widely used in genetics, molecular biology, and biochemistry. Radioisotopes are unstable atoms that return to a more stable state by a process called "decay," or "disintegration." For example, ^{14}C contains six protons and eight neutrons. Upon decay, one neutron is converted to a proton and a negatively charged beta particle is released (β^-). The new atom formed is ^{14}N, the normal isotope of nitrogen. The beta particle formed has a characteristic energy. In fact, each radioisotope is characterized by the energy released during decay. The most common unit of radioactivity used in biological applications of radioisotopes is the microcurie (μCi), which is equivalent to 2.2×10^6 disintegrations per minute (dpm). The ease in which radioactive decay events can be detected allows one to identify or trace the location of the radioactivity in a biological molecule or system.

In addition to their utility, radioisotopes also pose potential health hazards, and precautions must be taken when using them. Several different radioisotopes are commonly used in the biological sciences (see Table A13.1) and the health hazards vary with each. Due to the presence of phosphate in all nucleic acids, ^{32}P is probably the most commonly used radioisotope in molecular biology. It has a high energy emission and poses a significant health hazard if not properly used. ^{14}C, ^{35}S, and ^3H are also widely used but have much lower emission energies (see Table A13.1). These radioisotopes can be hazardous, however, if ingested, inhaled, or injected into the body.

TABLE A13.1 Commonly Used Radioisotopes

Isotope	Half-life	Type of radiation	Energy (MeV)
^3H	12.5 y	beta	0.018
^{14}C	5700 y	beta	0.155
^{35}S	87 d	beta	0.167
^{32}P	14.3 d	beta	1.72

Due to the potential health hazard, the use of radioisotopes is strictly regulated. In the United States, the regulations are established by the Nuclear Regulatory Commission. The precautions listed in this appendix are summarized from Federal regulations.

Handling of Radioisotopes

The goal of proper handling of radioisotopes is to keep all of the isotope where you want it and to avoid contaminating anything. Contamination can result in potential health hazards, and even extremely small nonhazardous levels of contamination can foul up experiments. All work with radioisotopes is done over impermeable, plastic-backed absorbent paper so that any spillage is contained. When handling radioisotopes, always wear plastic or latex gloves to avoid contamination of hands, but still maintain dexterity. Your work area and any glassware or equipment that becomes contaminated should be labeled with the type and amount of isotope present. In general, stay conscious of the fact that anything that has come in contact with a radioisotope will contaminate anything it touches. The use of radioisotopes is analogous to aseptic techniques, and if you have had experience in bacteriological techniques, you should adapt to using radioisotopes readily.

Particular precautions must be taken when working with high-energy radioisotopes such as ^{32}P. To minimize exposure, the user should be cognizant of four principles: (i) **time**; (ii) **distance**; (iii) **shielding**; and (iv) **monitoring**. The amount of dosage received will be a result of the length of time exposed to the isotope. Dry runs with protocols prior to using the radioisotope will help you become familiar with procedures and minimize exposure when the procedure is done with the radioisotope. The intensity of radiation decreases at a rate inversely proportional to the square of distance away from the source. Therefore, increasing the distance from a source is an effective way to minimize exposure.

Shielding is particularly important when using ^{32}P. All ^{32}P work should be done behind plexiglass shielding, and radioactive stocks, reagents, and waste must be shielded by plexiglass. When working with ^{32}P, you should continuously monitor your work area, gloves, and equipment with a Geiger counter. In addition, persons working with ^{32}P must wear radiation badges (dosimeters) to monitor whole body exposure. If high levels are used (>100 μCi) a ring badge should also be worn. These devices are used to monitor and record long-term exposure to radioactivity because the effects are cumulative. Federal standards govern the permissible radioactive exposure an individual can receive, and if these limits are exceeded, individuals will be prohibited from working with radioisotopes.

Disposal

All radioactive materials must be disposed of safely. The proper method depends on the energy and the half-life of the isotope. In general, disposal is done by: (i) letting the radioisotope decay to background levels; (ii) returning it to the environment by high dilution; or (iii) containment in a safe place. Because the half-life of ^{14}C is more than 5000 years, most users lack the time for the first method. Low levels of aqueous liquid waste can be disposed of by washing them down a sink with large volumes of water. Contaminated glassware is also rinsed (10 to 12 times) in a sink before being washed as regular glassware. Only designated sinks can be used for disposal of radioactive wastes and must be labeled as such. Solid wastes are generally disposed of by burial or sometimes, in the case of ^{14}C, by combustion. Nonaqueous liquid wastes and solid wastes are disposed of by burial. Gaseous radioactivity such as ^{14}CO$_2$ should be trapped by alkaline materials or liquids and disposed of as described above.

Monitoring Radioactive Contamination

High energy radioisotopes such as ^{32}P are easily monitored using a Geiger counter. Despite the potential health hazard associated with ^{32}P due to its high energy emission, its ease of detection makes it easy to work with and monitor. Geiger counters, however, are inefficient in detecting ^{14}C and ^{35}S, and incapable of detecting ^{3}H. Therefore, **swab tests** are routinely done to examine laboratories for contamination from these isotopes. Swab tests are done by using a 1-inch diameter absorbent filter paper to swab benchtops, floors, faucets, and so on. These filters are then counted in a scintillation counter. Cleaning up radioactive spills is often a difficult job, so every precaution should be taken to avoid such spills.

Counting Beta-Emitting Radioisotopes

Beta-emitting radioisotopes can be counted by gas ionization, excitation, or autoradiography. In gas ionization methods (Geiger counters and gas proportional counters), helium gas is ionized and the released electrons are quantified. Because the radiation must pass through the window on Geiger tubes, this type of counter is only useful for high-energy isotopes. Low-energy isotopes such as ^{3}H cannot penetrate the window, although thin-window Geiger counters may be used to monitor ^{14}C and ^{35}S. In gas proportional counters, the sample actually passes through the counting tube with a helium carrier gas, making this method useful for detecting low energy isotopes.

Scintillation counters are the most commonly used method of quantifying beta-emitting radioisotopes. The sample to be counted is placed in a cocktail that contains light-emitting organic compounds known as "fluors" dissolved in a solvent. Sometimes detergents are added to solubilize aqueous samples. When a beta particle strikes one of the fluors, light is emitted and is counted by a photomultiplier tube in the scintillation counter. For a variety of reasons, collectively known as "quenching," not every disintegration is counted, and the results are reported as **counts per minute** (cpm). This value can be converted to **disintegrations per minute** (dpm) by dividing by an efficiency factor. The counting efficiency of ^{14}C and ^{35}S is generally about 80–90%, whereas the efficiency for ^{3}H is usually about 50%.

Autoradiography is a third technique of measuring radioactivity in which the beta emissions react with a photographic emulsion, producing dark silver grains (similar to the effect of light on photographic film). This is an extremely sensitive technique because the emulsion may be kept in contact with the radioactive material for long periods of time (weeks). The location of the radioactivity can also be determined in this technique. For example, autoradiography may be used in combination with microscopy to determine the location of a given nucleic acid in a cell by *in situ* hybridization. Autoradiography is also used in molecular biology to detect labeled nucleic acids in procedures such as colony hybridization, hybridization of Southern and northern blots, and in nucleic acid sequencing.

Radiation Safety: Rules and Regulations

1. Eating, drinking, storing food, smoking, or applying cosmetics is forbidden in any area where radioisotopes are used or stored.

2. Work must be carried out only in the exact areas designated by your instructor.

3. **Direct contact** with radioactive materials must be avoided by wearing a protective lab coat, disposable gloves, safety glasses, and by using pipet aids. **Never** mouth pipet radioactive solutions.

4. Perform all procedures with radioactive materials over plastic-backed absorbent paper. Keep all contaminated glassware, pipet tips, and other materials on this paper or in appropriate disposal receptacles.

5. All containers with radioisotopes must be labeled with isotope warning labels showing what isotope is present and the amount of activity.

6. Contaminated wastes should be disposed of in appropriate receptacles. Liquids should be poured into specially marked containers designated by your instructor. Solids (e.g., towels, gloves, filters, pipet tips, and so on) are to be placed in designated dry-waste containers.

7. Place all contaminated glassware in the appropriate containers.

8. Report any spills **immediately** to your instructor. Mark any contaminated areas on your absorbent paper by circling with a marker and identify the radioisotope spilled. Mark spills on equipment or surfaces; these will be cleaned by the Radiation Safety Officer.

   If you contaminate an item of clothing, remove it and place it at designated sink areas. It will be cleaned for you later. Contaminated clothes must not leave the lab until the Radiation Safety Officer checks them for contamination.

9. If you cut yourself, thoroughly wash the wound with running water, allow some bleeding, and then bandage. Your instructor will monitor the wound for contamination when the bleeding stops.

10. Store radioactive materials **only** in areas designated by your instructor, and be sure they are properly labeled with a radioisotope warning label that includes the isotope and the amount of activity.

11. Monitor all work surfaces, micropipetters, and equipment at the end of each lab period as demonstrated by your instructor. Use a Geiger counter for ^{32}P work and swab tests for other radioisotopes.

12. **Always** wash your hands before leaving the laboratory if you have been working with radioactive materials. Check hands with the Geiger counter if you are working with ^{32}P. Procedures for monitoring will be demonstrated by your instructor.

A P P E N D I X   X I V
Equipment and Supplies

Basic Equipment

The following list includes the basic equipment needed to conduct the DNA isolation experiments, gene cloning and subcloning, and hybridization experiments. Additional equipment is required for PCR, DNA sequencing, and pulsed field gel electrophoresis.

Centrifuge, micro-. Should be able to spin 12–18 1.5-ml microcentrifuge tubes at 12–14,000 $\times$ g. Various models are available from general scientific suppliers.

Centrifuge, refrigerated. Should have rotors that spin 50-ml Oak Ridge tubes and 250-ml bottles. Available from general scientific suppliers, Sorval, or Beckman.

Horizontal electrophoresis chamber, mini gel format. Should have a 6- or 8- **and** 12- or 14-tooth combs.

Electrophoresis power source. Should have a constant voltage setting and be capable of producing 150–250 volts. Power sources with two or more sets of outlet jacks are more cost effective and can be used to run more than one gel at a time.

Heat sealer. For sealing hybridization bags. We use Seal-a-Meal™ sealers, which are available from discount department stores for about $20 to $25 each. This company also makes heat-sealable bags.

Incubator. Capable of maintaining 37°C for incubation of plates. Room temperature incubations are acceptable if an incubator is not available, but they will require longer times for bacterial colonies to develop.

Incubator, shaking. For growing bacterial cultures in shake flasks. A shaking water bath can be used, although a shaking incubator allows better culture agitation. These are available from Lab-Line or New Brunswick and are marketed through many general scientific suppliers.

Micropipetters. Adjustable 0.5–10 μl, 10–100 μl, and 100–1000 μl micropipetters are needed. Standard sterile glass or plastic pipets can be used instead of the 100–1000 μl pipetter if desired. Various brands are available from any general scientific supplier and "Pipetman" brand is available from Rainin.

Microwave oven. Used for melting agarose. An inexpensive, low-watt model is sufficient, but you should be sure that the chamber is high enough to accommodate the size flasks you wish to use.

Spectrophotometer, UV/Visible. Needed to quantify DNA (UV) and to determine culture density when making competent cells and preparing cells for pulsed field gel electrophoresis. Available from any general scientific supplier. Milton Roy (formerly Bausch and Lomb, manufacturer of the Spec 20) makes reasonably priced models.

Spreading turntable. These are optional, but make spreading plates much easier. They are available through Fisher Scientific.

Ultraviolet light transilluminator and camera. Required to photograph ethidium bromide stained gels. The transilluminator should have mid-range (300 nm) UV lamps to maximize visualization and minimize photo-nicking of DNA. Handheld hooded Polaroid cameras are easiest to use in the classroom as a dark room is not required. The patented FOTO/Phoresis™ system manufactured by FOTODYNE has a UV-blocking, interlocking safety cover that eliminates any chance of UV exposure and is ideal for student use. The FOTO/Phoresis™ transilluminator also accommodates a standard 100-mm Petri plate for photographing ethidium bromide plates used to quantify DNA (Exercise 9).

Note: A UV transilluminator is not necessary to visualize methylene blue stained gels, but a white light table (or overhead projector) and camera are required.

Vortex mixer. Available from any general scientific supplier.

Water bath. Small water bath for incubating restriction digestions, maintaining molten agarose, and so on. Scientific Products markets an inexpensive bath that we have found maintains excellent temperature control.

Water bath, shaking. For growing bacterial broth cultures and incubating membranes for hybridization. Scientific Products markets a model (YB-531) that has springs instead of separate platforms, which allows incubation of a variety of flasks, tubes, and so on, without changing platforms. This is also very convenient for submerging Seal-a-Meal™ bags for hybridizations. This model is also available from Carolina Biological Supply.

PCR, DNA Sequencing, and Pulsed Field Gel Electrophoresis Equipment

These procedures require the following more specialized (and more expensive) incubating and electrophoresis equipment.

Thermalcycler. This is required to generate the temperatures required for melting, annealing, and replicating DNA during the polymerase chain reaction. Available from Perkin Elmer and a variety of other suppliers. The PCR exercise in this manual was developed using the Perkin Elmer 2400, which is ideal for student use. Prices for thermalcyclers range from $3000 to $10,000.

DNA sequencer. Long vertical electrophoresis apparatus required to separate the nested DNA fragments generated during DNA sequencing reactions. These are generally available from any supplier that sells electrophoresis equipment and come in a variety of sizes. We recommend narrower units (for 18–20 cm wide gels) to minimize the number of students sharing one apparatus and to facilitate handling the gels. These will still accommodate 24 lanes per gel. Prices range from $600 to $1000.

High voltage power source. A 2000–3000 volt power source is required to run DNA sequencing gels and should have constant voltage or constant power settings. Available from any supplier that sells electrophoresis equipment. Prices range from $2000 to $2500.

Slab gel dryer. These are used to dry sequencing gels prior to autoradiography and cost about $1000 to $1200. Gels may be dried overnight in a 37°C incubator or by blowing warm air from a hair dryer on the gel. Gel dryers are available from companies that sell DNA sequencing equipment.

High vacuum pump. This is required to run a gel dryer and costs $1000 to $2000. Numerous models are available from any general scientific supplier.

Pulsed Field Gel Electrophoresis System. These systems are required to separate DNA fragments ranging from approximately 100 kilobases to several megabases in size. The exercise in this manual was developed for use with Bio-Rad CHEF™ systems that sell for $10,000 to $20,000 (depending on configuration and options). Other suppliers may market equipment suitable for this exercise, although Bio-Rad is currently the leader in pulsed field gel electrophoresis technology.

General Supplies

Most all of the supplies required in the course can be obtained from general scientific suppliers (VWR, Fisher, and so on). However, better prices are often found with specialty companies for such items as disposable plastic ware (tubes and tips) and certain chemicals. Addresses of suppliers are listed in Appendix XVII. Bacteriological media are generally purchased through DIFCO or BBL, which market through most general scientific suppliers.

APPENDIX XV
Media and Reagents

MEDIA RECIPES

Good Vibrio Medium (GVM) (per liter)

To 1000 ml deionized water add:

tryptone	10.0 g
casamino acids (Difco)	5.0 g
NaCl	25.0 g
MgCl$_2$	4.0 g
KCl	1.0 g

Adjust pH to 7.4 with 1 M NaOH. To prepare agar plates, add 15 grams agar per liter (1.5%) prior to autoclaving. Sterilize by autoclaving for 15 minutes (20 minutes for volumes greater than 250 ml). Swirl thoroughly after autoclaving to mix agar, and place flasks in a 55°C water bath until the plates are poured.

Photobacterium Broth

To 1000 ml deionized water add:

tryptone	5.0 g
yeast extract	2.5 g
ammonium chloride	0.3 g
magnesium sulfate	0.3 g
ferric chloride	0.01 g
calcium carbonate	1.0 g
monopotassium phosphate	3.0 g
sodium glycerol phosphate	23.5 g
sodium chloride	30.0 g

Adjust pH to 7.4 with 1 M NaOH and sterilize by autoclaving. To prepare agar plates, add 15 grams agar per liter (1.5%) prior to autoclaving. Sterilize by autoclaving for 15 minutes (20 minutes for volumes greater than 250 ml). Swirl thoroughly after autoclaving to mix agar, and place flasks in a 55°C water bath until the plates are poured. Placing a stir bar in the flask prior to autoclaving aids mixing after autoclaving. This medium forms a white precipitate, which is normal. Stir continually while pouring plates or when dispensing broth to containers. The medium will be cloudy.

LB (Luria-Bertani) Medium (available premixed from Difco)

To 1000 ml deionized water add:

Bacto-tryptone	10.0 g
Bacto yeast extract	5.0 g
NaCl	10.0 g

Dissolve all ingredients and pH to 7.2 with NaOH (about 1 ml of 4 M NaOH). Dispense into appropriate containers and sterilize by autoclaving for 15 minutes (20 minutes for volumes greater than 250 ml). For agar plates, add 15 g agar per liter (1.5%) prior to autoclaving. Swirl thoroughly after autoclaving to mix agar, and place flasks in a 55°C water bath until the plates are poured. If adding heat labile compounds after autoclaving (e.g., antibiotics, X-gal, and so on), place a stir bar in the flask prior to autoclaving.

Agar Stabs (0.1X LB in 0.3% agarose; for storage of *E. coli* strains)

To 1000 ml deionized water add:

Bacto-tryptone	1.0 g
Bacto yeast extract	0.5 g
NaCl	1.0 g
agarose	3.0 g (or 6.0 g agar)

Mix all ingredients until dissolved and adjust pH to 7.2. Heat with stirring on a hot plate (or in a microwave) until the agarose is molten (the medium should just start to boil). Dispense 3-ml aliquots into 1-dram (4-ml) vials and cap loosely. Sterilize by autoclaving 15 minutes. When vials are cool, tighten caps.

 Note: Vials **must** have rubber-lined lids to seal properly over long-term storage. We use Wheaton #224882.

Agar stabs for *Vibrio* strains

Prepare stabs as above except use 1X GVM with 0.3% agarose.

Agar stabs for *Photobacterium* strains

Prepare stabs as above except use 1X photobacterium broth with 0.3% agarose.

MEDIA ADDITIONS

The following sterile stocks should be added aseptically to autoclaved and cooled media (55°C). If additions are to be made to media after autoclaving, place a stir bar in the medium flask prior to autoclaving. Use the stir bar to mix the cooled (55°C) medium and slowly add the sterile stock down the inside surface of the flask to pre-warm it before it reaches the medium. It is best to aliquot stocks in small volumes to avoid repeated freeze/thaw cycles and contamination.

0.1 M IPTG

Dissolve 2.38 g of isopropylthio-β-D-galactoside (IPTG) in 8.0 ml deionized water. Make volume to 10.0 ml with deionized water. Sterilize through a 0.2-μm membrane filter and dispense in 2–3 ml aliquots in sterile vials or tubes. Store at −20°C. Add 2.5 ml per liter of **sterilized** medium that has cooled to 55°C (0.25 mM final concentration).

Note: IPTG can be left out of media used for α-complementation when using pUC or pGEM vectors. These are such high copy number vectors that there is enough expression of the α-fragment of β-galactosidase to obtain activity (i.e., blue colonies) without induction by IPTG.

40 mg/ml X-gal

Dissolve 0.4 g 5-bromo-4-chloro-3-indoyl-β-D-galactoside (X-gal) in 10 ml dimethylformamide (DMF) in a glass or polypropylene tube. **DMF will dissolve polycarbonate!** Store in 2–3 ml aliquots in foil-wrapped vials at −20°C. Do not sterilize, but treat as a sterile stock. Add 1.0 ml per liter of **sterilized** medium that has cooled to 55°C (40 μg/ml final concentration).

Note: The price of X-gal varies widely and it is most cost effective to purchase at least 1 g amounts. See Appendix XVII or the authors' Web Page for recommended suppliers.

50 mg/ml Ampicillin

Dissolve 0.50 g ampicillin (sodium salt) in 10 ml deionized water. Place ampicillin solution into a 10-cc syringe and sterilize through a sterile, disposable 0.2-μm membrane filter cartridge. Dispense in 2–3 ml aliquots into sterile 1-dram vials. Store at −20°C. Add 1.0 ml per liter of **sterilized** medium cooled to 55°C (50 μg/ml final concentration). Adjust volume accordingly for different concentrations.

REAGENT RECIPES

I. Commonly Used Stocks

The following stocks are commonly used for the preparation of many reagents used in these exercises. The virtue of such stock solutions is that they can be used to prepare many other reagents. To destroy nucleases, most stock solutions used in molecular biology are autoclaved. Be careful to use aseptic techniques when using sterile stocks, and minimize the number of times you enter a sterile stock. For example, if you need to aliquot 20 1-ml aliquots of TE buffer, remove 21 ml and place in a sterile tube. This tube may then be used to aliquot the 20 tubes. Use good quality deionized or glass distilled water to make reagents if possible. Most recipes give instructions to make one liter of reagent, although smaller volumes are often sufficient. Decrease volumes as appropriate for your needs. Additional recipes for commonly used stocks may be found in Table B-7 of Sambrook, et al. (1989).

β-mercaptoethanol (BME)

β-mercaptoethanol is purchased as a liquid at a concentration of 14.4 M. Store at 4°C in a dark bottle. **Do not autoclave** BME or buffers containing it, but treat it as a sterile stock.

50 mg/ml BCIP

Dissolve 0.5 g 5-bromo-4-chloro-3-indoyl phosphate (BCIP) in 10 ml dimethylformamide (DMF) in a **glass** vial or **polypropylene** tube. DMF will dissolve polycarbonate! A micro stir bar helps the BCIP to dissolve. BCIP can also be made as a 25 mg/ml stock. Store at 4°C in a foil-wrapped or brown vial. Do not sterilize.

1.0 M Calcium Chloride

Dissolve 14.7 g $CaCl_2 \cdot 2H_2O$ in 80 ml deionized water. Adjust the volume to 100 ml with deionized water. This may be stored at room temperature indefinitely without autoclaving.

100 mM Calcium Chloride (for preparation of competent cells)

Add 50.0 ml 1.0 M $CaCl_2$ to 450 ml deionized water and mix. Sterilize by autoclaving. Store at 4°C.

0.5 M EDTA, pH 8.0

Add 186.1 g of disodium ethylenediaminetetraacetate·$2H_2O$ to 800 ml deionized water that is being mixed with a magnetic stirrer. Adjust the pH to 8.0 with the addition of NaOH pellets (about 20 g). The EDTA will not completely dissolve until the pH has reached 8.0. Add the NaOH slowly so as not to adjust the pH above 8.0. Be patient; this is a rather time-consuming process (15 to 20 minutes). Make to 1000 ml with deionized water. Dispense into appropriate containers and sterilize by autoclaving. Store at room temperature.

95% Ethanol

May use 95% ethanol purchased in bulk or from any supplier. Store in autoclaved bottles, and prepare any dilutions of 95% ethanol (e.g., 70%) with sterile water.

10 mg/ml Ethidium Bromide

 Note: We strongly recommend purchasing this as a premixed 10 mg/ml solution (available from most suppliers). If you use powdered ethidium bromide, use extreme caution.

To prepare stock from powder, dissolve 0.25 g of ethidium bromide in 25.0 ml deionized water in a bottle with a lid. Stir at room temperature for several hours to ensure that the dye is completely dissolved. Store in a foil-wrapped or brown glass bottle at room temperature or 4°C.

 CAUTION: Ethidium bromide is a mutagen and a suspected carcinogen. Be sure to handle with gloves. Wear a mask when weighing dry powder. Be careful to avoid spills and immediately clean up any spilled dye. Label bottles:
**CAUTION, ETHIDIUM BROMIDE
TOXIC AND MUTAGENIC**

Isopropanol

This is available from any chemical supplier. Treat as a sterile stock.

50 mg/ml NBT

Dissolve 0.5 g nitro blue tetrazolium (NBT) in 10 ml of 70% dimethylformamide (DMF) in a glass or polypropylene tube. DMF will dissolve polycarbonate. Store at 4°C in a foil-wrapped or brown vial. Do not sterilize.

5.0 M Potassium Acetate

Dissolve 40.9 g of potassium acetate (anhydrous) in 75 ml deionized water. Make to 100 ml with deionized water. Sterilize by autoclaving and store at room temperature.

3.0 M Sodium Acetate (pH 5.2) (anhydrous MW = 82.03)

Add 24.6 g sodium acetate (anhydrous) to 70 ml deionized water and mix to dissolve. Adjust pH to 5.2 with glacial acetic acid. Make to 100 ml with deionized water. Sterilize by autoclaving and store at room temperature.

5.0 M Sodium Chloride

Dissolve 292.2 g NaCl in 700 ml deionized water. Heat may be necessary to completely dissolve the NaCl. Make to 1000 ml with deionized water and sterilize by autoclaving. Store at room temperature.

20% SDS (also available premixed from most suppliers)

Add 200 g of molecular biology grade sodium dodecyl sulfate (often called sodium lauryl sulfate) to 800 ml deionized water (heating to 50-60°C aids the SDS going into solution). Adjust the pH to 7.2 with a few drops of 6 M HCl. Make to 1000 ml with deionized water. Do not autoclave this stock. Store at room temperature. If the solution forms a precipitate, which is common in cool conditions, it may be resuspended by heating to 37°C for a few minutes.

 Note: Wear a mask when weighing out SDS powder.

4.0 M Sodium Hydroxide

Add 160 g NaOH pellets to 800 ml deionized water. Dissolve on a stir plate (solution will get hot). After all pellets are dissolved, allow solution to cool to room temperature and make to 1000 ml with deionized water. Store at room temperature in a **polypropylene** bottle (NaOH should **not** be stored in glass bottles!).

20X SSC (Standard Saline Citrate; 3 M NaCl, 0.3 M Na citrate)

To 800 ml deionized water add:

NaCl	175.3 g
sodium citrate·2H$_2$O	88.2 g

Adjust pH to 7.0 with 4 M NaOH and make to 1000 ml with deionized water. Sterilize by autoclaving. Store at room temperature.

1.0 M Tris, pH 8.0 (also available premixed from most suppliers)

Dissolve 121.1 g Tris base in 800 ml deionized water. Allow to cool to room temperature and adjust pH to 8.0 with concentrated HCl (about 42 ml).

CAUTION: Wear gloves, eye protection, and a mask when weighing Tris pow-

Make to 1000 ml with deionized water and mix. Dispense into appropriate containers and sterilize by autoclaving. Store at room temperature.

Note:
- The pH of Tris buffers is temperature dependent. Make all measurements at room temperature. The pH will decrease approximately 0.03 pH units for each °C increase in temperature.
- Many pH electrodes do not accurately measure the pH of Tris buffers. Be sure that the electrode you use is capable of accurately measuring the pH of Tris solutions. Sigma sells Tris electrodes and provides an excellent pamphlet on preparation and use of Tris buffers (Technical Bulletin #106B).
- Avoid inhaling Tris. Wear a mask when weighing out dry powder.

Triton X-100

Purchased as a liquid. Treat Triton X-100 as a sterile stock.

Deionized or Glass Distilled Water

Many recipes call for use of sterile deionized water. It is a good idea to have several bottles (100 to 1000-ml) of autoclaved deionized water available as a "stock."

II. DNA and Enzyme Stocks

DNA Solutions

DNAs (plasmids, lambda, and so on) are generally supplied at a known concentration in TE buffer (10 mM Tris, pH 8.0; 1 mM EDTA). All dilutions of DNA stocks should be made in sterile TE buffer. Wear gloves when handling DNA stocks, and exercise caution to avoid nuclease contamination. Store DNAs at 4°C if you are using them frequently, and at −20°C for long-term storage.

Restriction Enzymes, Ligase, and Polymerases

Restriction enzymes, ligase, and polymerases are supplied by manufacturers in buffers containing 50% glycerol and must be stored at −20°C. The concentration of commercially available enzymes varies (from 5–20 units/μl), but it is best not to dilute them for use. Dilution results in loss of activity and is time-consuming. Addition of an excess amount of enzyme makes reactions proceed faster in teaching laboratories. Refer to Appendix VIII (Care and Handling of Enzymes) prior to dispensing or using restriction enzymes, ligase, and polymerases.

All suppliers of these enzymes supply concentrated (10X) buffer solutions with each enzyme purchased.

Some enzymes are purchased as dry powders and are prepared as follows:

Lysozyme solution (10 mg/ml in 10 mM Tris, pH 8.0)

To 11.9 ml sterile deionized water add:

1.0 M Tris, pH 8.0	0.12 ml
lysozyme	0.12 g

Dissolve lysozyme and keep on ice. Make lysozyme solution **immediately** (1 to 4 hours) before use.

Proteinase K Stock (10 mg/ml)

Dissolve 25 mg proteinase K in 2.5 ml sterile deionized water. Dispense into sterile microcentrifuge tubes in 500 μl aliquots and store at −20°C.

RNase A Stock (10 mg/ml in 10 mM Tris, pH 7.6, 15 mM NaCl)

To 5 ml of deionized water add:

1.0 M Tris, pH 7.6	50.0 μl
5 M NaCl	15.0 μl
RNase A	50.0 mg

Dissolve RNase A and place in boiling water bath for 10 minutes to destroy DNase. Allow the solution to slowly cool to room temperature. Transfer to sterile microcentrifuge tubes in 500 μl aliquots and store at −20°C. Treat as a sterile stock.

III. Reagents and Buffers

A. Nucleic Acid Isolation

70% Ethanol

Add 74 ml 95% ethanol to 26 ml of sterile deionized water in a sterile bottle.

GTE/lysozyme

To about 80 ml deionized water add:

glucose	0.9 g
1.0 M Tris (pH 8.0)	2.5 ml
0.5 M EDTA (pH 8.0)	2.0 ml

Mix and make to 100 ml with deionized water. Sterilize this solution (GTE) by autoclaving and store at 4°C (for up to 2 months). Add powdered lysozyme **immediately** before use (5 mg/ml). Only add lysozyme to as much buffer as you will need for the day's lab—lysozyme is expensive. Many host strains of *Escherichia coli* will lyse without lysozyme.

Potassium Acetate Solution (for alkaline lysis, pH 4.8)

To make 150 ml combine:

5.0 M potassium acetate	90.0 ml
glacial acetic acid	17.2 ml
sterile deionized water	42.8 ml

(No need to check final pH; do not autoclave.) Be sure this is labeled: **"KOAc for Alkaline Lysis, pH 4.8"** to distinguish this solution from the 5 M potassium acetate stock solution. The potassium acetate for alkaline lysis solution is 3 M potassium and 5 M acetate.

1% SDS/0.2 M NaOH Stock (make fresh the day of use)

Add to 90 ml sterile deionized water in a sterile bottle:

20% SDS (molecular biology grade)	5.0 ml
4 M NaOH	5.0 ml

Add SDS to water first, mix, and add NaOH. DO NOT REFRIGERATE!

STET Buffer (for rapid boiling mini-preps)

To 17.8 ml sterile deionized water add:

1.0 M Tris (pH 8.0)	2.0 ml
0.5 M EDTA (pH 8.0)	4.0 ml
Triton X-100	0.2 ml
20% sucrose	16.0 ml

All stocks (except Triton X-100) should be sterilized by autoclaving and combined aseptically in a sterile bottle or tube. Dispense Triton X-100 with a sterile syringe or pipet.

TE Buffer (10 mM Tris, pH 8.0; 1 mM EDTA)

Stock	per 100 ml	per 500 ml	per 1000 ml
1.0 M Tris (pH 8.0)	1.0 ml	5.0 ml	10.0 ml
0.5 M EDTA (pH 8.0)	0.2 ml	1.0 ml	2.0 ml
deionized water	99.0 ml	494 ml	988 ml

Mix and sterilize by autoclaving. Dispense into sterile tubes if needed. Store at room temperature or 4°C.

TES Buffer (10 mM Tris, pH 8.0; 5 mM EDTA; 1.5% NaCl)

To 196 ml deionized water add:

1.0 M Tris (pH 8.0)	2.0 ml
0.5 M EDTA (pH 8.0)	2.0 ml
NaCl	3.0 g

Mix and sterilize by autoclaving. Dispense into sterile 15-ml tubes. Store at 4°C.

Chloroform:Isoamyl Alcohol (24:1)

To a 500 ml bottle of chloroform, add 21 ml of isoamyl alcohol. Mix and label "Chloroform:Isoamyl alcohol (24:1)." This solution may be stored at room temperature indefinitely.

"Phenol"

Phenol used in nucleic acid purifications must be highly purified as it readily oxidizes and the oxidized products can damage DNA. Historically, batches of inexpensive, commercially available phenol were redistilled at 160°C in a hazardous and time-consuming procedure to remove impurities. Many molecular biology suppliers sell redistilled phenol stored under an inert gas that is suitable for use with DNA. Before use, however, the phenol must be equilibrated with a buffer and mixed with antioxidants to prevent the formation of oxidation products that can damage DNA. Such equilibrated phenol is usually written with quotes around it ("phenol"). The preparation of "phenol" is a hazardous process that is no longer necessary. Many suppliers of reagents used in molecular biology market "phenol" that is equilibrated and ready to use. This equilibrated "phenol" may be aliquoted and stored at 4°C in a dark bottle for up to one month or indefinitely at −20°C.

"Phenol:Chloroform:Isoamyl alcohol" (25:24:1)

Mix equal volumes of "phenol" and chloroform:isoamyl alcohol (24:1) in a dark bottle. This is generally prepared immediately before use or may be stored overlayed with TE buffer in a dark bottle at 4°C for up to one month.

B. Restriction Digestion and Ligation of DNA

All suppliers of enzymes supply appropriate 10X buffers with each enzyme ordered. We recommend using these buffers rather than preparing your own. If you are performing digests with more than one enzyme that require different buffers, the enzyme supplier will provide a table indicating the activity of each enzyme in various buffers. Buffers that allow 50% or more activity of an enzyme are suitable for doing multiple digests.

Refer to Appendix VIII (Care and Handling of Enzymes) prior to dispensing or using restriction enzymes, ligase, and polymerases.

C. Agarose Gel Electrophoresis

Type I Electrophoresis Loading Dye

To 42 ml of warm (50-60°C) deionized water add:

bromphenol blue	0.25 g
xylene cyanol	0.25 g
sucrose	50.0 g
20% SDS	5.0 ml
0.5 M EDTA (pH 8.0)	20.0 ml

Mix to dissolve all ingredients; final volume will be 100 ml (sufficient for 100,000 digestions!). Do not autoclave. Store at room temperature or at 4°C.

This buffer stops restriction enzyme digestions (due to the SDS and EDTA) and makes the DNA solution very dense (from the sucrose) so that it will sink to the bottom of the well. The two dyes serve as tracking dyes; bromphenol blue migrates at approximately the same rate as a 0.3-0.5 kb fragment, and the xylene cyanol migrates at approximately the rate of a 4-5 kb fragment. Add 1-2 µl dye per 10 µl of DNA solution.

50X Tris Acetate EDTA (TAE) Buffer (for agarose gel electrophoresis)

To 600 ml of deionized water add:

Tris base	242.0 g
glacial acetic acid	57.1 ml
0.5 M EDTA (pH 8.0)	100.0 ml

Dissolve on a stir plate and adjust volume to 1000 ml with deionized water. Do not autoclave. Dilute 1/50 (20 ml stock per liter) as needed for electrophoresis. We find it convenient to mix 10 or 20 liters of 1X stock in a carboy for classes if you are doing numerous electrophoresis experiments.

10X Tris Borate EDTA (TBE) Buffer

TBE buffer may also be used for agarose gel electrophoresis. It has more buffering capacity than TAE, but results in slightly slower DNA migration and does not work well for some procedures such as DNA fragment isolation from agarose. The recipe for 10X TBE is given under the section on DNA sequencing.

D. Southern Blotting and Hybridization

0.4 M NaOH

Add 100 ml 4.0 M NaOH to 900 ml deionized water and mix. Store at room temperature in a **polypropylene** bottle. This does not need to be autoclaved.

5X SSC

Add 250 ml 20X SSC to 750 ml deionized water and mix. Sterilize by autoclaving.

Prehybridization Buffer (5X SSC, 1% SDS, 0.5% BSA)

To 700 ml deionized water add:

20X SSC	250.0 ml
20% SDS	50.0 ml
Bovine serum albumin (fraction V)	5.0 g

Mix gently on a stir plate and store at room temperature. Do not autoclave.

Note: If stored in a refrigerator or in a cool room, the SDS will precipitate. Precipitated SDS will readily go into solution with brief heating at 40–50°C.

Hybridization Buffer (with Alkaline Phosphatase Conjugated Probe)

Add the alkaline phosphatase conjugated oligonucleotide probe to prehybridization buffer to give a final concentration of 0.2 nM (the probe concentration of the stock is provided on the vial). Add probe **immediately** before use and store at room temperature.

Wash Solution I (0.5% SDS, 0.5% N-lauroylsarcosine, 2X SSC)

To 875 ml deionized water add:

20X SSC	100.0 ml
20% SDS	25.0 ml
N-lauroylsarcosine	5.0 g

Mix gently on a stir plate and store at room temperature (do not autoclave). Place in a 50°C water bath for several hours prior to use.

Wash Solution II (0.5% Triton X-100, 1X SSC)

To 945 ml deionized water add:

20X SSC	50.0 ml
Triton X-100	5.0 ml

Mix gently on a stir plate and store at room temperature (do not autoclave).

Alkaline Phosphatase Buffer

To 400 ml deionized water add:

Tris Base (Molecular Biology Grade)	6.05 g
NaCl	2.92 g
$MgCl_2 \cdot 6H_2O$	0.51 g

Adjust to pH 9.5 with HCl and make to 500 ml with deionized water. Be sure to use an electrode compatible with Tris buffers and calibrate the pH meter with pH 7.0 and 10.0 reference buffers.

To prepare the color development reagent add to each 10.0 ml of buffer:

50 mg/ml BCIP	25 µl
50 mg/ml NBT	50 µl

Mix and store at room temperature wrapped in foil. Use within **2 to 3 hours** of preparation.

E. DNA Sequencing

Polyacrylamide Gel Solution

Both acrylamide and bis-acrylamide are neurotoxins in the free form. To minimize potential hazards from contact with free acrylamide powder, most suppliers sell premixed solutions of acrylamide in commonly used ratios of acrylamide:bis. However, we recommend the purchase of premixed solutions formulated specifically for DNA sequencing gels. These solutions are available in 6 and 8% concentrations of acrylamide, and contain urea and TEMED in 1X TBE buffer. They are supplied in individual squeeze bottles with pouring spouts. Freshly prepared ammonium persulfate is added and the gels can be poured with minimal risk of contamination with toxic acrylamide.

10% ammonium persulfate (APS)

Dissolve 0.1 g ammonium persulfate in 1.0 ml deionized H_2O. Make fresh the day of use.

10X TBE Buffer

To 800 ml deionized water add:

Tris base	108.0 g
boric acid	55.0 g
0.5 M EDTA (pH 8.0)	40.0 ml

Mix until all ingredients are dissolved and make volume up to 1000 ml with deionized water. Store at room temperature.

0.5X TBE (upper gradient buffer)

Add 50 ml 10X TBE to 950 ml deionized water. Mix and store at room temperature.

3.0 M sodium acetate (for lower gradient buffer)

Add 246 g sodium acetate (anhydrous) to 700 ml deionized water. Mix to dissolve and make to 1000 ml with deionized water.

1X TBE, 1 M sodium acetate (lower gradient buffer)

To 850 ml deionized water add:

10X TBE	150.0 ml
3.0 M sodium acetate	500.0 ml

2 M NaOH, 2 mM EDTA

To 9.92 ml of sterile deionized water in a sterile polypropylene tube add:

4 M NaOH	10 ml
0.5 M EDTA (pH 8.0)	80 μl

Mix well on a vortex mixer. This solution is stable indefinitely at room temperature in polypropylene tubes.

2 M ammonium acetate, pH 4.6

To 80 ml deionized water, add 15.42 g ammonium acetate (anhydrous). Mix to dissolve and adjust pH to 4.6 with glacial acetic acid. Make to 100 ml with deionized water and sterilize by autoclaving. This is stable indefinitely at room temperature.

Gel Fixation Solution (5% acetic acid, 5% methanol)

To 3600 ml deionized or distilled water add:

glacial acetic acid	200 ml
methanol	200 ml

Stir to mix and store in tightly capped bottles at room temperature. This solution may be reused up to five times and should be handled as radioactive waste after the first use. Discard as low-level radioactive waste.

APPENDIX XVI
Bacterial Strains

The following strains are required to conduct the experiments in this course. Consult the World Wide Web Molecular Biology Home Page at the University of Wisconsin–La Crosse (http://www.uwlax.edu/MoBio) for information on obtaining these strains. As additional strains become available that are useful in this course, they will also be listed on the home page.

Vibrio fischeri MJ1. This is a bioluminescent strain of *Vibrio fischeri* isolated from the pinecone fish (*Monocentrus japonicus*). The entire *lux* operon in this strain is located on an 18 kb *Bam*H I restriction fragment and also on a 9 kb *Sal* I fragment (which is part of the 18 kb *Bam*H I fragment).

 Note: The *Vibrio fischeri* strain available from the American Type Culture Collection (ATCC 7744) is **not** suitable for these exercises due to a different restriction map.

***Photobacterium* sp.** This environmental isolate (from fresh shrimp) is a brightly glowing strain suitable for classroom demonstrations of bioluminescence.

Escherichia coli DH5α (F⁻,φ80d*lac*ZΔM15, *end*A1, *rec*A1, *hsd*R17 (r_k^-, m_k^+), *sup*E44, *thi*-1, *gyr*A96, *rel*A1, (Δ*lac*ZYA-*arg*F), U169, λ⁻. This is the host strain used in the genomic cloning and subcloning experiments. It is deficient in host restriction activity (*hsd*R17) and possesses a mutated *lacZ* gene (*lac*ZΔM15) whose gene product can form functional β-galactosidase in the presence of the α-fragment of β-galactosidase synthesized from a gene encoded on the cloning vector. This allows rapid identification of clones that contain insert DNA (see Exercises 10 and 11 for a detailed explanation). DH5α is also deficient in recombination (*rec*A1) that minimizes recombination of plasmid DNA with the host genome.

Escherichia coli DH5α (pJE202). This strain harbors the *V. fischeri lux* operon on the 18 kb *Bam*H I fragment cloned into pBR322. This plasmid may be used for subcloning the 9 kb *lux* fragment without generating a genomic library or used for restriction mapping.

Escherichia coli DH5α (pGEM™-3Zf[+]). This strain harbors the plasmid vector used in the genomic cloning.

Escherichia coli DH5α (pUC18) (or pUC19). These strains harbor the plasmid vector used in the *lux*AB subcloning exercises (17–22).

***Escherichia coli* DH5α (pUWL500) (or pUWL501).** These strains harbor the *V. fischeri lux* operon on the 9 kb *Sal* I fragment cloned into pGEM™-3Zf(+). These plasmids should be obtained by students successful in obtaining bioluminescent clones in Exercise 13.

***Escherichia coli* DH5α (pUWL506).** This strain harbors a plasmid containing a portion of the *V. fischeri lux*A gene subcloned into pGEM™-3Zf(+). This is used for DNA sequencing in Exercise 25.

A P P E N D I X   X V I I
Lists of Suppliers

The following are reliable companies that offer quality supplies and equipment at reasonable prices. These companies will be happy to place you on mailing lists so that you will receive their catalogs.

General Scientific Suppliers

The following companies carry a wide range of scientific supplies, equipment, and reagents. Many of these suppliers now carry molecular biology grade reagents and enzymes as well as equipment used in molecular biology. Many colleges and universities have state contracts with one or more of these companies that allow substantial discounts.

Carolina Biological Supply
2700 York Road
Burlington, NC 27215
(800) 334-5551

Fisher Scientific
1600 W. Glenlake Avenue
Itasca, IL 60143
(800) 766-7000

VWR Scientific
800 E. Fabyan Parkway
Batavia, IL 60510
(800) 932-5000
Note: VWR has purchased the Scientific
Products Division of Baxter Diagnostics, Inc.

Curtin Matheson Scientific, Inc.
P.O. Box 1546
Houston, TX 77251-1546
(713) 820-9898

Thomas Scientific
P.O. Box 99
Swedesboro, NJ 08085-6099
(800) 345-2100

Equipment

Beckman Instruments, Inc.
1050 Page Mill Road
Palo Alto, CA 94305
(415) 857-1150

Carries micro-, refrigerated, and ultracentrifuges.

Bio-Rad Laboratories
Life Science Group
2000 Alfred Nobel Drive
Hercules, CA 94547
(800) 424-6723

Carries electrophoresis equipment and supplies. Manufacturer of the CHEF™ pulsed field gel electrophoresis system used in this manual.

Integrated Separation Systems
Enprotech Corp.
21 Strathmore Road
Natick, MA 01760
(800) 433-6433

Carries a wide variety of reasonably priced electrophoresis equipment.

FOTODYNE, Inc.
940 Walnut Ridge Drive
Hartland, WI 53029
(800) 362-3686

Carries transilluminators, cameras, electrophoresis equipment, and supplies. Supplier of a transilluminator with an interlocking UV blocking lid. FOTODYNE has an Educational Products Division with a separate catalog.

Gibco BRL
Life Technologies, Inc.
P.O. Box 68
Grand Island, NY 14072-0068
(800) 828-6686

Carries electrophoresis equipment, membranes and supplies.

Owl Scientific, Inc.
10 Commerce Way
Woburn, MA 01801
(800) 242-5560

Carries a variety of electrophoresis equipment. Also sells stick-on thermometers used to monitor temperature of DNA sequencing gels.

Pharmacia Biotech
800 Centennial Avenue
Piscataway, NJ 08855
(800) 526-3593

Carries a wide variety of electrophoresis equipment and supplies. Supplier of Hoefer™ electrophoresis equipment.

Rainin Instrument Co., Inc.
Mack Road
Woburn, MA 01888
(617) 935-3050

Markets "Pipetman" micropipetters and motorized micropipets. These are not sold through any other catalogs.

Sorval
Du Pont Company
Biotechnology Systems
P.O. Box 80024
Wilmington, DE 19880-0024
(800) 551-2121

Carries micro-, refrigerated, and ultracentrifuges.

USA Scientific Plastics
P.O. Box 3565
Ocala, FL 34478
(800) 522-8477

Carries a variety of electrophoresis equipment and supplies.

Disposable Supplies

Disposables (tubes, micropipet tips, and so on) are carried by general suppliers, although they are often less expensive from some of the following suppliers. Hybridization membranes should be charged nylon and are supplied by several suppliers (Bio-Rad, Fisher, BRL, and so on). It is most cost effective to buy a 3 m roll, but smaller amounts can be purchased. We have found that Cuno, Inc., markets reasonably priced membranes, and we routinely use their Zetabind™ membranes.

Cuno, Inc.
Life Sciences Division
400 Research Pkwy
P.O. Box 1018
Meriden, CT 06450
(800) 231-2259

Carries Zetabind™ nylon membranes for hybridizations (#NM804-01-045SP for Southern hybridizations [20 cm × 3 m roll] and #NM908-01-045SP for colony hybridizations [82 mm discs])

FMC BioProducts
191 Thomaston St.
Rockland, ME 04841
(800) 341-1574

Carries excellent quality agarose. We use SeaKem LE agarose for routine agarose gel electrophoresis. More expensive grades of agarose are not necessary for the gels run in this manual.

International BioProducts, Inc.
P.O. Box 2728
Redmond, WA 98073
(800) 729-7611

Carries tubes, Petri plates, and so on. Best price we have found for sterile plastic Petri dishes.

Marsh Biomedical Products
565 Blossom Road
Rochester, NY 14610
(800) 445-2812 (outside New York)
(716) 271-7060 (in New York)

Carries microcentrifuge tubes, tips, gloves, and so on. Supplier of foam microcentrifuge tube floats (very handy at about $1 each!)

NEN Research Products
Du Pont Company
Customer Service
549 Albany Street
Boston, MA 02118
(800) 551-2121

Carries colony hybridization membranes with tabs and prepunched holes (NEF-978X, colony/plaque screen, 82 mm)

Phenix Research Products
3540 Arden Road
Hayward, CA 94545
(800) 767-0665

Carries microcentrifuge tubes, tips, and latex gloves.

USA Scientific Plastics
P.O. Box 3565
Ocala, FL 34478
(800) 522-8477

Carries microcentrifuge tubes, tips, storage boxes, and so on.

Enzymes, DNAs, Reagents and Biochemicals

All of the following companies provide 10X buffers with enzymes purchased. The catalogs also offer excellent and up-to-date information on use and characteristics of enzymes, maps of vectors, and so on.

BIO 101, Inc.
P.O. Box 2284
La Jolla, CA 92038-2284
(800) 454-6101

Supplier of the GENECLEAN™ Kit used for purification of DNA from agarose gels.

Boehringer Mannheim Biochemicals
P.O. Box 50414
Indianapolis, IN 46250
(800) 428-5433

Carries a complete line of restriction and other enzymes, molecular biology grade reagents, and vectors. Best source we have found for proteinase K, which can vary widely in quality from various suppliers.

Gibco BRL
Life Technologies, Inc.
P.O. Box 68
Grand Island, NY 14072-0068
(800) 828-6686

Carries a complete line of restriction and other enzymes, molecular biology grade reagents, vectors, and the 1 kb (BRL) ladder used as a size standard. Best source we have found for *Sal* I, which can vary widely in quality from various suppliers.

Gold BioTechnology
8620 Pennel Drive
St. Louis, MO 63114
(800) 248-7609

Cheapest source we have found for X-gal ($80.00/gram).

New England Biolabs
32 Tozer Road
Beverly, MA 01915-5510
(800) 248-7609

Carries a complete line of restriction and other enzymes, molecular biology grade reagents, and vectors.

Promega Corporation
2800 Woods Hollow Road
Madison, WI 53711-5399
(800) 356-9526

Carries a complete line of restriction and other enzymes, molecular biology grade reagents, and vectors.

Sigma Chemical Company
P.O. Box 14508
St. Louis, MO 63178
(800) 325-3010

Carries virtually all chemicals and biochemicals used in molecular biology. Sells molecular biology grade reagents and restriction and other enzymes.

United States Biochemical Corporation
26111 Miles Road
Cleveland, OH 44128
(800) 321-9322

Carries a complete line of restriction and other enzymes, molecular biology grade reagents, and vectors. Has reasonably priced molecular biology grade reagents and a good price on LE agarose.

Radioisotopes (for DNA sequencing or labeling radioactive probes)

Amersham Corporation
2636 S. Clearbrook Drive
Arlington Heights, IL 60005
(800) 323-9750

NEN Research Products
Du Pont Company
Customer Service
549 Albany Street
Boston, MA 02118
(800) 551-2121

APPENDIX XVIII
Recommended References

Alberts, B., D. Bray, J. Lewis, M. Raff, K. Roberts, and J. D. Watson. 1994. *Molecular biology of the cell.* 3rd ed. New York: Garland Publishing, Inc. This is an excellent cell biology text with a molecular emphasis.

Ausubel, F., R. Brent, R. E. Kingston, D. D. Morre, J. G. Seidman, J. A. Smith, and K. Struhl. 1995. *Short protocols in molecular biology.* 3rd ed. New York: John Wiley and Sons, Inc. This is a shortened version of *Current protocols in molecular biology* which is the most comprehensive compilation of molecular biology procedures published. *Short protocols* includes basic procedures in molecular biology as well as many procedures in immunology and biochemistry.

Beckwith, J., and T. J. Silhavy. 1992. *The power of bacterial genetics.* New York: Cold Spring Harbor Laboratory Press. This is a great book based on classic papers from the literature covering the history and principles of bacterial genetics. The book is divided into 13 sessions (each consisting of 5 to 8 papers) with the final session dealing with the social impact of science.

Brock, T. D. 1990. *The emergence of bacterial genetics.* New York: Cold Spring Harbor Laboratory Press. This is an excellent historical overview of classical and bacterial genetics, including the role of bacterial genetics in the development of recombinant DNA techniques. A superb reference if you like to go back to the original experiments that led to current concepts in genetics.

Davies, J., and W. S. Reznikoff. 1992. *Milestones in biotechnology.* Boston: Butterworth-Heinemann. This is a collection of classic papers related to the development and applications of modern molecular biology.

Freifelder, D. 1987. *Molecular biology.* 2d ed. Boston: Jones and Bartlett Publishers, Inc. A good but somewhat dated general text in molecular biology.

Glick, B. R., and J. J. Pasternak. 1994. *Molecular biotechnology: Principles and applications of recombinant DNA.* Washington, D. C.: American Society for Microbiology. This is an excellent and clearly written text covering the principles and applications of recombinant DNA technology.

Gonick, L., and M. Wheelis. 1991. *The cartoon guide to genetics.* New York: Harper Perennial. On the lighter side, this is a humorous look at the development of genetics. Great reading for late nights prepping for the next day's lab.

Lodish, H., D. Baltimore, A. Berk, S. L. Zipursky, P. Matsudaria, and J. Darnell. 1995. *Molecular cell biology.* 3rd ed. New York: W. H. Freeman and Co. A good general text with a molecular emphasis; a Scientific American Book.

Neidhardt, F. C., ed. 1996. *Escherichia coli* and *Salmonella typhimurium*. Cellular and molecular biology. 2d ed. Washington, D. C.: American Society for Microbiology. A superb and current two-volume collection of reviews on the genetics, molecular biology, and physiology of these two important species.

Sambrook, J., E. F. Fritsch, and T. Maniatis. 1989. *Molecular cloning: A laboratory manual*. 2d ed. New York: Cold Spring Harbor Laboratory Press. This is often considered the cloning bible and is a superb three-volume reference on most protocols used in cloning. The manual also includes excellent discussions on basic molecular biol-ogy and the theory and applications of the techniques presented. It also contains excellent appendices on preparation of media, reagents, suppliers, and basic techniques.

Singer, M., and P. Berg. 1991. *Genes and genomes*. Mill Valley, CA: University Science Books. This is an excellent up-to-date molecular biology text. It is divided into four sections on: (i) Molecular Basis of Heredity; (ii) The Recombinant DNA Breakthrough; (iii) The Molecular Anatomy, Expression, and Regulation of Eukaryotic Genes; and (iv) Understanding and Manipulating Biological Systems.

Watson, J. D., M. Gilman, J. Witkowski, and M. Zoller. 1992. *Recombinant DNA*. 2d ed. New York: W. H. Freeman and Co. This is an excellent text on the development and uses of recombinant DNA techniques. Although written at an introductory level, it is also suitable for advanced courses.

Watson, J. D., N. H. Hopkins, J. W. Roberts, J. A. Steitz, and A. M. Weiner. 1987. *Molecular biology of the gene*. 4th ed. Menlo Park, CA: The Benjamin/Cummings Co. This is a superb and well-written text in molecular biology. Although a bit dated, it contains an excellent discussion of DNA structure and chemistry. It can also be purchased separately as two volumes: one on general molecular biology and a second on specialized aspects such as development, immunology, cancer, and evolution.

APPENDIX XIX
Restriction Mapping Problems

The sizes of fragments generated by digestion of a given DNA molecule with restriction endonucleases can be used to construct a **restriction map** of the DNA. Restriction maps show the relative positions of the restriction enzyme recognition sites and can be used to devise subcloning strategies, to study evolutionary relationships between different species, and to provide physical maps of virtually any piece of DNA.

The basic strategy in restriction mapping is to use the fragments generated by one enzyme (preferably one that makes only one cut in a molecule of DNA) to map the location that the enzyme cuts the molecule. By then using the sizes of fragments generated from other enzymes singly and in combination, the sites of the other enzymes may be deduced. When examining the results, look to see what fragments present in a single digest disappear when a double digest is done—this indicates that the second enzyme cleaved internally in the original fragment. Often you will not initially be able to definitively map the location of a restriction cleavage site but only know that it cuts at one of two possible sites. Additional digestions should eliminate all but one of the possibilities, allowing you to identify the site on your map. The following simple examples illustrate the logic behind restriction mapping and are followed by a series of problems of increasing difficulty.

Example 1. Mapping a Linear DNA Molecule

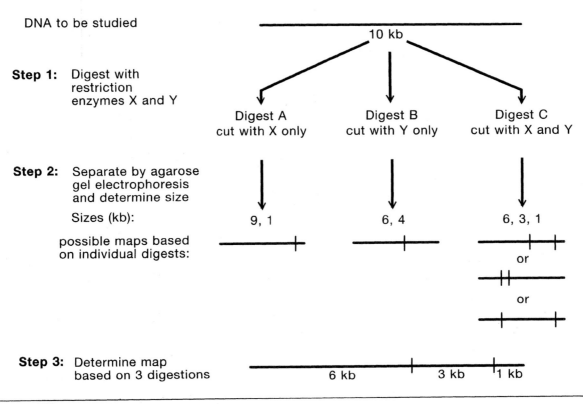

Example 2. Mapping a Circular DNA Molecule

Many of the DNA molecules used in molecular biology are circular (e.g., plasmids). The same general logic used in mapping linear molecules applies to constructing a circular restriction map. However, a single cut in a plasmid provides one linear piece of DNA that can be distinguished from the uncut plasmid on a gel by differential mobility (see Appendix IV). Similarly, two cuts produce two fragments, three cuts produce three fragments, and so on.

In the following example, a 12 kb plasmid (pUWL11) is digested with three restriction enzymes in single and combination digests. Use the circle drawn below the table to map the restriction sites of each of the three enzymes. Start by designating the *Eco*R I site as 12 o'clock on the circle. Sometimes it is useful to draw circular molecules as a linear molecule with the understanding that the ends represent a common point on the circular molecule (note the possible maps in the table). By making each kb a defined length (e.g., 1 kb/cm), the location of specific restriction sites may be more accurately placed on a line than on a circle.

Enzyme(s)	Fragments produced (kb)			Possible maps
*Eco*R I	12			E _____ E
*Hin*d III	12			E ____ H? ____ E
Pst I	12			E ____ P? ____ E
*Eco*R I + *Hin*d III	11	1		E H _____ E
*Eco*R I + *Pst* I	6	6		E ____ P ____ E
*Hin*d III + *Pst* I	7	5		H ____ P? ____ E H
*Eco*R I + *Pst* I + *Hin*d III	6	5	1	E _____ E

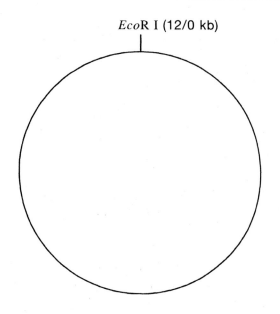

*Eco*R I (12/0 kb)

Restriction Mapping Problems

The following five problems are more challenging than the examples, and in some cases illustrate how information other than the sizes of the restriction fragments in a digest can be used to aid in the construction of a map. In each problem, first map the cleavage sites of enzymes that cut the DNA only once. Look for a way to approach each problem that allows you to get a start with the information provided and then build upon this by examining the results of other digests.

Problem 1

If two enzymes cut in the same gene, this information is often useful in constructing a map. This strategy can be used in the following problem. pUWL22 is a circular, double-stranded, 10.5 kb plasmid. It contains a gene encoding an enzyme that confers ampicillin resistance in the host bacterium. Cloning into the *Kpn* I and *Sst* I sites abolishes ampicillin resistance, whereas cloning into other sites on the plasmid does not. Digestion with the following restriction enzymes singly and in combination results in the fragments listed.

a. Map the cleavage sites for these enzymes on the circular map of pUWL22 provided on the following page. Designate the *Hin*d III site as 12 o'clock (10.5/0 kb) on the map. After you have completed your map, indicate the relative position of the ampicillin gene.

Enzyme(s)	Fragments produced (kb)		Possible maps
*Hin*d III	10.5		
Kpn I	10.5		
Not I	10.5		
Pvu II	10.5		
Sst I	10.5		
Stu	10.5		
*Hin*d III + *Kpn* I	8.5	2.0	
*Hin*d III + *Not* I	8.0	2.5	
*Hin*d III + *Pvu* II	6.0	4.5	
*Hin*d III + *Sst* I	7.5	3.0	
*Hin*d III + *Stu* I	7.5	3.0	
Stu I + *Not* I	10.0	0.5	
Stu I + *Pvu* II	7.5	3.0	

b. Propose two other double digests using the enzymes listed in the table above that could be done to confirm your map. Predict the sizes of the fragments generated from these digests.

Map of pUWL22

*Hin*d III (10.5/0 kb)

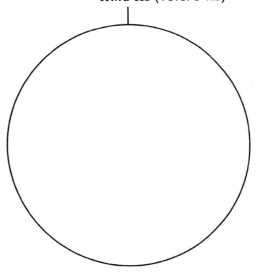

Problem 2

You purify a 15 kb plasmid from an environmental bacterial isolate (because it is a natural isolate, it does not contain a known cloning vector sequence) and name it pMUD101. Digestion with the following restriction enzymes singly and in combination results in the fragments listed in the following table. Map the cleavage sites for these enzymes on the circular map of pMUD101 provided. Designate the *Bam*H I site as 12 o'clock (15/0 kb) on the map. The data provided should allow you to map each enzyme with the use of only the single and double digests. Use the data from the triple digest to confirm your map.

Enzyme(s)	Fragments produced (kb)				Possible maps
*Bam*H I	15				
*Hin*d III	7	5	3		
Pst I	15				
*Bam*H I + *Pst* I	8	7			
*Bam*H I + *Hin*d III	6	5	3	1	
Pst I + *Hin*d III	7	4	3	1	
*Bam*H I + *Pst* I + *Hin*d III	6	4	3	1 (doublet band)	

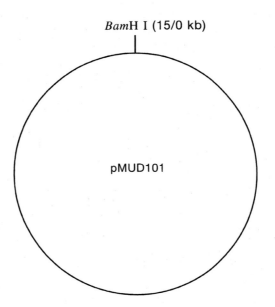

BamH I (15/0 kb)

pMUD101

Problem 3

When DNA is cloned into a vector, preparing a restriction map of the cloned DNA is facilitated by knowing the location of restriction sites in the vector. In the following problem, you are given a recombinant plasmid (pDNA44) that has an insert cloned into the *EcoR* I site of the vector pUWL33. Use the restriction map of the vector (provided below) and the data from the restriction digests to create a map for the recombinant plasmid pDNA44 on the following page. Designate one of the *EcoR* I sites as 12 o'clock on the map and clearly distinguish the vector and insert regions. The data provided should allow you to map the cleavage site of each enzyme with the use of only the single and double digests. Use the data from the triple digest to confirm your map.

Enzyme(s)	Fragments produced (kb)						
EcoR I	5.0	4.2					
BamH I	7.3	1.9					
Pst I	7.4	1.1	0.7				
EcoR I + *Pst* I	4.5	2.9	1.1	0.5	0.2		
BamH I + *Pst* I	3.5	2.0	1.9	1.1	0.7		
EcoR I + *Pst* I + *BamH* I	3.5	2.0	1.1	1.0	0.9	0.5	0.2

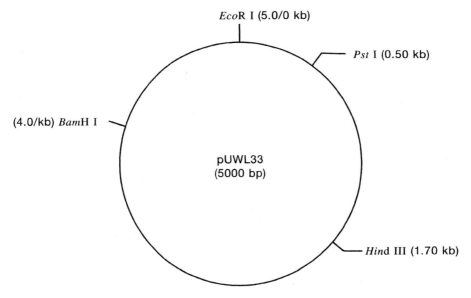

Restriction map of the vector pUWL33

Map of pDNA44

*Eco*R I (_____/0 kb)

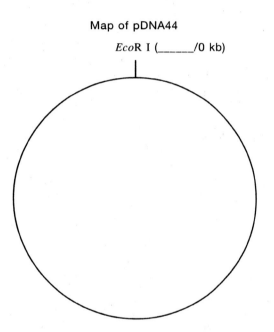

Problem 4

You have cloned a *Bam*H I—*Sal* I fragment into the vector pUC19 (a map of pUC19 is provided in Appendix XI) and named your new recombinant plasmid pDNA99. You now wish to prepare a detailed restriction map of the insert DNA for further sub-cloning. To accomplish this, use the map of the vector and the following information to construct an accurately scaled linear map of the insert DNA (include and identify the vector sequence on your map). Use a scale of 0.5 kb/cm.

Enzyme(s)	Fragments produced (kb)			
*Eco*R I	5.4	3.5		
*Hin*d III	5.3	3.6		
Kpn I	7.4	1.5		
Xba I	6.2	2.7		
*Bam*H I + *Sal* I	6.2	2.7		
*Bam*H I + *Xba* I	4.0	2.7	2.2	
*Eco*R I + *Kpn* I	3.9	3.5	1.5	
*Eco*R I + *Hin*d III	4.5	2.7	0.9	0.8
*Hin*d III + *Xba* I	3.6	2.7	1.3 (doublet)	

Map of pDNA 99:

Problem 5

You have constructed a recombinant plasmid (pDNA199) by directionally inserting an *Eco*R I—*Hin*d III fragment of *Plasmodium yoellii* DNA into pGEM™-3Zf(+) (a map of this vector is provided in Appendix XI). The vector and insert DNA were both digested with *Eco*R I and *Hin*d III prior to ligation. You wish to create a restriction map of the cloned DNA, but you also want to identify the region of the insert that codes for a specific surface antigen (*sag*A gene). To assist in the identification of the coding region, a 600 bp probe for the *sag*A gene from a closely related species is available. After the coding region for the *sag*A gene has been identified, it then will be subcloned into another vector for sequencing and expression.

You perform a number of restriction digests and resolve the fragments by agarose gel electrophoresis. After photographing the gel, the restriction fragments are transferred to a nylon membrane by Southern blot, then hybridized to the 600 bp probe. The results are given in the table that follows. The bands that hybridized to the probe are indicated with an asterisk(*). Not all enzyme combinations were done, but they are not needed to determine the map.

Diagram an accurately scaled linear map of pDNA199 (include and identify the vector sequence on your map). Use a scale of 0.5 kb/cm.

Note: The amount of data makes this mapping problem look forbidding at first. The best way to start building the map is to first identify the location of the four enzymes that make single cuts (don't forget to use the map of the vector). Then use these restriction sites to place the remaining sites. It is not possible to map all the sites using only the restriction data; you will need to use the hybridization data to map all the restriction sites. In fact, use of the hybridization data throughout the map-building process will simplify your task.

Enzyme(s)	Fragments produced (kb)					
EcoR I	10.4*					
Hind III	10.4*					
Hpa I	10.4*					
EcoR V	10.4*					
Cla I	6.85	3.55*				
Xho I	9.5*	0.5*	0.4			
EcoR I + Hind III	7.25*	3.15				
EcoR I + Hpa I	9.8*	0.6				
EcoR I + EcoR V	5.3	5.1*				
EcoR I + Cla I	4.35	3.55*	2.5			
EcoR I + Xho I	6.7	2.8*	0.5*	0.4		
Hind III + Hpa I	6.65*	3.75				
Hind III + EcoR V	8.25*	2.15				
Hind III + Cla I	5.65	3.55*	1.2			
Hind III + Xho I	5.95*	3.55	0.5*	0.4		
Hpa I + EcoR V	5.9	4.5*				
Hpa I + Cla I	4.95	3.55*	1.9			
Hpa I + Xho I	7.3	2.2*	0.5*	0.4		
EcoR V + Cla I	6.85	2.6*	0.95			
EcoR V + Xho I	8.1*	1.4	0.5*	0.4		
Cla I + Xho I	6.85	2.35	0.5*	0.4	0.3*	
EcoR V + Hpa I + Cla I	4.95	2.6*	1.9	0.95		
EcoR V + Hpa I + Xho I	5.9	2.2*	1.4	0.5*	0.4	
EcoR V + Cla I + Xho I	6.85	1.4	0.95	0.5*	0.4	0.3*
EcoR V + Cla I + EcoR I	4.35	2.6*	2.5	0.95		
Hpa I + Cla I + Xho I	4.95	2.35	1.9	0.5*	0.4	0.3*

*Indicates restriction fragments that hybridize with the sagA probe.

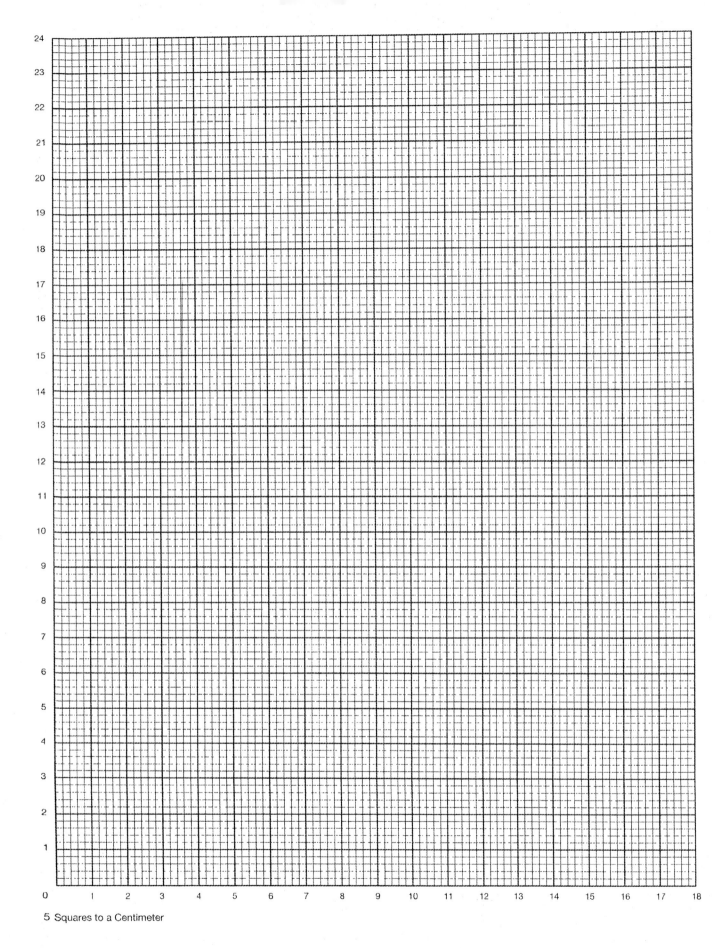

5 Squares to a Centimeter